Byzantine Military Tactics
in Syria and Mesopotamia
in the Tenth Century

BYZANTINE MILITARY TACTICS IN SYRIA AND MESOPOTAMIA IN THE TENTH CENTURY

A Comparative Study

Georgios Theotokis

EDINBURGH
University Press

Edinburgh University Press is one of the leading university presses in the UK. We publish academic books and journals in our selected subject areas across the humanities and social sciences, combining cutting-edge scholarship with high editorial and production values to produce academic works of lasting importance. For more information visit our website: edinburghuniversitypress.com

© Georgios Theotokis, 2018, 2020

First published in hardback by Edinburgh University Press 2018

Edinburgh University Press Ltd
The Tun – Holyrood Road
12 (2f) Jackson's Entry
Edinburgh EH8 8PJ

Typeset in 11/ 13 JaghbUni Regular by
IDSUK (DataConnection) Ltd

A CIP record for this book is available from the British Library

ISBN 978 1 4744 3103 3 (hardback)
ISBN 978 1 4744 3104 0 (paperback)
ISBN 978 1 4744 3105 7 (webready PDF)
ISBN 978 1 4744 3106 4 (epub)

The right of Georgios Theotokis to be identified as author of this work has been asserted in accordance with the Copyright, Designs and Patents Act 1988 and the Copyright and Related Rights Regulations 2003 (SI No. 2498).

Contents

Acknowledgements	vi
List of Rulers	vii
Map 1 Anatolia and Upper Mesopotamia	viii
Map 2 Armenian Themes and Principalities	ix
Introduction	1
1 The 'Grand Strategy' of the Byzantine Empire	23
2 Byzantine and Arab Strategies and Campaigning Tactics in Cilicia and Anatolia (Eighth–Tenth Centuries)	52
3 The Empire's Foreign Policy in the East and the Key Role of Armenia (c. 870–965)	69
4 The Byzantine View of their Enemies on the Battlefield: The Arabs	105
5 Methods of Transmission of (Military) Knowledge (I): Reconnaissance, Intelligence	128
6 Methods of Transmission of (Military) Knowledge (II): Espionage	147
7 Tactical Changes in the Byzantine Armies of the Tenth Century: Theory and Practice on the Battlefields of the East	192
8 Tactical Changes in the Byzantine Armies of the Tenth Century: Investigating the Root Causes	219
9 Byzantine–Arab Battles of the Tenth Century: Evidence of Innovation and Adaptation in the Chronicler Sources	236
10 Tactical Innovation and Adaptation in the Byzantine Army of the Tenth Century: The Study of the Battles	276
Summaries and Conclusions	298
Primary Bibliography	308
Secondary Bibliography	313
Index	341

Acknowledgements

This monograph has been eight years in the making, from the time it was first conceived as an original idea while discussing over a glass of wine with colleagues from the University of Glasgow, during which time I was still struggling to finish my doctoral thesis, to the completion of the final draft in the study cubicles of the National Library of Latvia in Riga. In this process of endless research, reading and writing, I benefitted from the support of several institutions, most notably the Medieval Institute of the University of Notre Dame, which provided me with both an ideal working environment and the financial support to get my idea off the ground and develop it into a viable project. I am grateful to the late Olivia Remie Constable, Robert M. Conway Director of the Medieval Institute at the University of Notre Dame, who gave me this rare opportunity to work across the Atlantic. I also spent six fruitful months in the Institute of Arab and Islamic Studies at the University of Exeter where, with the help and guidance of Professor Dionisius Agius, my monograph acquired its final form. Finally, I am grateful to the three most important people of my life, my parents and my wife, who truly helped this 'ship sail safely through stormy seas' with their endless love and encouragement.

Boğaziçi, February 2018

Rulers

Byzantine Emperors of the Macedonian Dynasty

Basil I (867–86)
Leo VI (886–912)
Alexander (912–13)
Romanus I Lecapenus (920–44)
Constantine VII (913–59)
Romanus II (959–63)
Nicephorus II Phocas (963–9)
John I Tzimiskes (969–73)
Basil II (976–1025)

Hamdanid Rulers of Mosul and Aleppo

Hamdan ibn Hamdun
Abu al-Hayjaʾ ʿAbdallah (Mosul 905–13, 914, 925–9)
Abu al-Saraya Nasr (Mosul 929–34)
Abu al-Ula Saʿid (Mosul 934–5)
Abu Muhammad 'Nasir ad-Dawla' al-Hasan (Mosul 929, 934, 935–67)
Abu al-Hasan 'Sayf ad-Dawla' (Aleppo 944–67)
'Abu Taghlib' Uddat ad-Dawla Fadlallah (Mosul 967–79)
'Abu al-Maʿali' Saʿad ad-Dawla Sharif I (Aleppo 967–91)

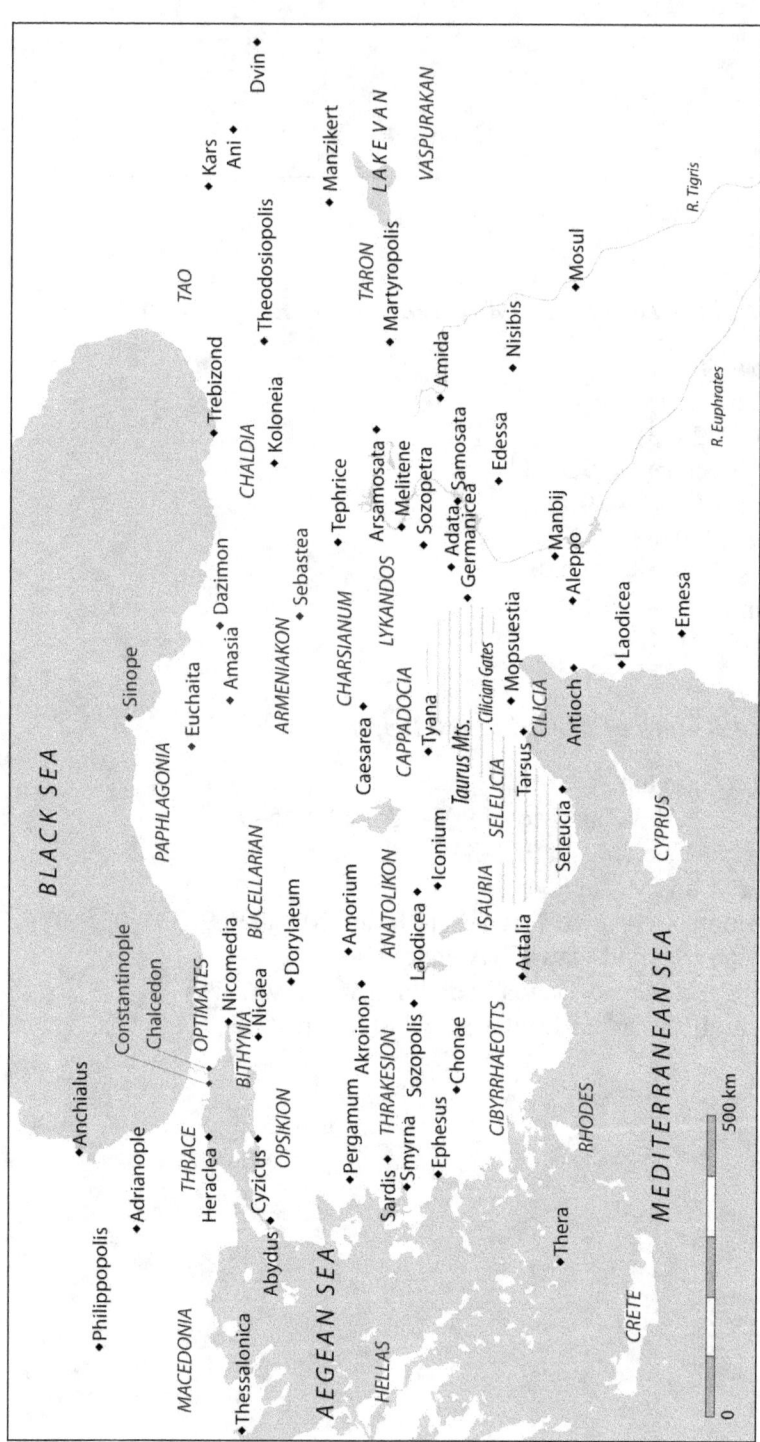

Map 1 Anatolia and Upper Mesopotamia (Tenth Century AD): Major Themes, Principalities and Towns

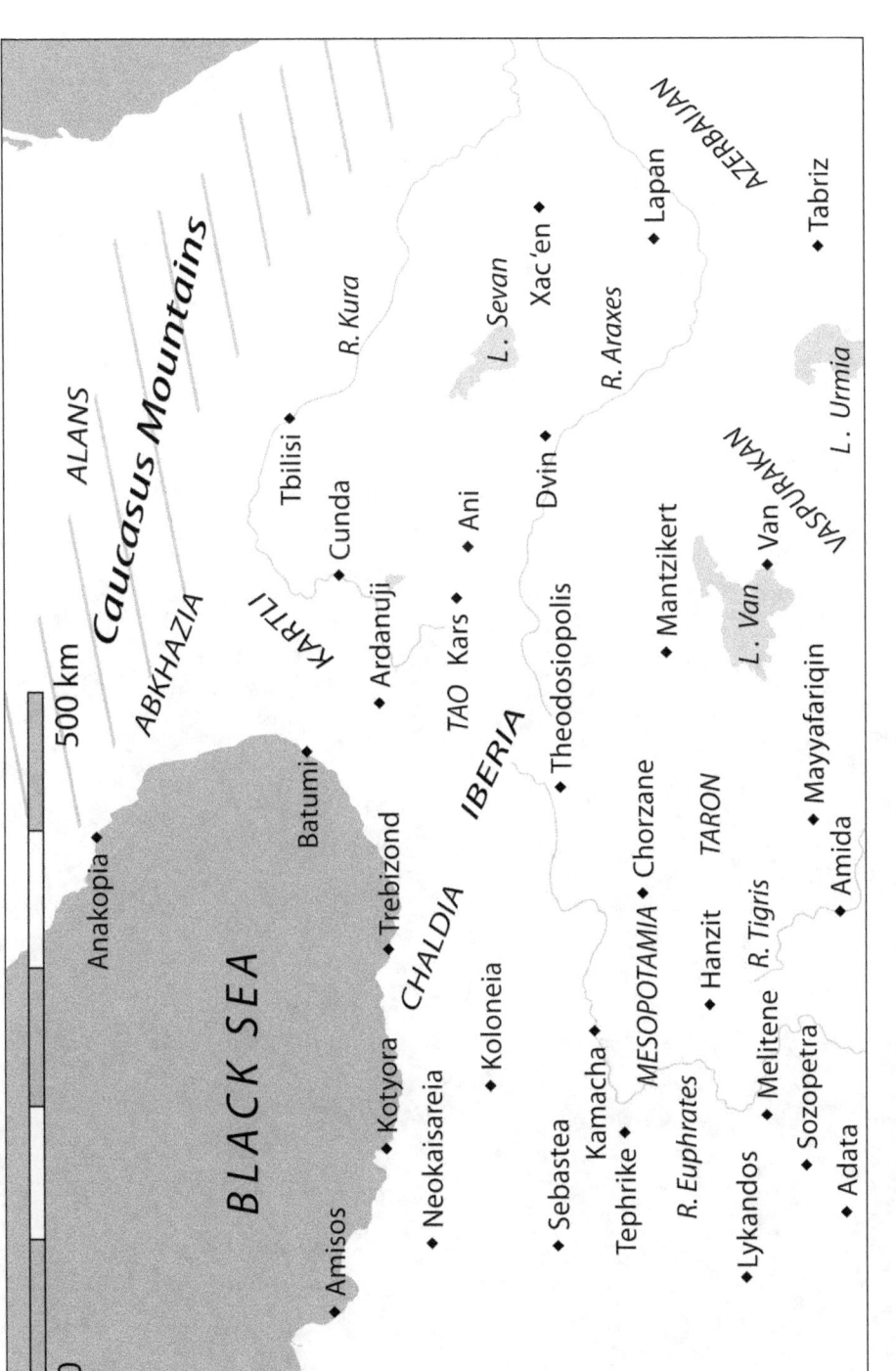

Map 2 Armenian Themes and Principalities (Tenth Century)

Introduction

Fas est et ab hoste doceri
(It is right to learn, even from the enemy)
—Ovid, *Metamorphoses*, 4.428

As a doctoral student, I was interested in the raids of the Norman dukes of southern Italy on the Byzantine provinces in the Balkans between 1081 and 1108. Foremost, I conducted a study of the military organisation of the Norman and Byzantine states in that period, their overall strategies and their military tactics on the battlefield. In that time, I had the chance to venture into the world of warfare in the eastern Mediterranean, from Italy and the Balkans to Asia Minor and the Middle East, examining the military organisation, tactics and strategies of the Byzantines and the Seljuk Turks, the Arabs of Egypt and the Crusaders. It was while studying battles in the same geographical area that I could identify the numerous tactical innovations and adaptations between different armies in their battle tactics after a pitched battle or a skirmish with the enemy. Therefore, a key question was quickly raised: can it be said that the general who shows the most willingness to adapt to the tactics of the enemy has significantly better chances of winning the battle and, perhaps, even the war? Thus, the main aim of the present study is to examine in detail the way each state adapted to the strategies and tactics of its enemies in a specific operational theatre: the region that is bordered by Antioch and Aleppo to the south, Taron and Vaspourakan around Lake Van to the east, and the mountain ranges of the Taurus and Anti-Taurus to the north and west, between the second and third quarters of the tenth century.

The period that I have chosen to study should be considered within the political context of the Byzantine wars of expansion that dominated the eastern frontiers of the empire for the best part of the tenth century. Since 927–8, when the threat from the Bulgarian tsar Symeon had disappeared, the empire's foreign policy had already shifted to the preservation of a pro-Byzantine Armenia and the establishment of control over the strategic cantons of Taron and Vaspourakan, around Lake Van – an area that controlled the invasion routes into Byzantine Chaldea through north-eastern

Anatolia.¹ This period comes in complete contrast to the previous decades of incessant raiding in eastern and central Anatolia; the second and third quarters of the third century witnessed an increasingly aggressive Byzantine foreign policy during which the need for more professional units of heavy infantry and cavalry became pressing. If Armenia, however, was strategically far more important to the Byzantine government than Cilicia and Syria, then how can the empire's extensive territorial gains in Cilicia in the third quarter of the tenth century be explained?

It all comes down to the personal and political image of the Byzantine emperor as a sovereign chosen by God to protect His people. The struggle between the Emperor Constantine VII (945–59) and his successors and Sayf ad-Dawla of Aleppo (944–67) was titanic, and by the close of the 950s had escalated into an all-out conflict where no one could afford (politically) to succumb. In the end, it was the vast resources Byzantium poured into the wars in the East that turned the tide in their favour by 962. After the conquest of Antioch seven years later, Nicephorus II Phocas (r. 963–9) wrecked the Emirate of Aleppo, swept away the bases for Arab raids in Anatolia and replaced them with an impregnable wall of Byzantine themes which was to withstand foreign invasions for another hundred years. After his death, the Emirate of Aleppo was to become a Byzantine dependency, thus enabling the Byzantines to come into direct contact with the Fatimids of Egypt, who held southern Palestine.

As the title suggests, this is a comparative study of the military cultures that clashed in Cilicia, Syria and northern Mesopotamia in the tenth century. For the purposes of this comparison, I examine two pools of primary material: first, the accounts of the largest and most important raids, sieges and pitched battles of this period, since they could have had a decisive outcome on the course of a campaign or even a war, through whatever information can be deduced from the contemporary historical accounts; second, the Byzantine and Arab military manuals, which, being prescriptive and not descriptive in nature, provide crucial information on how armies should have been organised and deployed in the battlefield up to the period when they were compiled, thus reflecting decades and even centuries of experience in fighting.

My strategy is twofold: first, I focus on the tactical changes that took place in the units of the imperial army in the tenth century. The richest and most useful – but ostensibly underutilised – sources for identifying these changes are military treatises such as the *Praecepta Militaria* (*Military Precepts*) of Nicephorus Phocas (c. 969), the anonymous

Introduction

Sylloge Taktikorum (*Collection of Tactics*) (c. 930), the *Taktika* attributed to Emperor Leo VI (written c. 895–908) and numerous others. I discuss the recommendations of the authors regarding the marching and battle formations, the armament and battlefield tactics of the Byzantine army units, and I ask whether they reflect any kind of innovation or tactical adaptation to the strategic situation in the East. Then I explain how far we can say that 'theory translated into practice' in the campaigns and battlefields of this period, such as Hadath (954), Tarsus (965), Dorystolo (971), Alexandretta (971), Orontes (994) and Apamea (998), according to the accounts of contemporary historians from both sides of the political, religious and cultural spectrums.

I try to understand the mechanisms that lie behind this diffusion of 'military knowledge', and I approach the issue by asking several related questions:

- What were the tactics of each state in the region under consideration and what basic *similarities* can we trace? How suitable were they for the warfare in the region?
- Can we identify any similarities in the tactics used against different enemies in the same region or do we notice a change depending on the enemy faced in the field? Are we able to trace the origins of a tactic and what does that say about the influence each culture had on its neighbours?
- What is the connection between adaptability in the battlefield and the overall strategy of a state? Do we see nations that pursue a more defensive strategy adapting more easily to the changing tactics of their enemies?
- Can we say that certain cultures are more susceptible to tactical changes than others and, if so, what are the deeper reasons behind this phenomenon? In what way does this reflect their social structure and any changes in it? What was the role of religion and religious enthusiasm in this process?

The standard work on the political and social history of the Hamdanids of Aleppo remains Canard's *Histoire de la dynastie des Hamdanides*,[2] a remarkable survey of the naissance, heyday and decline of the Muslim dynasty, through the study of Arab and non-Arab, literary and non-literary sources, supplemented by a rich bibliography and more than 200 pages of material on the historical geography of the Jazira (Upper Mesopotamia), Mesopotamia, Syria and Armenia-Azerbaijan. Other works include

Bikhazi's PhD thesis that revised Canard's narrative and analysis in a number of areas, especially concerning the emirate in Mosul.[3] The towns and populations of the *thughūr* (frontiers), the Muslim frontier districts that formed the bulwark between the Byzantine Empire and the provinces of the Abbasid Caliphate, have been thoroughly studied by Asa Eger, Bonner, Bosworth, Holmes, and Haldon and Kennedy.[4] Finally, Shepard has produced two influential papers on the Byzantine notion of frontiers and imperial expansionism, and on the empire's foreign policy in the East, with a particular focus on Armenia.[5]

Of direct relevance are the works by Haldon and Treadgold that examine the structure, consistency, battle tactics and formations of the Byzantine army up to the eleventh century.[6] Although these editions are invaluable in providing a general overview of the transformation of the imperial army through the centuries, McGeer's laborious and detailed work on Byzantine warfare in the tenth century is more helpful for the purposes of this study.[7] His monograph forms a contribution to the understanding not simply of two of the most important military treatises of tenth-century Byzantine military literature, but also very skilfully places them within the political and social context of the period's expansionist policies against the Arabs in Cilicia, Syria and Armenia, thus revealing the sophistication of the imperial military system at the time.

Crossing the political frontier, many rich and engaging studies have been produced on the military organisation of the Muslim states bordering the empire in the East and, especially, the Fatimids of Egypt. I would argue that the works by Lev have set the bar very high for the quality of work produced on both Mediterranean and Islamic history during a period (tenth to twelfth centuries) of profound changes in the eastern Mediterranean.[8] Kennedy's study on the armies of the caliphs provides a competent analysis of the history, organisation and equipment of the Muslim armies for the first three centuries of the Muslim expansion.[9] Other useful monographs are Beshir's 'Fatimid Military Organization',[10] Hamblin's PhD dissertation titled 'The Fatimid Army during the Early Crusades',[11] followed by Bosworth's articles on the 'Military Organization under the Buyids of Persia and Iraq' and 'Ghaznavid Military Organization',[12] a collection of studies edited by Parry and Yapp,[13] and a number of general overviews on the topic with rich endnotes by Nicolle.[14] Regrettably, the only study on the military organisation and tactics of the Hamdanid armies is a dense but well-written chapter in McGeer's *Sowing the Dragon's Teeth*.

Much of what I touch upon in this monograph has been covered in reasonable detail within the last fifty years. In many of the chapters I review

Introduction

the findings and breakthroughs of scholars like McGeer and Haldon concerning the theory and practice of tenth-century warfare in the empire, and the political, cultural and social environment that imposed a deep divide between the empire's centre and its provinces. Therefore, what I am suggesting is a synthesis of what we know of tenth-century Byzantine warfare, but within a very focused thematic (campaign strategy, comparative tactics, raids, siege warfare), geographic (Armenia, Cilicia, Syria) and chronological (tenth century) scope. I offer a fresh and critical perspective on warfare in the region by delving deeper into well-known, but ostensibly underutilised, sources. For that purpose, my intention is to compare tactics with military cultures found in the tenth-century military manuals and place them as a sort of 'benchmark' against which to test narrative accounts. As a consequence, the analysis will delve further into the issue of 'innovation' and identify developments within and between military cultures.

Two conventions of conflict that need to be explained and formulated within the context of the medieval strategy of campaigning are the theory of Vegetian warfare and the role of battle as a component or tool in a commander's repertoire.[15] The former theory describes a cross-cultural[16] method of waging war in pre-modern economic and technological conditions, which places a particular emphasis on the avoidance of battle at all costs and underlines the logistical and geographical constraints imposed upon the medieval and early modern commander. This theory, of course, dates back many centuries – even further back than the derogatory words put by Herodotus into the mouth of a bewildered Mardonius commenting on the fighting habits of the ancient Greeks, a characteristic example of an outsider reviewing intracultural warfare tactics.[17]

The use of the term Vegetian (warfare/theory/strategy), however, to describe the principle of avoiding an engagement by using 'other means', should not be taken as a misnomer, giving the false impression that I am creating an artificial problem by importing a research question pertaining to the medieval West that is utterly irrelevant to Byzantium. As I will show in the second part of Chapter 1, the Byzantines had very similar attitudes towards pitched battles: they did not risk taking to the battlefield unless the odds were overwhelmingly stacked in their favour. And it was not just European commanders that followed Vegetius' dicta well into the eighteenth century. Many of Byzantium's Muslim enemies, including their Ottoman successors, complied with the same theory of battle avoidance as a principle for conducting military operations.[18] Due to the lack of any equivalent modern historical term to describe the Byzantine culture of war – something in the spirit of 'Maurician' (referring to the sixth-century *Strategikon* attributed to Emperor Maurice) or 'Leonid' (referring to the *Taktika* written by Emperor Leo VI c. 895–908) strategy in Byzantium – I had no choice but to borrow the term 'Vegetian' as a convenient reference to a specific set of principles of war that defined European strategy for almost fifteen centuries.

Thus far we have established that battle is a hazardous course of action only to be followed when the odds are overwhelmingly in one's favour or as a last resort. Pitched battles were risky because they might lead to waste of life and resources and undo the arduous work of months or even years in just a few hours. That is precisely the reason why battles were rare: because they could, potentially, be decisive. Rather, it was diplomacy, bribery, tricks, stratagems, sieges, raiding and plundering that took precedence in the Byzantine officers' campaigning repertoire against the empire's multitudes of enemies in East and West. Armies, therefore, were built as much for ravaging and siege work as for battle, which affected how they fought on the battlefield. But how decisive could a battle be and what exactly do we mean by the term 'decisive battle'?

In a detailed article on the different categories of wars the Byzantines fought in their long history, W. Treadgold argued:

> My point is not that the Byzantines defended themselves badly, because for the most part they defended themselves very well. After all, the empire lasted longer than all its enemies. It outlived the Persian Empire, two Muslim caliphates, two Bulgarian empires, and the Seljuk Turks; even the Ottoman Turks, who brought Byzantium down, founded an empire that was to have a much shorter life. The Byzantines were probably wise to avoid major pitched battles, especially because they lost most of the few they did fight, like the battle of Adrianople against the Goths, the battle of the Yarmouk during the initial Muslim expansion, and the battles of Manzikert and Myriocephalum against the Seljuk Turks.[19]

Thus, according to Treadgold's very aptly put argument on their military abilities, the Byzantines were not very good at taking their chances on the battlefield against their Christian and Muslim enemies. But then how can we explain the paradox of an 'empire that would not die'? Surely, occasional crises united the empire by forcing the imperial court, the provincial ruling classes and the church closer together,[20] but there is an additional factor here, one that takes us back to the previous question regarding the decisiveness of battles. If the Byzantines were so hopeless at fighting

Introduction

external enemies in pitched battles – an oversimplification based solely on their track record – then were these battles decisive or not?

What makes a battle decisive? From Sir Edward Shepherd Creasy's well-known nineteenth-century monograph to more contemporary works by John Keegan, Victor Davis Hanson and Stephen Morillo, the answer is straightforward: impact.[21] A decisive battle should have long-term socio-political implications between adversaries and profoundly affect the balance of power on more than just the local level. Inevitably, many battles spring to mind: Salamis, appraised as having saved the Western world and democracy from oriental despotism and having provided the necessary impetus for the rise of Athenian (naval) power; Lechfeld, effectively ending the raiding of the Magyars in central Europe and seeing the rise of the eastern Frankish power in Europe; Stanford Bridge and Hastings, starting a new chapter in the socio-political history of England; Talas River, putting a halt on Tang China's expansionist policies towards the west, resulting in Muslim control of Transoxiana for the next 400 years; Manzikert, although not the cataclysmic event it was once considered to be, still viewed as an engagement that decisively lost central and eastern Anatolia to the Turks; Antioch (1098), when the future of the Crusading movement hung in the balance; Bouvines, securing Normandy for the Capetians and turning France into the most powerful monarchy in Europe; and the second battle of Kosovo, where the Ottomans managed to annihilate the Serbian army, thus forcing the Serbian principalities that were not already Ottoman vassals to become so in the following years.

Let us consider in more detail the concept of a *decisive* battle through a series of examples that involved the empire in a long-term conflict on its eastern frontiers. I am referring to the four-century long war with the Sassanid Empire that went on from the 230s to Heraclius' campaigns in the 620s, involving several smaller campaigns and peace treaties lasting for years at a time. Despite the inevitable religious-ideological element of the conflict that was omnipresent, both parties were constantly vying for control of the strategic buffer zones of Iberia and Lazica, in the Caucasus, and northern Mesopotamia; the Persians were dominant in the south and the Byzantines were more successful in the north, nearer their bases in Asia Minor. Military activity in the region depended on a delicate balance between the Romano-Byzantine use of fortifications and heavy infantry, and the Persian advantage on mobile heavy cavalry forces (*clibanarii*). A key aspect of these wars is that they 'tended toward indecisiveness and often ended in a truce by mutual agreement to avoid fiscal crisis'.[22] What Morillo aptly points out is the correlation between the decisiveness, the strategic aim/outcome and the logistical constraints of a campaign and/or war.

This conflict reached its climax in the first quarter of the seventh century. The Persians proved largely successful during the first stage of the war from 602 to 622, conquering much of the Levant, Egypt and parts of Anatolia, while Heraclius' campaigns in Persian lands from 622 to 626 forced the Persians onto the defensive, allowing his armies to regain momentum that culminated in the famous Battle of Nineveh in 627. It is important to note, however, that until 619 'Rome was on the verge of extinction, and the empire of Darius on the verge of being re-established';[23] this was no longer a conflict regarding regional buffer zones, but rather an operation for the liquidation of the Byzantine Empire. Heraclius followed up the triumph of Nineveh with a direct strike to Ctesiphon that delivered the crushing blow to Khusro's regime, and that resulted in the installation of his son Kavadh as king, who returned all Persian conquests to Roman rule. This extraordinary reversal of fortunes after 622 owes much to the determination, brilliant generalship and leadership, diplomacy and financial management of one man, Heraclius.[24]

Heraclius' achievements were briefly undermined by the advent of a new enemy from the south-east, who stroke a devastating blow against Byzantine arms at Yarmouk in 636. This was one of the rare occasions when the Byzantines allowed themselves to be drawn into a pitched battle, as they had a sound strategic objective: to flush the Arabs out of Syria.[25] The Arabs had an equally sound strategic objective and that was luring as many Byzantine forces as possible into Syria so that they could inflict a decisive defeat.[26] The magnitude of the Arab victory at Yarmouk can be encapsulated not only in the number of fatalities inflicted on the Byzantine armies but, most importantly, in its immediate aftermath:

> They [the Arabs] concentrated on sound military goals, the destruction of the remaining Byzantine forces as organized armies, and only then worried about conquering and organizing rich lands and towns. These actions transformed what was merely a great victory into a very decisive one, and one of the worst of all Byzantine military disasters.[27]

The survival of the empire might not, realistically and with the benefit of hindsight, have been at stake, but Yarmouk did provide a crucial window of opportunity that inevitably led to permanent changes in the balance of power in Syria.

Introduction

Thus a great victory – to use Kaegi's term – could be considered a decisive one according to the strategic aims of the battle-seeking army's commander. A detailed analysis of the Byzantine and Arab strategies and campaigning tactics in Cilicia and Anatolia (eighth to tenth centuries) is the subject of the second chapter in this study, but what I should mention here are the so-called 'basic forms of strategy': annihilation, exhaustion and attrition.[28] A key difference between the strategies of annihilation and exhaustion is the target; the strategy of annihilation targets the enemy's armed forces, while the strategy of exhaustion aims at the people's will to continue fighting. The latter is known in Western medieval history under the French term *chevauchée*, although historians of the eastern Mediterranean would better identify it with the razzia: limited warfare verging on brigandage that avoided head-on confrontations and instead emphasised raiding and looting, usually of livestock.[29] Although both the *chevauchée* and the razzia served the same strategic purposes, the key difference between them was the religious element that was dominant in the Fertile Crescent.

Based on what I have examined so far about the decisiveness of a pitched battle in medieval history and the significance of (1) the impact of the battle, and (2) the strategic aims of the battle-seeking commander, let me return to the perplexing conclusion put forward by Treadgold about the survival of the Byzantine Empire. The Byzantines defended themselves brilliantly, despite their numerous defeats in pitched battles, because rarely was their centre of power – the imperial court – seriously endangered, and no defeat proved as devastating for the fighting capabilities of the Byzantine armed forces as Adrianople (378) or Yarmouk (636).

A battle can be said to be decisive because of its impact, but in no way does this form conclusive evidence of the superiority of one military culture over another: the interpretation of battles as proof of historical superiority, bearing in mind cultural, social, economic and technological factors, is unfounded.[30] The superiority one side may have before the battle certainly provides it with the best chances of prevailing, but that does not consider a hugely influential factor: chance. An accidental arrow, unexpected rainfall, fog or a royal horse running astray on the battlefield could upset the turn of events. This is what Clausewitz calls 'friction':

> The only concept that more or less corresponds to the factors that distinguish real war from war on paper . . . This tremendous friction, which cannot, as in mechanics, be reduced to a few points, is everywhere in contact with chance, and brings about effects that cannot be measured, just because they are largely due to chance.[31]

This factor offers battle the ability to change the balance of power between two forces in a truly chaotic manner, simply because 'small inputs can create very large perturbations'.[32] Thus battles were important because failure to employ correct tactics could have a profound political impact, in a period when national leaders often fought in the front ranks.

During the wars in the East between the Byzantine and the Hamdanid armies that dominated the best part of the tenth century, both sides fought numerous battles and competed for control over an equally large number of castles and mountain and river crossings. At no time, however, did the Byzantine Empire feel threatened for its very existence, as it had some three centuries earlier, since we know that the Arab threat to the capital had already ceased by 718. This, however, does not negate the decisive nature of some of the battles of the period. Clearing out the Muslim outposts west of the Taurus and Anti-Taurus Mountains, pushing the eastern marches of the empire further east to include the strategic cities of Tarsus and Mopsuestia, and neutralising the city of Aleppo by turning it into a buffer zone between the empire and the emerging Fatimids of Egypt, who controlled southern Palestine, were the main strategic objectives of the period. In pursuing an expansionist regional strategy in three consecutive phases that involved a change from a strategy of exhaustion into one of annihilation, with the final phase (963–5) focusing on the conquest of Cilicia despite the logistical challenge, pitched battles assumed an overriding – although not exclusive – role on the socio-political shaping of the region in the second half of the tenth century.[33]

A final concept that needs to be defined and formulated within the context of warfare in medieval Anatolia and the Middle East is 'military culture'. First and foremost, we need to underline the distinction between social structure and culture:

> Social structure comprises the relationships among groups, institutions, and individuals within a society; in contrast, culture (ideas, norms, rituals, codes of behaviour) provides . . . [a] 'web of meaning' shared by members of a particular society or group within a society.[34]

Culture is a system of beliefs and behavioural norms that influence what people think is (morally) right and wrong, how they make judgements and how they categorise things. Therefore, 'military social structure' can be defined as the structure that consists of the arrangement into military groups, such as divisions, battalions and companies; 'military culture', conversely, can be understood as the 'operational code of war' that is

Introduction

followed by an entire nation or people or, to give a more straightforward definition, it could mean 'a way of understanding why an army acts as it does in war'.[35]

As different cultures perceive war differently,[36] the Byzantines had their own military culture that was transmitted to them through the centuries from ancient Greece and classical Rome, partly through the written tradition of what we have already identified as military manuals, also called (*biblia*) *Strategika* or (*biblia*) *Taktika*.[37] These manuals carried on a late Roman tradition of theoretical and practical writing about military organisation, structures and hierarchies, battle tactics and war ethos/ideology, placing war at the epicentre of the imperial foreign policy's repertoire complemented by other tools such as bribery, tricks, diplomacy and so on.

This book, then, is a contribution to the study of the military culture of the Byzantines in the tenth century, comparing it with the received wisdom on warfare from earlier periods (for instance, ancient Greece and Rome) and contrasting it with the contemporary and equally developed and detailed accounts surviving from Byzantium's Muslim neighbours. It focuses on behaviour in war and battle, identifying the norms and expectations of both warriors and others who observed them concerning the conduct of war. It also provides a comprehensive comparison with the military institutions and systems, methods of recruitment and ideology of the Byzantine adversaries in the region during the period in question.

This book does not go into detail concerning the institutional framework of the polities of the eastern Mediterranean region, nor does it break new ground in the logistics of the wars and campaigns of the period; it does not talk about the experience of the common soldier in battle or the impact of war on the border societies of eastern Anatolia and Mesopotamia. This would certainly raise some eyebrows as to why I have deviated from the 'fashionable' (an inappropriate term, still used by some historians, that pertains conformity and adherence to popular norms) narratives of the so-called 'new military history' school that has had a dominant influence in historical output since the 1980s. This school aspires to bring closer together the field of military history and the socio-economic analyses of Karl Marx, in stark contrast to the academic model before the 1950s that largely focused on the *Art of War* and the study of campaigns and battles as models and exemplars of military history.

'New military history' focuses on three main contexts: (1) the political-institutional context that covers the relation between the political and the military institutions within a state and the degree to which an army could be used as an instrument of politics; (2) the socio-economic context, an area that includes the impact of war on societies (economic productivity, logistics, recruitment, technology and so on) and that of societies on war; and (3) the cultural context that encompasses the interaction of warrior values with the cultural values of societies in general (glorification or condemnation of warrior values through epic poems, folk songs and tales and so forth). [38]

Keeping this in mind, I am not disputing the importance of matters such as administration, the institutional framework for warfare, supply systems and logistics, and society during war; after all, it was the vast logistical and economic resources of the Byzantine Empire that brought the Emirate of Aleppo to its knees through a relentless series of campaigns, conducted both in the summer and winter months between 963 and 965.[39] In the same vein, I do not attempt to reshape the contemporary view that requires sieges, raids, skirmishes and ambushes to dominate medieval warfare;[40] historians note that in the aforementioned period between 963 and 965 'most of the military activity [in northern Syria] took the form of sieges and pillaging rather than pitched battle',[41] because Aleppo could no longer offer any significant resistance to the war of exhaustion that had been waged against it since 959.

As mentioned already, this is a study with a very focused thematic (campaign strategy, comparative tactics, raids, siege warfare), geographic (Armenia, Cilicia, Syria) and chronological (tenth century) scope. It is my intention to reintegrate the operational, tactical, technical and equipment aspects of the conduct of warfare, whilst incorporating into the discussion its impact on wider society. This is because, regardless of whether battles are trustworthy or untrustworthy assessments of historical entities and movements, they are rare events and they form the ultimate 'Darwinian test' for two sides facing each other in a frenzied and violent interaction that would provide history with a winner. 'For it is not through what armies *are* but by what they *do* that the lives of nations and of individuals are changed',[42] especially considering that this was a period when the emperor or the emir were at the forefront of fighting and their units often bore the brunt of an enemy attack.

An Overview of the Book

The purpose of Chapter 1 is to provide a condensed view of the 'grand strategy' of the Byzantine Empire and the different perceptions of strategy and tactics present in the military manuals of the ancient Greek and Roman authors of *strategika* like Aeneas Tacticus (writing in 357–6 BC),

Introduction

Onasander (writing in AD 59), Aelian (writing in AD 106), Polyaenus (writing in AD 163–5) and the anonymous author of the *Strategikon* attributed to Emperor Maurice (probably written in AD 591–610). I ask in what way the Byzantine theorists and generals of the period up to the tenth century established their theoretical and practical ideas about warfare, and in what way can the Byzantine emperors be said to have pursued a 'grand strategy'?

Next, I compare the notion of 'cultures of bravery' (and therefore of cowardice) in the Byzantine, Western European and Islamic worlds in an attempt to show the degree to which the acceptability of, for instance, feigned flights, other sorts of ruses, ambushes and so on varies between cultures. For some, such tactics were indeed construed as unmanly and as signs of cowardice, since for them bravery was constructed around notions of how one fought, with the 'how' usually centred on the honour to be gained in face-to-face combat with melee weapons. For others, such tactics were signs of cleverness – bravery and manliness having been constructed more around whether one won a battle than how one fought it. This will show how willing different cultures were to adopt military tactics and strategies from their enemies.

Chapter 2 focuses on a different set of questions: what is the kind of warfare that dominated the geographical area under consideration and what does it reveal about the strategy and strategic goals of the Arabs in the region? In view of the wider debate between modern scholars like C. J. Rogers, J. Gillingham and S. Morillo about the term 'Vegetian strategy', I will ask whether historians can characterise any of the strategies applied by the Arabs and the Byzantines in the operational theatres of the East leading up to the 960s as 'Vegetian'? What key role did the three Muslim bastions for razzias (Tarsus, Melitene and Theodosiopolis) play in this conflict? How did the regional geography of eastern and central Anatolia shape the kind of warfare that was waged in the region? What were the invasion routes taken by the Arab raiding parties that led them over the Taurus and Anti-Taurus into Anatolia?

The fact that these three Arab bases on the frontiers of the Christian–Muslim world in Anatolia were such a thorn in the side of the empire would be made abundantly clear. These razzias over the Taurus and Anti-Taurus contributed to the political instability, militarisation, and economic, commercial, agricultural and demographic decline of central and eastern Asia Minor. The breaking of the power of these emirates and, in effect, their neutralisation would be at the top of the priorities of the governments of Romanus I Lecapenus, Constantine VII and their successors, and it would dominate their foreign policy from the 920s onwards.

The main aim of Chapters 3 and 4 is to bring the imperial policy in the East into the spotlight and provide some perspective on the empire's frontier wars against its Muslim enemies, which reached a climax in the middle of the tenth century. Few studies have been produced about the Byzantine Empire's foreign policy in Armenia and northern Mesopotamia during the reign of the first three emperors of the Macedonian dynasty, even though the empire considered this region as one of its most significant and fragile territories that required careful diplomatic negotiations and the show of brute force to prevent it from falling under the sphere of influence of the Abbasid Caliphate. Hence, Chapter 3 will consider several interrelated aspects of Byzantium's strategic management of the eastern frontier regions: the political reasons behind its involvement into this operational theatre; the wars with the Muslims; the emperors' delicate diplomatic negotiations with the Armenian princes; the imperial campaigns in the Armenian and Mesopotamian frontiers and the emergence of a new enemy – the brothers Nasir and Sayf ad-Dawla.

There is, however, a paradox in Byzantium's expansionist wars in the first half of the tenth century: in none of the cases were the Byzantines contemplating any kind of permanent territorial expansion – these remained just raids to capture prisoners that would later be exchanged for ransom money and to enhance the emperor's influence and popularity. If that was the case, then why did the empire decide to crush the dynasty of the Hamdanids of Aleppo in a series of expansionist wars that placed huge sums of money and materiel at the disposal of the imperial generals in the East? What were the attractions that drew Byzantium and the Hamdanid emir to the region of Armenia, Taron, Vaspourakan and northern Mesopotamia? Was it considered to be a fight to the bitter end for Byzantium and the Hamdanids of Aleppo? These are some of the issues that I attempt to formulate in the second part of Chapter 3.

In Chapter 4, I wish to concentrate on the methods of exchange of information and intelligence between cultures. In this case I focus on the Byzantine view of their enemies on the battlefield – the Arabs: 'What are they really like? What weapons do they make use of in military campaigns? What are their practices? How does one arm oneself and campaign against them and thus carry out operations against them?' These are the kind of questions Leo VI was asking about the Arabs.

What I have been able to discern by looking through lay and ecclesiastical sources of the period provided me with a rather contradictory view of the Arabs as warriors. Leo the Deacon offers the most stereotypic view of the Arabs as warriors of jihad: they are an enemy unsophisticated

Introduction

in warfare and technologically inferior. This notion of an opportunistic soldier can also be found in Ioannes Kaminiates' history of the siege and conquest of Thessaloniki in 904. Kaminiates despised them, but he was also highly impressed by their fighting abilities: they were intelligent, furiously resilient, with high morale and ready to die for their cause. Leo VI's *Taktika* reveals a feeling of respect for the empire's Muslim adversaries on the battlefield. He describes them as formidable enemies who excel all foreign nations in intelligence and who have adopted Roman weapons and often copied Roman tactics.

The most striking difference of opinion can be found in the works of Emperor Constantine VII. In his oration to his troops in the autumn of 950 he portrays the Arabs of Aleppo as feeble women, while in his magisterial manual on kingcraft he highlights their prowess in battle and the splendour of their weapons and armour. Constantine VII's oration is a good example of Byzantine propaganda literature. A small victory against the Hamdanids was exploited for propaganda purposes rather than for its real strategic value. The purpose was simple: to restore some much-needed prestige to the regime of Constantine Porphyrogenitus after the humiliation of the Cretan expedition the year before. His true thoughts about the Arabs as warriors can be seen in his *De Administrando Imperio* and they are much more pragmatic. This is also the case for the military treatises of the period, such as the *Praecepta Militaria*; they paint a clearer and more refined picture of their enemies as ingenious and brave soldiers capable of injecting fear and confusion into their adversaries, and likely to stand their ground and even fight a losing battle rather than strike camp and retreat. But it is the Byzantines who have the moral high ground; it is they who will achieve eternal glory with the help of God.

Crossing over to the methods of transmission of (military) knowledge between cultures, the focal point of Chapter 5 is intelligence and the different methods of procuring accurate and reliable information, which could have a potentially lethal effect on the battlefield. In this section, I concentrate on a series of questions: what types of intelligence does a modern general have at his disposal before a great battle and what did a medieval general have to cope with in order to obtain all the necessary information that would shape his strategy? Why is intelligence a fundamental aspect of warfare in any period and what was its role in medieval Islamic and Christian armies?

Knowing as much as possible about your enemy is paramount for the successful outcome of a battle or even a war. Some fundamental questions include the state of the sovereign's army, its numbers, armament and morale/loyalty, and how many (and which) officers can be found in each fortress and town in the border areas – obviously hinting at their battle worthiness and loyalty to the regime. Another crucial question is the state of the country's economy, thus considering whether an invasion army can be logistically supported, if there are rich pickings to be had by an invasion or if there is any public feeling of discontent against the central government to be taken advantage of.

Intelligence on the enemy could be acquired in two ways: by reconnaissance (or tactical intelligence), where a commander openly sent scouts (either light infantry, cavalry or swift scouting ships) to observe the enemy army and collect information about its numbers, composition and the general's intentions, or by espionage, where disguised or covert agents operated in secret in enemy territory collecting information about the enemy. I ask what do the primary sources tell us about intelligence in this period and, more specifically, how could information be passed on or procured in the interests of war? Chapter 6 follows with an investigation of the role of espionage in Byzantine foreign relations, and the official and unofficial channels that provided the Byzantines with the necessary information that shaped their foreign policy.

Chapters 7 and 8 of this study focus on the battlefield deployment of the Byzantine armies in the tenth century, and the changes and adaptations that took place according to the military manuals of the period. As the Byzantines encountered many enemies in different operational theatres of war during their long history, their numerous military treatises amply illustrate their willingness not only to produce works that describe in detail the fighting habits and customs of their enemies, but also to learn from them and adapt their methods of war to those of their opponents. With the study of war energetically renewed in tenth-century Byzantium, the number of important manuscripts and texts dating from this period also proliferated – there are six extant treatises on tactics that date from this century. It has always been a difficult problem for historians, however, to establish whether the theory of tactical change in the Byzantine army of the tenth century, as described by the contemporary military manuals, translated into practice on the battlefields of Cilicia and Syria.

The most useful primary sources for identifying these changes are the military treatises, such as the *Praecepta Militaria* (*Military Precepts*) of Nicephorus Phocas (c. 969) or the *Sylloge Taktikorum* (*Collection of Tactics*) by an anonymous author (c. 930), which among others provide crucial information on how armies should have been organised and deployed on the battlefield up to the period when they were compiled.

Introduction

I discuss the recommendations of the authors of the treatises regarding the marching, battle formations, armament and battlefield tactics of the Byzantine cavalry units, and I ask whether they reflect any kind of innovation or tactical adaptation to the strategic situation in the East. This section is immediately followed by Chapters 9 and 10 that scrutinise the evidence of innovation and adaptation found in the contemporary historical sources in regard to the battles between the Byzantines and the Arabs in the East during the better part of the tenth century.

The objective of Chapter 9 is to examine the most detailed primary sources for the period of the Byzantine expansion in the tenth century. These include two Byzantine sources, namely Leo the Deacon and John Skylitzes, whose accounts of the Byzantine wars in the Balkans are considered the best and most detailed that modern historians have to hand, a local Syriac one, Yahya ibn Said al-Antaki from Antioch, and three Muslim, al-Mutanabbi, Abu Firas and Ibn Zafir, who provide us with invaluable information about the Byzantine–Arab conflicts of the 940s–60s in Cilicia, Syria and northern Mesopotamia. I direct my attention to the chroniclers' social, religious and educational backgrounds, the dates and places of the compilation of their works, their own sources and the ways they collected information from them, their biases and sympathies and, thus, the extent of their impartiality as historians. This section is followed by a comparative analysis of the aforementioned sources strictly from a military perspective, reaching significant conclusions regarding their value as 'military historians'.

Finally, to determine whether theory translated into practice, in Chapter 10 I examine the largest and most important campaigns and pitched battles of this period: Hadath (954), Tarsus (965), Dorystolo (971), Alexandretta (971), Apamea (994) and Orontes (998), through the accounts of contemporary lay and ecclesiastical sources. In the process, I give answers to such questions as: How successful were the Byzantines at adapting to the changing military threats posed by their enemies in the East? How far can we see the Byzantines responding to the tactical and strategic threats of enemies in ways not anticipated by the manuals? These questions give me the opportunity to discern the place of literacy in the Byzantine military command structure and the training of the officer class, and to reach some conclusions on the question of professionalism in the Byzantine army. This will be coupled with the whole idea of 'adaptability' to the new and innovative elements of warfare in the East, shedding light on the (military) culture, ideology and – once again – the level of professionalism in the imperial armies of the tenth century.

Notes

1. The best study on the Byzantine Empire's foreign policy of the period remains: S. Runciman (1929), *The Emperor Romanus Lecapenus and his Reign: A Study of Tenth-Century Byzantium*, Cambridge: Cambridge University Press.
2. M. Canard (1953), *Histoire de la dynastie des Hamdanides*, Paris: Presses Universitaires de France. See also M. Canard (1948), 'Les Hamdanides et l'Armenie', *Annales de l'Institut d'études orientales d'Alger*, 7, pp. 77–94.
3. R. J. Bikhazi (1981), 'The Hamdanid Dynasty of Mesopotamia and North Syria 868–1014', PhD dissertation, University of Michigan. For the history of Aleppo at a later period, see J. S. Nielsen (1991), 'Between Arab and Turk: Aleppo from the 11th till the 13th centuries', *Byzantinische Forschungen*, 16, pp. 323–40; R. Burns (2017), *Aleppo: A History*, London: Routledge.
4. A. Asa Eger (2015), *The Islamic–Byzantine Frontier*, London: I. B. Tauris; M. D. Bonner (2006), *Jihad in Islamic History, Doctrines and Practice*, Princeton, NJ: Princeton University Press, especially chapters 6–8; M. D. Bonner (1996), *Aristocratic Violence and Holy War: Studies in the Jihad and the Arab–Byzantine Frontier*, New Haven, CT: American Oriental Society; M. D. Bonner (1994), 'The Naming of the Frontier: Awāṣim, Thughūr, and the Arab Geographers', *Bulletin of the School of Oriental and African Studies*, 57, pp. 17–24; M. D. Bonner (1987), 'The Emergence of the *Thugur*: The Arab–Byzantine Frontier in the Early Abbasid Age', PhD dissertation, Princeton University; C. E. Bosworth (1992), 'The City of Tarsus and the Arab–Byzantine Frontiers in Early and Middle Abbasid Times', *Oriens*, 33, pp. 268–86; C. Holmes (2002), 'The Byzantine Eastern Frontier in the Tenth and Eleventh Centuries', in D. Abulafia and N. Berend (eds), *Medieval Frontiers: Concepts and Practices*, Aldershot: Ashgate, pp. 83–104; J. F. Haldon and H. Kennedy (1980), 'The Arab–Byzantine Frontier in the Eighth and Ninth Centuries: Military Organization and Society in the Borderlands', *Zbornik Radova Vizantoloski Institut*, 19, pp. 79–116.
5. J. Shepard (2002), 'Emperors and Expansionism: From Rome to Middle Byzantium', in D. Abulafia and N. Berend (eds), *Medieval Frontiers: Concepts and Practices*, Aldershot: Ashgate, pp. 55–82; J. Shepard (2001), 'Constantine VII, Caucasian Openings and the Road to Aleppo', in A. Eastmond (ed.), *Eastern Approaches to Byzantium*, Aldershot: Ashgate, pp. 19–40.
6. W. Treadgold (1995), *Byzantium and its Army, 284–1081*, Stanford, CA: Stanford University Press; J. F. Haldon (1999), *Warfare, State and Society in the Byzantine World, 565–1204*, London: UCL Press.
7. E. McGeer (1995), *Sowing the Dragon's Teeth: Byzantine Warfare in the Tenth Century*, Washington, DC: Dumbarton Oaks.
8. Y. Lev (1997), *War and Society in the Eastern Mediterranean, 7th–15th Centuries*, Leiden: Brill; Y. Lev (1991), *State and Society in Fatimid Egypt*, Leiden: Brill; Y. Lev (2006), 'Infantry in Muslim Armies during

the Crusades', in J. H. Pryor (ed.), *Logistics of Warfare in the Age of the Crusades*, Aldershot: Ashgate, pp. 185–207; Y. Lev (1988), 'The Fatimids and Egypt 301–358/914–969', *Arabica*, 35, pp. 186–96; Y. Lev (1987), 'Army, Regime, and Society in Fatimid Egypt, 358–487/968–1094', *Journal of Middle East Studies*, 19, pp. 337–65; Y. Lev (1982), 'The Fatimids and the *Aḥdāth* of Damascus 386/996–411/1021', *Die Welt des Orients*, 13, pp. 97–106; Y. Lev (1980), 'The Fatimid Army, A.H. 358–427/968–1036 C.E., Military and Social Aspects', *Asian and African Studies*, 14, pp. 165–92. Also of direct relevance is J. L. Bacharach (1981), 'African Military Slaves in the Medieval Middle East: The Cases of Iraq (869–955) and Egypt (868–1171)', *International Journal of Middle East Studies*, 13, pp. 471–95.

9. H. Kennedy (2001), *The Armies of the Caliphs: Military and Society in the Early Islamic State Warfare and History*, London: Routledge.
10. B. J. Beshir (1978), 'Fatimid Military Organization', *Islam*, 55, pp. 37–56.
11. W. J. Hamblin (1985), 'The Fatimid Army during the Early Crusades', PhD dissertation, University of Michigan.
12. C. E. Bosworth (1965–6), 'Military Organization under the Buyids of Persia and Iraq', *Oriens*, 18–19, pp. 143–67; C. E. Bosworth (1961), 'Ghaznavid Military Organization', *Der Islam*, 36, pp. 37–77.
13. V. J. Parry and M. E. Yapp (eds) (1975), *War, Technology and Society in the Middle East*, London: Oxford University Press.
14. D. Nicolle (2007), *Crusader Warfare, Muslims, Mongols and the Struggle against the Crusades*, London: Hambledon Continuum; D. Nicolle (1993), *Armies of the Muslim Conquests*, London: Osprey.
15. J. Gillingham (1989), 'William the Bastard at War', in C. Harper-Hill et al. (eds), *Studies in Medieval History Presented to R. Allen Brown*, Woodbridge: Boydell & Brewer, pp. 141–58; J. Gillingham (1984), 'Richard I and the Science of War in the Middle Ages', in J. Gillingham and J. C. Holt (eds), *War and Government in the Middle Ages*, Woodbridge: Boydell & Brewer, pp. 78–91; J. Gillingham (2004), '"Up with Orthodoxy!" In Defence of Vegetian Warfare', *Journal of Medieval Military History*, 2, pp. 149–58; S. Morillo (2003), 'Battle Seeking: The Context and Limits of Vegetian Strategy', *Journal of Medieval Military History*, 1, pp. 21–41; C. J. Rogers (2003), 'The Vegetian "Science of Warfare" in the Middle Ages', *Journal of Medieval Military History*, 1, pp. 1–19.
16. For a comprehensive explanation of the distinctions between transcultural (intercultural and subcultural) and intracultural warfare, and cross-cultural norms as prerequisites for the use of Vegetian strategy in war, see S. Morillo (2006), 'A General Typology of Transcultural Wars: The Early Middle Ages and Beyond', in H. H. Kortüm (ed.), *Transcultural Wars: From the Middle Ages to the 21st Century*, Berlin: Akademie Verlag, pp. 29–42. See also R. Cox (2012), 'Asymmetric Warfare and Military Conduct in the Middle Ages', *Journal of Medieval History*, 38, pp. 100–25.

17. Herodotus, *History*, 7.9–10; Morillo, 'General Typology of Transcultural Wars', pp. 33–4.
18. Morillo, 'Battle Seeking', pp. 23–8; R. Murphy (1999), *Ottoman Warfare, 1500–1700*, London: UCL Press, ch. 2 and pp. 127–9.
19. W. Treadgold (2006), 'Byzantium, the Reluctant Warrior', in N. Christie and M. Yazigi (eds), *Noble Ideals and Bloody Realities: Warfare in the Middle Ages, 378–1492*, Leiden: Brill, p. 215.
20. J. F. Haldon (2016), *The Empire That Would Not Die: The Paradox of Eastern Roman Survival, 640–740*, Cambridge, MA: Harvard University Press; J. F. Haldon (1990), *Byzantium in the Seventh Century*, Cambridge: Cambridge University Press; M. Whittow (1996), *The Making of Byzantium, 600–1025*, Los Angeles: University of California Press, especially ch. 5.
21. S. Morillo (1996), *The Battle of Hastings: Sources and Interpretations*, Woodbridge: Boydell & Brewer, pp. xv–xx; J. Keegan (1993), *A History of Warfare*, New York: Vintage, p. 67; E. Shepherd Creasy (1863), *The Fifteen Decisive Battles of the World: From Marathon to Waterloo*, New York: A. L. Burt, p. iv. See also V. D. Hanson (1989), *The Western Way of War: Infantry Battle in Classical Greece*, New York: A. Knopf; V. D. Hanson (2001), *Carnage and Culture: Landmark Battles in the Rise of Western Power*, New York: First Anchor; J. Dahmus (1983), *Seven Decisive Battles of the Middle Ages*, Chicago, IL: Nelson-Hall; J. MacDonald (1988), *Great Battlefields of the World*, New York: Michael Joseph; R. A. Gabriel and D. W. Boose (1994), *The Great Battles of Antiquity: A Strategic and Tactical Guide to Great Battles that Shaped the Development of War*, Westport, CT: Greenwood.
22. S. Morillo, J. Black and P. Lococo (2009), *War in World History: Society, Technology and War from Ancient Times to the Present*, 2 vols, New York: McGraw-Hill Education, I, p. 142. See also M. H. Dodgeon and S. N. C. Lieu (eds) (2002), *The Roman Eastern Frontier and the Persian Wars, AD 226–363: A Documentary History*, London: Routledge; G. Greatrex and S. N. C. Lieu (eds) (2002), *The Roman Eastern Frontier and the Persian Wars, AD 363–628: A Narrative Sourcebook*, London: Routledge.
23. Morillo et al., *War in World History*, I, p. 142.
24. J. Howard-Johnston (1999), 'Heraclius' Persian Campaigns and the Revival of the East Roman Empire, 622–630', *War in History*, 6, pp. 1–44.
25. W. E. Kaegi (2000), *Byzantium and the Early Islamic Conquests*, Cambridge: Cambridge University Press, p. 122.
26. Kaegi, *Early Islamic Conquests*, p. 128.
27. Ibid., p. 140.
28. C. J. Rogers (2006), 'Strategy, Operational Design, and Tactics', in J. C. Bradford (ed.), *International Encyclopaedia of Military History*, New York: Routledge.
29. H. J. Hewitt (1966), *The Organization of War under Edward III, 1338–62*, Manchester: Manchester University Press, p. 99; C. Allmand (1988), *The Hundred Years War*, Cambridge: Cambridge University Press, pp. 54–5.

Introduction

Rogers has argued strongly that the *chevauchée* is a battle-seeking rather than a battle-avoiding strategy, basing his assumptions on the meticulous analysis of Edward III's 1346 campaign in northern France that culminated in his victory at Crecy. See C. J. Rogers (1999), 'Edward III and the Dialectics of Strategy, 1327–1360', in C. J. Rogers (ed.), *The Wars of Edward III: Sources and Interpretations*, Woodbridge: Boydell & Brewer, pp. 265–84; C. J. Rogers (2000), *War Cruel and Sharp: English Strategy under Edward III, 1327–1360*, Woodbridge: Boydell & Brewer, pp. 238–72. As I argue in the second chapter of this study, a campaign of this type – whether we choose to call it a *chevauchée* or a razzia – could be either battle-avoiding or battle-seeking, depending on the campaign strategy of the invading force.

30. See the very interesting discussion in Y. N. Harari (2007), 'The Concept of "Decisive Battles" in World History', *Journal of World History*, 18, pp. 251–66.
31. C. von Clausewitz (1984), *On War*, ed. and trans. M. Howard and P. Paret, Princeton, NJ: Princeton University Press, pp. 119–20.
32. Morillo, *Battle of Hastings*, pp. xix–xx; G. A. Reisch (1991), 'Chaos, History, and Narrative', *History and Theory*, 30, pp. 1–20.
33. W. Garrood (2008), 'The Byzantine Conquest of Cilicia and the Hamdanids of Aleppo, 959–965', *Anatolian Studies*, 58, pp. 127–40.
34. *The U.S. Army & Marine Corps Counterinsurgency Field Manual, U.S. Army Field Manual No. 3–24, Marine Corps Warfighting Publication No. 3–33.5* (2007), Chicago, IL: University of Chicago Press, pp. 85–94. For an analysis of the terms 'military culture' and 'militarism' and their distinctive features, see P. H. Wilson (2008), 'Defining Military Culture', *Journal of Military History*, 72, pp. 11–41; S. Morillo (2013), 'Justifications, Theories and Customs of War', in D. Graff (ed.), *The Cambridge History of War*, Cambridge: Cambridge University Press, pp. 1–24.
35. I. V. Hull (2005), *Absolute Destruction, Military Culture and the Practices of War in Imperial Germany*, Ithaca, NY: Cornell University Press, pp. 93–4; J. L. Soeters, D. J. Winslow and A. Wibull (2003), 'Military Culture', in G. Caforio (ed.), *Handbook of the Sociology of the Military*, New York: Springer, pp. 237–54.
36. For the perception of war in different cultures, see B. Heuser (2010), *The Evolution of Strategy: Thinking War from Antiquity to the Present*, Cambridge: Cambridge University Press, pp. 18–20; M. van Creveld (1991), *The Transformation of War*, London: Free Press, chs 1, 2 and 4; M. van Creveld (2008), *The Culture of War*, New York: History Press.
37. G. Theotokis (2014), 'From Ancient Greece to Byzantium: Strategic Innovation or Continuity of Military Thinking?', in B. Kukjalko, I. Rūmniece and O. Lāms (eds), *Antiquitas Viva 4: Studia Classica*, Riga: University of Latvia Press, pp. 106–18.
38. S. Morillo (2013), *What is Military History?*, Cambridge: Cambridge University Press, p. 4; Morillo et al., *War in World History*, I, p. x. See also the

collection of essays in H. H. Kortüm (ed.) (2006), *Transcultural Wars: From the Middle Ages to the 21st Century*, Berlin: Akademie Verlag, especially part 2.
39. Explained in detail in Garrood, 'Byzantine Conquest of Cilicia', pp. 127–40. See also M. Whittow (1996), *The Making of Orthodox Byzantium, 600–1025*, New Studies in Medieval History, London: Palgrave, pp. 310–34.
40. The bibliography on this view is vast but an exhaustive footnote can be found in Harari, 'Decisive Battles', p. 253 n. 6.
41. Garrood, 'Byzantine Conquest of Cilicia', p. 135.
42. J. Keegan (2004), *The Face of Battle: A Study of Agincourt, Waterloo and the Somme*, London: Pimlico, p. 18.

1

The 'Grand Strategy' of the Byzantine Empire

> Strategy is how good commanders put their military training
> into practice, their drilling with stratagems, and putting
> together ways of defeating [the enemy].[1]

The Meaning of the Terms Strategy, Tactics and Stratagems in the Pre-modern World

According to one of the greatest theorists on warfare of modern times, Carl von Clausewitz (1780–1830), the conduct of war consists of the planning and organisation of fighting in a greater or lesser number of single acts, each complete in itself, identified by the term 'engagements'. We thus arrive at the distinction between 'the use of armed forces in the engagement', identified as tactics, and 'the use of engagements for the object of war', defined as strategy.[2] In 1814, the Archduke Charles (1771–1847), the Habsburg commander in the wars against Napoleon, defined *strategy* as the 'science of war: it designs the plan, circumscribes and determines the development of military operations; it is the particular science of the military commander'. He defined *tactics*, however, as 'the art of war': 'It teaches the way in which strategic designs are to be executed; it is the necessary skill of each leader of troops.'[3] Therefore, strategy ends and tactics begin where opposing forces clash – on the battlefield.[4] Although this distinction between the two meanings of the conduct of war may seem reasonable to a modern expert, as 'it is now almost universal',[5] most authors before the French Revolution wrote about neither strategy nor tactics but about military matters in the tradition of the Roman author Flavius Vegetius Renatus (writing c. AD 400), or about the 'art of war' like Machiavelli did eleven centuries later.[6]

The term strategy (στρατηγεία or στρατηγική) had a different meaning in ancient Greece. It derives from the noun *strategos* (στρατηγός) and meant the office or the skills of a general in command of an army during the war.[7] In order to understand how different the meaning of strategy was from that of tactics in ancient Greece, however, we must read Xenophon's

Cyropaedia, where the author launches an attack on the narrow tactical conception of the military education of his time, making the case that tactics represent only a small part of generalship.[8] When Xenophon imagines Cyrus asking his father to pay for a tutor who had promised to teach him generalship, Cyrus' father retorted: 'The man to whom you are taking the pay has given you instruction in domestic economy as a part of the duties of a general, has he not?' As Cyrus replied that he had not received any such instructions, his father insisted by asking whether 'he [the tutor] had said anything to [Cyrus] about health or strength, inasmuch as it would be requisite for the general to take thought for these matters as well as for the conduct of his campaign'. Receiving another negative reply, Cyrus' father finally asked his son if his tutor

> had put [Cyrus] through any training so that [Cyrus] might be able to inspire [his] soldiers with enthusiasm ... And when this also appeared not to have been discussed at all, [the father] finally asked [Cyrus] what in the world he [the tutor] had been teaching [Cyrus] that he professed to have been teaching ... generalship (στρατηγίαν). And thereupon [Cyrus] answered, 'tactics' (τακτικά). And [his father] laughed and went through it all, explaining point by point, as [he] asked of what conceivable use tactics could be to an army, without provisions and health, and of what use it could be without the knowledge of the arts invented for warfare and without obedience.[9]

The first attempt to make a formal distinction between the two meanings of strategy and tactics is found in the military treatise Περί Στρατηγίας (*On Strategy*), no longer believed to have been compiled by an anonymous author but rather belonging to the compendium of Syrianus Magistrus written, most likely but not conclusively, sometime in the third quarter of the ninth century.[10] According to the anonymous author, 'Strategy (στρατηγική) is the means by which a commander may defend his own lands and defeat his enemies. The general is the one who practises strategy.'[11] Indeed, the author goes as far as to differentiate between two kinds of strategy, the defensive by which the general acts to protect his own people and their property, and the offensive by which he retaliates against his opponents. It is this particular division of the meaning of strategy into two categories – offensive and defensive – that will play a key role in the following chapters, where I present the strategic thinking of the Byzantine Empire in the eastern theatre of war. Conversely,

> tactics (τακτική) is a science which enables one to organize and manoeuvre a body of armed men in an orderly manner. They may be divided into four

parts: (a) proper organization of men for combat, (b) distribution of weapons according to the needs of each man, (c) movement of an armed body of troops in a manner appropriate to the occasion, and (d) the management of war, of personnel and materials, including an examination of ways and causes as well as of what is advantageous.[12]

The terms strategy and tactics have not been systematically distinguished in the military treatises of previous centuries, including the works of Aeneas Tacticus (writing in 357–6 BC), Onasander (writing in AD 59), Aelian (writing in AD 106), Polyaenus (writing in AD 163–5) and the anonymous author of the *Strategikon* (writing probably in AD 591–610). Perhaps the sole exception is the work *Strategemata* by the Roman Sextus Julius Frontinus (end of first century AD), which is a collection of *strategemata* – acts of a general, tricks or devices – drawn from the author's personal experience on the battlefields of the period, but also taken from the ancient Greek and Roman historiography and oral tradition.[13] Frontinus clearly differentiates between the two terms, strategy and tactics, 'which are by nature extremely similar', already in the preface of his work. He also uses the Greek words since none of the terms has an exact equivalent in Latin. Hence, everything achieved by a commander, be it characterised by foresight, advantage, enterprise or resolution, belongs under the heading of 'strategy' (στρατηγικά), while those deeds, particularly the successful ones, employed under a special type of 'strategy' are called 'stratagems' (στρατηγήματα).[14]

Emperor Leo VI 'the Wise' (886–912), drawing extensively on the works of ancient authors like Onasander, Aelian, Polyaenus and, of course, the author of the *Strategikon*, whom he paraphrased in several places in his constitutions, used the terms strategy and tactics in the same hierarchical way as the ninth-century treatise *On Strategy*: 'Strategy is how good commanders put their military training into practice, their drilling with stratagems, and putting together ways of defeating [the enemy]', while 'tactics is the science of movement in warfare . . . Tactics is the military skill [that is concerned with] battle formations, armament, and troops movements. Its aim is to defeat the enemy by all possible means of assaults and actions.'[15] Therefore, the Byzantine theorists and generals of the period up to the tenth century had well-established theoretical and practical ideas about warfare – from strategies and battle tactics, to army formations, armament, logistical organisation and tricks to deceive the enemy, all of which were coupled with the intimate notion of a Christian empire where nothing could be achieved without God's favour.[16]

Byzantine Strategic Thinking and the Factors that Shaped It

Focusing on the concept of strategy and its evolution during the centuries, the main elements of the political ideology that shaped the empire's foreign policy were (1) the defence of the Christian Roman *Oikumene*,[17] (2) the propagation of the true faith among the infidels[18] and (3) the recovery of former imperial lands.[19] This ideology, expressed in theological and even apocalyptic terms,[20] assumed the status of a world order and its realisation became the destiny of Byzantine emperors.[21] They considered themselves the successors of the Roman *Domini* and the protectors of the Orthodox Christian faith against heretics (Arians, Paulicians), the Persian Zoroastrians and the Muslims:

> Be well aware, therefore, o' general, that it is not you alone who ought to be a serious promoter and lover of the fatherland (σπουδαίος και φιλών την πατρίδα) and defender of the correct faith of Christians, ready, if it so transpires, to lay down your very life, but also all the officers under your command and the entire body of soldiers should be ready to do the same. May those who share the same noble [ideal] remain such.[22]

The wars of the *Reconquest*[23] by Justinian in the West harboured the notion of the restoration of the Roman *Oikumene*, with Procopius writing in his *De bello Gothico*: 'Γότθοι Ιταλίαν την ημετέραν βία ελόντες ουχ όσον αυτήν αποδιδόναι ουδαμή έγνωσαν' (The Goths, having seized by violence *our* Italy, they refuse to give it back).[24] Next, there is Heraclius' recovery of Syria, Palestine and Egypt, and even though these series of campaigns in the 620s have been portrayed as a sort of 'proto-crusade' due to the religious enthusiasm infused upon the soldiers and people of the empire by the government's propaganda, this war was, indeed, a defensive one;[25] with the balance of resources and military manpower shifting dramatically in favour of the Persians, it is not an exaggeration to say that the empire was fighting for its very existence.[26] Finally, whether the emperors of the first half of the tenth century and their immediate predecessors in the second half of the ninth century had any grand plans to conquer the Muslim lands of Syria, Palestine and northern Mesopotamia, or whether they were simply interested in stopping the Muslim raids from Melitene, Tarsos and Crete that were a serious threat to the socio-economic stability of Anatolia and the Aegean, will be examined in detail in the following chapters of this study.

If the Byzantine emperors did pursue a grand strategy, what form did it take? For the emperors and high officials there was no succinct concept of 'grand strategy', at least not in a way scholars would have understood it in the twentieth century,[27] but rather a reaction to the socio-political events in the world that surrounded the empire – a sort of 'crisis management on a grand scale' as very aptly put by Haldon.[28] Yet we can identify the basic strategic considerations (or factors) that determined the empire's strategic thinking and planning. These interrelated factors were the following:[29] (1) the position of the empire in the wider geostrategic context of the Balkans, Asia Minor and the Middle East; (2) the economy and manpower of Byzantium in relation to warfare; (3) several cultural approaches that affected the attitude of the Byzantines towards warfare. To begin with, the strategic position of the empire played a prominent role in its military organisation and the shaping of its attitude towards its neighbours and warfare in general. In order to understand its history and strategic thinking, one must realise the geopolitical significance of Asia Minor and, especially, Constantinople to the wider region of the eastern Mediterranean.[30] With its capital situated at the crossroads of Asia and Europe, the Byzantine Empire controlled vast territories in the Balkans and Asia Minor and the Middle East. Inevitably, it had to face different enemies – not only from outside the empire but internally as well in terms of religion, economy and socio-political and military organisation – in two geographical areas that were culturally and otherwise as far apart as they could be.

Different geographical areas meant different operational theatres of war for the Byzantine armed forces and, in certain cases, distinct organisation of the *themata* (themes) and varying distribution of troops.[31] Hence, there is a clear correlation between the two major operational theatres – the Balkans and Asia Minor – of the Byzantine army, the resources that were available in each area at any given time and the ability of the empire to move adequate reinforcements from one place to another to deal with an external threat. The sad reality that the emperors in Constantinople had to face was that the limited resources in money and manpower constituted the waging of war in more than one theatre an almost inconceivable prospect.[32]

A number of factors raised Asia Minor to the top of the strategic priorities of the emperors, including the region's strategic importance and proximity to the capital, its economic importance for the state due to its rich resources in agriculture and minerals, and the high number of recruits for the army.[33] It was only when the East had been pacified that the Byzantines were able to concentrate their efforts in the Balkans (it is to this reason, for example, that we can attribute the success of the Bulgars in establishing themselves south of the Danube in the last two decades of the seventh century), while any kind of insurrection or military activity in Bulgaria or Macedonia could inhibit or postpone military operations in the East.[34]

Up until the final two centuries of the empire's existence, the largest share of the state's income came from agrarian production. Trade might have made an important contribution to a state's coffers, but only agriculture could provide the resources to support a major political power. Resources flowed into Constantinople through a comprehensive system of land taxation, while customs duties and several other taxes were imposed on commercial transactions. In addition, imperial estates were another source of revenue for the government, critical for the land reforms of the late ninth and late eleventh centuries (*pronoia* system). Since a fundamental principle of the Byzantine state was to maximise its revenues, this entailed the intensification of the exploitation of the land. Until the end of the eleventh century, the armed forces must have been the largest employer in the empire, numbering some 645,000 men and sailors in the fourth century according to the historian Zosimus, while recent estimates put the total number of soldiers in the year of Basil II's death (1025) to 247,800.[35] Since the mid-eighth century, a distinction had been made between the full-time soldiers of the *tagmata* units that were maintained by the state with salaries (*rogai*), and the part-time (in social and economic terms, not in fighting ability) soldiers of the *themata* who drew their income from the military lands. This mass of people and their dependants (families, servants and people who traded with them in the local communities) naturally had a great impact on the economy and society in general, as we read in the late ninth-century *On Strategy*:

> The financial system was set up to take care of matters of public importance that arise in on occasion, such as the building of ships and of walls. But it is principally concerned with paying the soldiers. Each year, most of the public revenues are spent for this purpose.[36]

No pre-modern economy based on agrarian production reacted well to raids and the destruction of farmland, and the ensuing loss of state income resulted in the limited ability of the government to raise sufficient numbers of troops. Furthermore, prolonged military activity, and the attendant pestilence and famine, was one of the major factors for the decline of the population of the empire and, consequently, of the effectives conscripted into the army, as recorded for the period between c. 550 and c. 775 when the estimated population was reduced from 19.5 million to 7 million.[37] Therefore, the type of warfare that involved regular raiding by small- or medium-sized parties proved to be highly destructive to the local communities of eastern Asia Minor. Indeed, the author of the mid-tenth-century Byzantine treatise *On Skirmishing* provides us with an excellent description of the strategies and tactics applied by raiders in the central and eastern parts of Anatolia, but also of the types of raids and the measures that should have been taken by local commanders to fend off those threats. As I will show in a following chapter of this study in more detail, there were three main effects of this type of warfare for the empire and the local communities: (1) economic – the living off the land, the pillaging and the taking of prisoners resulted in economic destruction, either short-term or long-term depending on the local communities themselves and the frequency of the raids;[38] (2) demographic – these raids caused the seasonal migration or immigration of large parts of the population of eastern Asia Minor either further inland into the Anatolian plateau or into Arab-held territories;[39] and (3) political – these raids challenged the authority of the Byzantine emperor as ruler of these lands, and in the eastern theatre of war satisfied the desire of volunteers from the *thughūr* to participate in the jihad.

Finally, one of the greatest differences between the West and the Christian and Islamic East throughout the Middle Ages was in the perception of warfare, and more specifically the notion of chivalric and honourable battle. It is also important to emphasise the plural in 'cultures of bravery' (and therefore of cowardice), as different cultures constructed the central characteristics of bravery and cowardice differently.[40] The acceptance of feigned flights, other sorts of ruses, ambushes and so on, for example, varied widely. For some cultures, such tactics were indeed construed as unmanly and signs of cowardice, since bravery was constructed around notions of how one fought, with the 'how' usually centred on the honour to be gained in face-to-face combat with melee weapons. For others, such tactics were signs of cleverness, bravery and manliness having been constructed more around whether one won a battle than how one fought it. Similar divisions separated warrior classes, some of which disdained the use of long-range weapons, especially the bow, and others for which it was the weapon par excellence.

A sense of honour dominated the behaviour of the Western knight in his life, forbidding any flight before the enemy as the utmost shame and cowardice. Despite the restrictions on the use of violence enforced upon the arms-bearers by the Church's teachings on sin, penance and atonement, the warrior aristocracy of Western Europe was equally constrained by the notions of honour and shame.[41] In order to be regarded as a prud'homme – the knightly ideal of medieval chivalry – one had to display many qualities, including *prouesse* (prowess) on the battlefield and in performing feats of arms. It was considered a great honour to strike the first blow. Attacking an enemy whose forces were overwhelmingly larger was eagerly sought out by knights as the ultimate chivalric performance on the field of battle. We need to bear in mind, however, that the notion of chivalry presented knightly conduct in a rather idealised light, and that the reality of war was not far removed from the brutalities that struck at the most defenceless elements of society.

Numerous texts serve to illustrate the fundamental connection between the fighting upper classes of Western Europe and physical valour and reputation, from Geoffrey de Charny, the renowned fourteenth-century French knight and author of the *Livre de chevalerie*,[42] the *Song of Roland*,[43] the *Histoire de la guerre sainte*[44] and the *Histoire de Guillaume le Maréchal*.[45] Hence, when the author of the *Histoire de Guillaume le Maréchal* has William exhort the royalist army to resist the invading army of Prince Louis of France prior to the Battle of Lincoln in 1217, William's priorities are most revealing: the safeguarding of reputation comes at the top of the list, before the defence of the family, the *patria* and the Holy Church:

> Hear me, true and noble knights ... since we're about to take up arms in defence of our names, our land, our lives and the lives of those we love, our wives, our children, and to win the greatest honour, and to restore the peace of Holy Church which our enemies have shattered and violated, and to earn redemption and forgiveness for all our sins, be sure that none of you lacks courage this day![46]

Another good example that illustrates the fundamental link between honour, reputation and martial prowess and the fighting upper classes on the field of battle comes from the *Chansons de geste* – Old French for 'Songs of Heroic Deeds'.[47] These were epic poems that appeared in French literature between the late eleventh and early twelfth centuries and narrate legendary incidents (sometimes based on real events) in the history of France during the eighth, ninth and tenth centuries. Clearly, these poems describe an overwhelmingly masculine, aristocratic and martial world where a man's standing rested directly on his feats of arms in war. Yet this was a world where notions of chivalry set the model to which men aspired, but of which many – no doubt – fell short. We read in the *Chanson d'Antioche*[48] a poem inspired by the events of 1097–9 and said to have been composed by a French (or Flemish) eyewitness, a jongleur named Richard le Pèlerin:

[261] When they saw the fight, each Frenchman angrily shouted: 'Holy Sepulchre! Barons, clearly he will never have honour who does not perform here – let us grant to each what he will win – things will go well for those barons who fight well.'

[263] The duke of Boulogne said: 'We have nothing to be proud of as long as these false wretches stand up against us. Better to lose one's life in this encounter than not to drive them out of this place. Holy Sepulchre!' he shouted, rallying the French.

The battle at Brémule in 1119 is a fine example of intracultural warfare characterised by mutual comprehension, adherence to the agreed norms and respect for the fellowship in arms, notwithstanding the inherent hostility between adversaries in a war (Orderic Vitalis' *notitiaque contubernii*). All these are concepts which typified closed cultural systems like the Anglo-Norman and the French at the beginning of the twelfth century.[49] Orderic Vitalis explains the eagerness with which King Louis VI of France was pursuing a pitched battle with the defending army of King Henry I of England 'making frequent complaints to his attendants that they could not meet with the king of England in an open field'.[50] The same source also reports that, despite several of his officers trying to dissuade Louis from taking the field against the English, 'at last, it was generally understood, by the exchange of messengers ... that both kings were in presence at the head of their armies, and, if they wished, battle might be joined'. This was, according to Orderic, what the French king 'had long desired'.

A battle could, potentially, be a ceremonial of exhibiting valour and prowess on a much grander scale than a tournament, a judicial operation where everything unfolded according to specific rules of engagement and from where no one was able to withdraw without shame. Hence, before confronting Thierry of Alsace on the morning of 20 June 1128, at Axpoel, Count William Cliton of Flanders proclaimed that he preferred to 'die rather than suffer such a great shame'.[51] The Count of Boulogne is also claimed to have said to Otto IV of Germany before the Battle of Bouvines in 1214 that 'the custom of the people of France is never to flee but to die or win in battle'.[52] Honour and reputation lay at the heart of the self-perception of Western European knighthood, and it is within this context that adherence to specific conventions of conduct in war should be considered.

The great contrast between the military cultures of the Christian East and West is revealed through Emperor Leo VI's words in his *Taktika*:

'You should not endanger yourself and your army if it is not of utmost need or if you are not to have major gains. Because these people who do this, they greatly resemble those who have been deceived by gold.'[53] These words neatly sum up the Byzantine attitude towards warfare – Byzantine officers were professionals who saw battle as the chance to achieve their objectives using every means possible, fair or unfair, chivalric or unchivalric.[54] Undoubtedly influenced to a large extent by Christian ethics and the Roman imperial tradition,[55] the prevailing attitude of the Byzantines, or at least that of the dominant cultural elite, as attested by their own writings from the sixth to the eleventh centuries on the subject, was to praise the use of diplomacy,[56] the paying of subsidies,[57] and the employment of stratagems,[58] craft, wiles, bribery and 'other means' to deceive the enemy and bring back the army with as few casualties as possible. It was, after all, considered absurd to lose experienced soldiers and money to draw a campaign to a violent and uncertain end. This strategy of non-engagement may have been sensible in military terms, but it ran contrary to the chivalric ideals of honourable combat of Western knighthood.

> Warfare is like hunting. Wild animals are taken by scouting, by nets, by laying in wait, by stalking, by circling around, and by other such stratagems rather than by sheer force. In waging war we should proceed in the same way, whether the enemy be many or few. To try simply to overpower the enemy in the open, hand to hand and face to face, even though you might appear to win, is an enterprise which is very risky and can result in serious harm. Apart from extreme emergency, it is ridiculous to try to gain victory which is too costly and brings only empty glory.[59]
>
> It is good if your enemies are harmed either by deception or raids, or by famine; and continue to harass them more and more, but do not challenge them in open war, because luck plays as a major role as valour in battle.[60]
>
> We are recalling these matters, General, for your own protection and that of your men, if the army under your command is really quite small and very much inferior to that of the enemy. If the fighting men under your command number about five or six thousand, then you should hasten to draw them up in formation directly facing the enemy. Make use, then, of devices, stratagems, special operations, and, when necessary, surprise attacks against them.[61]
>
> If the enemy force far outnumbers our own both in cavalry and infantry, avoid a general engagement or close combats and strive to injure the enemy with stratagems and ambushes . . . Avoid not only an enemy force of superior strength but also one of equal strength, until the might and power of God restore and fortify the oppressed hearts and souls of our host and their resolve with His mighty hand and power.[62]

And only when you know everything about your enemy, only then you must stand and fight them, but do not let your army perish for no reason. Fight in such a way by applying tricks and machinations and ambushes to humiliate your enemy, and only when it is the last choice of all, and in the utmost need, only then stand and fight.[63]

For wars are usually won not so much by a pitched battle, as by cautious planning, and victories won with cunning at the opportune moment.[64]

In a letter to Emperor Isaac Comnenus, Michael Psellus alludes to the futility of squandering men in a conflict with 'barbarians' – specifically referring here to the nomadic peoples that inhabited the northern borders of the empire. As their numbers seemed, literally, endless to any contemporary observer, there is a clear distinction in Psellus' writings between the empire's 'finite' numbers of troops and the, seemingly, 'infinite' resources in manpower that could be fielded by the empire's enemies, especially the nomads from the north whose numbers were swollen by their dependants – women and children – who regularly accompanied them even on lesser raids. Hence, Psellus expressed the following hope: 'How much better is it that not one of our men should fall in battle, and that all the barbarians should be overcome by peaceful means?'[65] To add to Psellus' hopes, the anonymous author of the ninth-century treatise *On Strategy*, sums up the basic motives behind the Byzantine attitude to military activity in times of apprehended war: 'When faced with two evils, the lesser is to be chosen. Negotiating for peace may be chosen before other means, since it might very well offer the best prospect for protecting our own interests.'[66]

In his work, widely known as *De Administrando Imperio*, Emperor Constantine VII details how his son and heir – and privileged reader of the work – should avoid the costly and laborious process of going to war with neighbouring peoples. It was common knowledge that a system of alliances could enable the empire to activate a network of friendly states and people that would attack a potential enemy from the rear, thus saving the Byzantine government the cost of mobilising for war, which could – as we saw in the Introduction – have an unexpected outcome. This is more frequently seen on the Danube front, where successive Byzantine governments would invoke the threat of other nomads to the east against the Bulgars, Magyars or Pechenegs:

> So long as the Emperor of the Romans is at peace with the Pechenegs, neither Russians nor Turks can come upon the Roman dominions by force of arms ... for they fear the strength of this nation [Pechenegs] which the emperor can turn against them while they are campaigning against the Romans. For the

Pechenegs, if they are leagued in friendship with the emperor and won over by him through letters and gifts, can easily come upon the country both of the Russians and of the Turks.[67]

Anna Comnena also praises her father's peaceful nature and his placing of a positive value on the minimal use of force. The princess propagates on multiple occasions in her work that her father was an ardent proponent of peace, using force of arms only when peaceful methods had failed, thus highlighting the Aristotelian perception of war as only a means to a good end.[68] The reader, of course, would have held no doubt that this was no other than the peace of the Christian *Oikumene* ruled by the Roman emperors, God's appointees on earth:

> And at a time, when the Emperor had not yet overcome the difficulties at home, all the world outside burst into a blaze just as if Fortune were making the barbarians abroad and the pretenders at home spring up simultaneously like the self-grown Giants. And this in spite of the Emperor's administrating and managing the government in a very peaceful and humane way, and overwhelming everybody with kindnesses. For some he gladdened with honours and promotions, and never ceased enriching by handsome gifts; while as for the barbarians of whatever country they were, he never gave them any pretext for war nor enforced the necessity of it upon them, but when they made a tumult he checked them; for it is bad generals who in a time of universal peace purposely excite their neighbours to war. For peace is the end of every war.[69]

At the same time, Anna Comnena condemns the purposeful provocation of an enemy into battle or armed conflict as bad generalship:

> The general (I think) should not invariably seek victory by drawing the sword; there are times when he should be prepared to use finesse . . . and so achieve a complete triumph. So far as we know, a general's supreme task is to win, not merely by force of arms; sometimes, when the chance offers itself, an enemy can be beaten by fraud.[70]

This attitude of the Byzantines in subduing their enemies by 'other means' is discussed with a mixture of bewilderment and disdain by contemporary foreign authors, such as the tenth-century Saxon chronicler Widukind of Corvey. In his *Res gestae Saxonicae*, he reported that 'the Greeks had been the lords of very many peoples, and they prevailed by tricks (*artibus*) over those whom they could not defeat by courage (*virtute*)'.[71] Amatus of Montecassino, one of the three main historians of the expansion of the Normans in Italy in the eleventh century, comments in the same manner about the Byzantine methods of defeating their enemies: 'Since the Greeks had the habit of defeating their enemies through malicious ratiocination and subtle treachery.'[72] William of Apulia, the second of the historians of the Norman expansion in the south, also reports the following after the imperial armies had been thrice defeated on the field of battle in 1041, and had suffered a catastrophic reversal of fortune in their campaign to recover Sicily from the Kalbite Muslims between 1038 and 1041 – a campaign that led to the rebellion of the general in command, George Maniakes:

> Since he [Emperor Constantine IX] knew them [the Normans] to be expert at war and unconquerable by force, he hoped to trick them with promises. He had heard that the Norman people (*gens*) was always prone to avarice, loving greatly that which greatly benefited them. He ordered Argyros [the Byzantine general] to bring them great sums of money, silver, precious vestments, and gold, that the Normans might be persuaded to leave the frontiers of Italy (*Hesperiae*), hasten across the sea and mightily enrich themselves in imperial service. He also ordered that if they refused to depart then those bribes destined for them should be given to others, with whom he should launch a savage attack on the Gauls.[73]

The author of the *Russian Primary Chronicle* expresses similar feelings of disdain against the cunning methods employed by the Byzantines in fighting their enemies when referring to Svyatoslav's campaign in 971. In an attempt to discover the numbers under Svyatoslav's command, the Byzantines pretended they could not offer resistance to the advancing Rus' and asked for a precise number of troops in order to pay tribute. This was a simple but ingenious trick by the Byzantines to turn the tables in their favour:

> The Greeks made this proposition to deceive the Russes, for the Greeks are crafty even to the present day. Svyatoslav replied that his force numbered twenty thousand, adding ten thousand to the actual number, for there were really but ten thousand Russes. So, the Greeks armed one hundred thousand men to attack Svyatoslav, and paid no tribute.[74]

The same principles of avoiding war unless the chances were overwhelmingly in their favour, and the defeat of an enemy army by tricks, machinations, stratagems and ambushes, also characterise the attitude to war in medieval Islam. This attitude is made clear in the writings of the Muslim treatises on the art of war, which proliferated during the

Mamluk period but reflect earlier traditions in warfare, as is the case with the Byzantine military manuals already mentioned. According to the *Muqqadimah* (1377) of Ibn Khaldun:

> There is no certain victory in war, even when the equipment and the numerical [strength] that cause victory, exist. Victory and superiority in war come from luck and chance. There are external factors [that decide victory], such as the number of soldiers, the perfection and good quality weapons, the number of brave men, [the skilful] arrangement of the line formation, the proper tactics, and similar things. Then, there are hidden factors. [These] may be the result of human ruse and trickery, such as spreading alarming news and rumours to cause defections [in the enemy ranks] . . . hiding in thickets or depressions and concealing oneself from the enemy in rocky terrain, so that the armies suddenly appear when [the enemy] is in a precarious situation and he must then flee to safety, and similar things.[75]

Here Ibn Khaldun raises a crucial point: 'superiority in war is, as a rule, the result of hidden causes, not of external ones', and it is in these hidden causes – and more specifically in 'the terror that God threw in the hearts of the unbelievers' – that he attributes the victories of the Muslim armies in the early period of the expansion of Islam. Both Ibn Khaldun and al-Ansari, the author of a military treatise from thirteenth-century Egypt, underline a verse from the Quran: 'War is deception.'[76] Adding to that, we read the following principles in the fourth book of the *Tafrij al-kurub* 'about deception and stratagems which obviate war':

> There is no disputing that deception and stratagems in war are required by law and by reason . . . As for reason: there is no disagreement among men of intelligence that victories which have occurred through excellence of stratagem and grace of ingenuity, with the self safe and the armies preserved and with no expenditure of effort, are the best, more salutary and higher in value and degree.[77]

This paragraph perfectly encapsulates the essence of Islamic warfare and the attitude of Muslim tacticians and other theorists of the art of war: turn the odds in your favour, it is preferable to win a battle by using trickery and machination in order to inflict the most casualties on your enemy while your own army can escape with as few as possible, than face them in an honourable and fair battle where chance plays a significant role in the outcome and where you see your chances of winning diminishing.

This is not to suggest, of course, that the Muslim and Byzantine fighting upper classes did not have a sense of honour or shame in battle, as the heroic poetry of the era makes clear. Examples of heroic action in Greek literature, such as duels between champions, go as far back as the *Iliad* and extend to those more contemporaneous with our period of study, such as Leo the Deacon's description of Peter the *stratopedarches* accepting the challenge of a 'Scythian' commander to single combat, or the dash made by Anemas – commander of the imperial bodyguard at the Battle of Dorystolon (971) – against the second-in-command of the Rus' army.[78] Muslim tacticians also placed the daredevil warriors or champions (*mubarizan*), who sought fame in battle, in the front ranks of the Muslim armies of the period, while epic battles are a common feature in Muslim literature.[79] For a commander of a Muslim army, however, throwing his troops against an enemy force without making sure that he had turned the odds in his favour through every means necessary, chivalric or otherwise, was considered careless and unwise.

Finally, I wish to return to the Western European 'culture of bravery' and its notion of chivalric and honourable battle, as encapsulated in the heroic poetry of the *Chansons de geste* (Songs of Heroic Deeds). As Strickland has indicated in his meticulous study of war and chivalry in England and Normandy in the eleventh and twelfth centuries, 'in many instances . . . there was a happy and far from accidental correlation between honourable conduct in war and pragmatic self-interest'.[80] Practically, what this means is that many 'Western' fighters were not any further removed from notions of trickery and cunning behaviour in war than their 'Eastern' counterparts.

In order to illustrate this fact, I will present a few examples from the histories of the Normans in the Mediterranean, in an attempt to explain the paradox between notions of honourable conduct and tactical realism on the field of battle. Historians such as Loud, Davies and Albu have long recognised that cunning and deceit were essential features of the *Normanitas*, the unique character claimed for the *gens Normannorum* by their own historians, who appear to have embraced and promoted this reputation. During the siege of the Sicilian city of Messina by Roger Hauteville in May 1061, Geoffrey Malaterra reported:

> Seeing their enemies facing their army on the other shore and no prospect of doing anything, Count Roger resorted as was his custom to cunning proposals, as if he had read, 'What is to be done? Success falls to the crafty weapons'. He gave this advice to the duke [Robert 'Guiscard' (the Cunning) Hauteville],

that the latter should remain there with his army and show himself to the enemy; meanwhile he himself with a hundred and fifty knights would go to Reggio [the capital of Calabria], there board their ships under the cover of darkness, cross the sea while the enemy was unaware [of their presence] and invade Sicily.[81]

When the citizens of Iato in western Sicily – who, as Malaterra reports, were quite numerous with some 13,000 families – refused to pay their taxes to Roger Hauteville, his reaction shows his preference for 'other means' in subduing his enemies:

> The count sent envoys to them [the citizens of Iato], first soothing them with honeyed words, then concentrating their minds with threats, to stop them rebelling against him. When this accomplished nothing, he brought up his troops and attacked those whom he was unable to subdue through bribes or menaces, to make them submit through the use of force.[82]

Another example comes from William of Apulia and his *Deeds of Robert Guiscard*, which he wrote around the end of the eleventh century:

> While he [Robert 'Guiscard' Hauteville] plundered hither and thither [in Calabria], he was unable to capture any *castrum* or city, and so he resorted to a stratagem to enter a certain place . . . The cunning [Robert] thought up an ingenious trick. He told his people to announce that one of their number had died. The latter was placed on a bier as though he were dead. Swords were hidden on the bier under the 'body's' back . . . While a simple funeral service was being conducted the man who was about to be buried suddenly sprang up; his companions seized their swords and threw themselves on the inhabitants of the place who had been deceived by this trick. What could those stupid people do? They could neither fight nor flee, and all were captured.[83]

We understand that some Western leaders were indeed prepared to sacrifice tactical, strategic or political advantage for honour, in compliance with their notions of chivalry. We should also appreciate, however, that others were equally prepared to violate such norms for military advantage, especially when the enemy was not playing by the rules either.[84] In fact, guile and surprise were regarded as fundamental aspects of war in the West, and were not viewed in any sense as dishonourable, especially if no truces or prearranged agreements were broken. Rather, what brought shame was lying under oath and promising to abstain from such acts.[85] As Strickland notes, 'pragmatism, self-interest and even profit were the powerful dynamics that frequently lay behind the development and acceptance of such [surprise, guile and ravaging] usages', something that should warn modern historians against 'anachronistic assumptions about the nature of chivalry'.[86]

The Norman warriors in the Mediterranean were bands of men with a special set of drivers, and their image-makers (or propagandists?) ascribe the Norman success to a series of psychological characteristics, their energy (*strenuitas*) being particularly prominent in Malaterra's account, which portrays the protagonists of the expansion as the epitome of manliness. They were also courageous, fighting bravely to gain fame on the battlefield, often against high odds, as we see in Roger Hauteville's nephew Serlo's charge with only thirty-six knights against a Muslim army of 3,000 cavalrymen and many infantrymen at Cerami in 1063.[87] Further, Malaterra commends Robert Guiscard's decision not to withdraw from fighting in Calabria 'like a coward who retreats to avoid his enemies', despite the fact that his actions against the locals immediately afterwards made him look like a common thief.[88] Amatus of Montecassino underlines that 'the Normans were prepared to die before they would flee', referring to the three major battles fought against the Byzantines in 1017, from where the rebel Lombards emerged victorious owing to the decisive help of the Norman knights.[89]

A characteristic example of the youthful sense of eagerness to win honour and fame prevailing over the cautious and calculating nature of a commander in battle is the case of the Norman victory at Cerami in 1063. Roger Hauteville's nephew, Serlo, leading a reconnaissance party of thirty-six knights, beat back a much larger Muslim force in a surprise attack on their camp that overlooked the River Cerami. Finally, the main force of a hundred knights under Roger arrived and contemplated what to do next:

> They [Roger and his officers] urged that the victory ... was sufficient, and if they were to continue the pursuit then their luck might change and disaster might ensue. But when asked by the count, Roussel de Bailleul replied fiercely that he would never again help him unless he brought the enemy to battle. When the count heard this, he was angry and sternly reprimanded the faint-hearted. He marched in haste towards the enemy's camp [where they had taken refuge] to offer battle to them.[90]

The Normans were also resourceful, able to take command of the situation, and particularly distinguished for their craftiness and cunning spirit: 'What is to be done? Success falls to the crafty weapons.'[91] Like the

Byzantines and their Muslim enemies, they too were willing to take over a city by 'other means' regardless of the glory of conquering it by the sword, as Malaterra writes of Roger Hauteville's conquest of castles in Calabria: 'Roger ... led the army wisely, and in a very short space of time, partly by threats, and in part by diplomacy, he had gained eleven of the most important castra'.[92]

Thus the idea that the Byzantine Empire was a non-bellicose state, based on its socio-political approach to war and peace and the extensive influence of Christian ethics, must be discounted.[93] The foreign policies formulated by successive governments in Constantinople, which were based on the extensive use of non-bellicose means before resorting to conflict, were a product of what we may call 'political pragmatism' in the medieval Roman Empire.[94] In short, any means that guaranteed the empire's status quo – including diplomacy, bribery, trickery and every 'other means' mentioned thus far – was preferable and, in a cold calculating way, cheaper and less risky than military action. War, then, should be understood as the penultimate means of political negotiation, a true political instrument and, in a very Clausewitzian manner, a continuation of political intercourse.[95] Therefore, war was the political escalation that successive Byzantine governments would consider only as the last resort.[96]

As wars happened, the Byzantines had a specific mentality when facing their enemies in battle, one which is attested clearly and in detail, not just in the works of contemporary historians and commentators, but also in the numerous military treatises that proliferated in the tenth century. Expansionist wars, such as the ones conducted by Nicephorus II Phocas, John Tzimiskes and Basil II, were the result of an unexpectedly favourable strategic situation and prove that the imperial governments were capable of understanding when the equilibrium of power favoured the conduct of war in a specific operational theatre such as the Balkans or Anatolia.[97]

Different political aims and military capabilities, especially after the empire's territorial contraction in the seventh century, dictated different perceptions of peace and war and mixed attitudes towards military confrontations with the empire's neighbours. Understandably enough, it is to be expected that any expansion of the frontiers would not have been achieved solely by diplomatic means, hence the (ethically) legitimising mechanisms of the recovery of the Roman *Oikumene* (which usually meant a military response to an aggressor) and the propagation of the Christian faith among neighbouring peoples, introducing what modern historians call *dikaios polemos* (just war) into the empire's political ideology.[98]

Finally, the basic considerations that shaped the empire's strategic thinking and planning (or 'reacting') should be paid due attention before any in-depth analysis of the wars in eastern Anatolia, Syria and Mesopotamia in the middle of the tenth century. These included the empire's and, most importantly, the capital's, geopolitical location in the confluence of two continents; the state's reliance on agriculture and the economy's reaction to warfare; and the Byzantines' cultural approaches to warfare. All three were interrelated and helped define and develop a sort of strategic thinking for the empire that raised awareness over material considerations and the state's limited ability to face enemies in different operational theatres at the same time, with Asia Minor always taking precedence over the Balkans and Italy in the state's strategic priorities. The concern to avoid battles, minimise loss of life and prevent the destruction of the local economy was the reflection of a long-standing cultural tradition, as appears frequently in treatises of the Hellenistic and Roman period, which was coupled with the early Christian distaste for the shedding of blood.

The aim of this chapter was to show the ways in which Byzantine military culture justified and theorised war, and developed particular customs that shaped the conduct of warfare. The second part of the discussion was, in essence, a comparative study of how the warrior classes (or elites) of the Christian West, Byzantium and the Muslim East constructed the central characteristics of bravery and cowardice around notions of 'honourable' battle, and how attitudes towards battle were construed as manly or cowardly by contemporary and later commentators. Since medieval cultures of war developed within the contested and ever-changing intersections of war with culture, the state and socio-economic structures, it is hoped that in the chapters to follow the reader will be able to appreciate the Byzantine officers' preference for craft, intelligence, wiles, bribery and 'other means', and discern their willingness to adapt to the changing tactics of their enemies on the battlefields of the East.

Notes

1. Leo VI (2010), *The Taktika of Leo VI* [hereafter *Taktika*], Corpus Fontium Historiae Byzantinae, trans. G. T. Dennis, Washington, DC: Dumbarton Oaks, I.3, p. 12.
2. Clausewitz, *On War*, pp. 127–8.

3. Charles, Archduke of Austria (1893–4), *Ausgewählte Schriften weiland seiner kaiserlichen Hoheit des Erzherzogs Carl von Oesterreich*, Vienna: W. Braumüller, p. 57.
4. A. T. Mahan (1892), *The Influence of Sea Power Upon History, 1660–1783*, London: Little, Brown, p. 8.
5. Clausewitz, *On War*, p. 128.
6. B. Heuser (2016), 'Theory and Practice, Art and Science in Warfare: An Etymological Note', in D. Marston and T. Leahy (eds), *War, Strategy and History: Essays in Honour of Professor Robert O'Neill*, Canberra: Australian National University Press, pp. 179–96; Heuser, *Evolution of Strategy*, pp. 3–9.
7. *A Greek–English Lexicon* (1953), ed. H. G. Liddell and R. Scott, Oxford: Clarendon Press, p. 1652.
8. Xenophon (1914), *Cyropaedia*, trans. W. Miller, 2 vols, London: Heinemann, I.6, pp. 12–14.
9. Ibid.
10. *The Anonymous Byzantine Treatise On Strategy* [hereafter *On Strategy*] (2008), in *Three Byzantine Military Treatises*, trans. G. T. Dennis, Washington, DC: Dumbarton Oaks; P. Rance (2007), 'The Date of the Military Compendium of Syrianus Magister (Formerly the Sixth-Century Anonymous Byzantinus)', *Byzantinische Zeitschrift*, 100, pp. 701–37; S. Cosentino (2009), 'Writing about War in Byzantium', *Revista de Història das Ideias*, 30, pp. 83–99; S. Cosentino (2000), '"Syrianos" Strategikon – A Ninth-Century Source?', *Byzantinistica*, 2, pp. 243–80; C. Zuckerman (1990), 'The Military Compendium of Syrianus Magister', *Jahrbuch der Österreichischen Byzantinistik*, 40, pp. 209–24.
11. *On Strategy*, 4–5, pp. 20–2.
12. Ibid., 14, p. 44.
13. Σέξτος Ιούλιος Φροντίνος [Sextus Julius Frontinus] (2015), *Στρατηγήματα* [*Strategemata*], ed. G. Theotokis, trans. V. Pappas, Athens: Hellenic Army Press; E. L. Wheeler (1988), 'The Modern Legality of Frontinus' Stratagems', *Militärgeschichtliche Mitteilungen*, 44, pp. 7–29.
14. Sextus Julius Frontinus (1950), *The Stratagems and the Aqueducts of Rome*, ed. M. B. McElwain, trans. C. E. Bennett, London: Heinemann, p. 6; Wheeler, 'Frontinus' Stratagems', p. 12.
15. Leo VI, *Taktika*, I.1–4, p. 12.
16. Ibid., prologue 9, p. 8; *Maurice's Strategikon: Handbook of Byzantine Military Strategy* [hereafter *Strategikon*] (1984), trans. G. T. Dennis, Philadelphia: University of Pennsylvania Press, preface, p. 9; II.1, p. 23; VII, p. 64.
17. In the context of Byzantine historiography, the word *Oikumene* (Οικουμένη) usually denotes the world under Roman control. On the meaning of the term *Oikumene*, see M. Mann (2012), *The Sources of Social Power: A History of Power from the Beginning to AD 1760*, Cambridge: Cambridge University

Press, chapter 10, pp. 301–38. On the Byzantine claims of 'world domination' (ecumenicity) and how these were projected in the outside world, see A. Kaldellis (2017), 'Did the Byzantine Empire have 'Ecumenical' or 'Universal' Aspirations?', in C. Ando and S. Richardson (eds), *Ancient States and Infrastructural Power: Europe, Asia, and America*, Philadelphia: University of Pennsylvania Press, pp. 272–300. Chrysos has pointed out that after the fourth century there is more frequent use in the Latin and Greek sources of the term *orbis Romanus* (or καθ' ἡμᾶς οἰκουμένην) in the place of *orbis terrarum* to express Byzantine ecumenicity and 'universal aspirations'; see E. Chrysos (2005), 'Το Βυζάντιο και η διεθνής κοινωνία του Μεσαίωνα' [Byzantium and the International Community of the Middle Ages], in E. Chrysos (ed.), *Το Βυζάντιο ως Οικουμένη* [Byzantium as *Oikumene*], Athens: National Research Institute, pp. 59–78. See also E. Chrysos (1992), 'Byzantine Diplomacy, A.D. 300–800: Means and Ends', in J. Shepard and S. Franklin (eds), *Byzantine Diplomacy: Papers from the Twenty-Fourth Spring Symposium of Byzantine Studies*, Aldershot: Ashgate, pp. 25–39; S. Patoura-Spanou (2005), 'Όψεις της Βυζαντινής διπλωματίας' [Facets of Byzantine Diplomacy], in S. Patoura-Spanou (ed.), *Διπλωματία και Πολιτική, Ιστορικές Προσεγγίσεις* [Diplomacy and Politics, Historical Approaches], Athens: National Research Institute, pp. 131–64; H. Ahrweiler (1965), *L'idéologie de l'empire byzantine*, Paris: Presses Universitaires de France, pp. 9–24. On the ideology of the 'limited *Oikumene*' that was promoted by the Macedonian dynasty, see T. Lounghis (1993), *Η Ιδεολογία της Βυζαντινής Ιστοριογραφίας* [The Ideology of Byzantine Historiography], Athens: Herodotos, pp. 48–9, 75, 77, 129; T. Lounghis (2010), *Byzantium in the Eastern Mediterranean: Safeguarding East Roman Identity (407–1204)*, Nicosia: Cyprus Research Center, chs 3 and 5, pp. 77–114 and 147–86; T. Lounghis (1995), 'Die byzantinische Ideologie der "begrenzten Ökumene" und die römische Frage im ausgehenden 10. Jahrhundert', *Byzantinoslavica*, 56, pp. 117–28.

18. 'The Christian Roman emperor had the divine task of spreading Orthodoxy on a world-wide scale': L. V. Simeonova (2000), 'Foreigners in Tenth-Century Byzantium: A Contribution to the History of Cultural Encounter', in D. C. Smythe (ed.), *Strangers to Themselves: The Byzantine Outside*, Aldershot: Ashgate, pp. 229–44 (quote from p. 230). This view has been modified in the last decade, because historians now believe that the missionary policy of the Byzantine governments after 800 was mostly passive and opportunistic without any specific infrastructure for such activity: S. Ivanov (2002), 'Casting Pearls before Circe's Swine: The Byzantine View of Mission', *Travaux et mémoires*, 14, pp. 295–301; W. Treadgold (2015), 'The Formation of a Byzantine Identity', in M. B. P. Maleon and A. E. Maleon, *Studies in Byzantine Cultural History*, Bucharest: Editura Academiei Române, pp. 315–37, especially pp. 330–3.

19. Kaldellis, '"Ecumenical" or "Universal" Aspirations?', pp. 285–91. Here, Kaldellis explains the ways in which ecumenical ideologies promoted imperial projects.
20. C. Mango (1965), 'Byzantinism and Romantic Hellenism', *Journal of the Warburg and Courtauld Institutes*, 28, pp. 29–43. Here, Mango traces the line of continuity from imperial to messianic Byzantinism.
21. J. Chrysostomides (2001), 'Byzantine Concepts of War and Peace', in A. V. Hartmann and B. Heuser (eds), *War, Peace and World Orders in European History*, London: Routledge, pp. 91–102; Haldon, *Warfare, State and Society*, pp. 34–46.
22. Leo VI, *Taktika*, XVIII.16, p. 442.
23. For a useful division of the Byzantine wars into wars of expansion, wars of reconquest and civil wars, see Treadgold, 'Byzantium, the Reluctant Warrior', pp. 209–33.
24. Procopius (1924), *History of the Wars: Book VI, The Gothic War* [hereafter *De Bello Gothico*], trans. H. B. Dewing, London: Heinemann, V.5, p. 26. This is what Kaldellis calls 'a useful rhetorical trope for normalizing the acquisition of new territories and putting them on the path of full absorption into Romania' (Kaldellis, '"Ecumenical" or "Universal" Aspirations?', pp. 286–7).
25. Howard-Johnston, 'Heraclius' Persian Campaigns', pp. 1–44; Haldon, *Warfare, State and Society*, pp. 18–21.
26. For more on Heraclius' strategy and propaganda, see W. E. Kaegi (2003), *Heraclius, Emperor of Byzantium*, Cambridge: Cambridge University Press; E. Luttwak (2009), *The Grand Strategy of the Byzantine Empire*, Cambridge, MA: Harvard University Press, pp. 393–421.
27. On the use of a state's 'grand strategy' as a tool for its overall political aims, first postulated by Clausewitz, Rühle and Jomini in the early nineteenth century, but reaching universal consensus in the twentieth century, see Heuser, *Evolution of Strategy*, pp. 1–28.
28. J. F. Haldon (1992), 'Blood and Ink: Some Observations on Byzantine Attitudes towards Warfare and Diplomacy', in J. Shepard and S. Franklin (eds) *Byzantine Diplomacy: Papers from the Twenty-Fourth Spring Symposium of Byzantine Studies*, Aldershot: Ashgate, pp. 281–95; Haldon, *Warfare, State and Society*, p. 43.
29. Anthropologists such as Snyder have identified three sets of variables that determine war: material (natural environment, technology, etc.), social (institutions that shape peoples' relationships within social formations) and cultural (collectively shared beliefs and patterns of behaviour). See J. Snyder (2002), 'Anarchy and Culture: Insights from the Anthropology of War', *International Organization*, 56, pp. 7–45. A detailed list of variables is provided by Baron A. H. de Jomini (2008), *The Art of War*, restored edition, Kingston, ON: Legacy Books Press, pp. 21–43.

30. Selected works on the topic include M. Whittow (1996), *The Making of Byzantium, 600–1025*, Berkeley: University of California Press, pp. 15–37; M. Whittow (2009), 'The Political Geography of the Byzantine World: Geographical Survey', *OHBS*, pp. 219–31; Haldon, *Warfare, State and Society*, pp. 46–60; G. L. Huxley (1982), 'Topics in Byzantine Historical Geography', *Proceedings of the Royal Irish Academy. Section C: Archaeology, Celtic Studies, History, Linguistics, Literature*, 82C, pp. 89–110.
31. Compare the establishment of frontier districts in the Lower Danube with those in Syria and Mesopotamia in the late tenth century: A. Madgearu (2013), *Byzantine Military Organization on the Danube, 10th–12th Centuries*, Leiden: Brill; Haldon, *Warfare, State and Society*, pp. 65–6.
32. N. Koutrakou (2005), 'Βυζαντινή διπλωματική παράδοση και πρακτικές. Μια προσέγγιση μέσω της ορολογίας' [Byzantine Diplomatic Tradition and Practices: A Terminology Approach], in S. Patoura-Spanou (ed.), *Διπλωματία και Πολιτική, Ιστορικές Προσεγγίσεις* [Diplomacy and Politics: Historical Approaches], Athens: National Research Institute, pp. 92–5.
33. Neatly summed up by H. Ahrweiler (1962), 'L'Asie mineure et les invasions arabes', *Revue historique*, 227, p. 2; P. Charanis (1975), 'Cultural Diversity and the Breakdown of Byzantine Power in Asia Minor', *Dumbarton Oaks Papers*, 29, pp. 1–20. For a more detailed analysis of the economic importance of Asia Minor, see Haldon, *The Empire That Would Not Die*, pp. 215–48.
34. To give just a few examples: 'When the emperor Justinian considered the situation was as favourable as possible, both domestically and in his relations with Persia, he turned his attention to affairs in Libya' (Procopius, *Wars*, III.10); 'While our forces were engaged against the Saracens, divine Providence led the Turks, in place of the Romans, to campaign against the Bulgarians' (Leo VI, *Taktika*, XVIII.40, p. 452); Theophanes Confessor, *The Chronicle of Theophanes Confessor* [hereafter *Chronicle*] (1997), ed. C. Mango and R. Scott, Oxford: Oxford University Press, pp. 434–5; John Skylitzes (2010), *A Synopsis of Byzantine History, 811–1057*, trans. J. Wortley, Cambridge: Cambridge University Press, pp. 172, 197; Symeon Magister and Logothete (2006), *Chronicon*, Corpus Fontium Historiae Byzantinae, vol. 44.1, ed. Staffan Wahlgren, Berlin: Walter de Gruyter, p. 304; *Theophanes Continuates* (1838), in *Theophanes Continuates*, Ioannes Cameniata, Symeon Magister, Georgius Monachus, *Οι μετά Θεοφάνην* [The Continuators of Theophanes], Corpus Scriptorum Historiae Byzantinae, vol. 33, ed. I. Bekker, Bonn: Webber, p. 181.
35. Treadgold, *Byzantium and its Army*, pp. 53–85; Whittow, *Orthodox Byzantium*, pp. 181–93; Haldon, *Warfare, State and Society*, pp. 99–103; J.-C. Cheynet (1995), 'Les effectifs de l'armée byzantine aux Xe–XIIe siècles', *Cahiers de civilisation médiévale*, 152, pp. 319–35.
36. *On Strategy*, 2.18–21, p. 12.

37. A. Laiou and C. Morrisson (2007), *The Byzantine Economy*, Cambridge: Cambridge University Press, pp. 38–63; W. Treadgold (1997), *A History of the Byzantine State and Society*, Stanford, CA: Stanford University Press, pp. 371–413; Haldon, *The Empire That Would Not Die*, pp. 232–9.
38. J. F. Haldon (1997), *Byzantium in the Seventh Century: The Transformation of a Culture*, Cambridge: Cambridge University Press, pp. 99–114. Paleo-environmental data, such as the analysis of pollen sequences, can confirm the impact of raiding and warfare at a regional level in Anatolia; see Haldon, *The Empire That Would Not Die*, pp. 232–9; J. Haldon et al. (2014), 'The Climate and Environment of Byzantine Anatolia: Integrating Science, History, and Archaeology', *Journal of Interdisciplinary History*, 45, pp. 113–61.
39. Asa Eger's study makes a useful distinction between the different types of Byzantine settlements that withstood the worst of the Muslim raids, such as the *kataphygia* or *ochyromata* (fortified villages or refuge settlements), the *phrouria* (garrison stations) and the *aplekta* (fortified camps). See Asa Eger, *Islamic–Byzantine Frontier*, pp. 248–63.
40. Morillo has published several articles on the topic of military cultures: S. Morillo (2013), 'Justifications, Theories and Customs of War', in D. Graff (ed.), *The Cambridge History of War*, Cambridge: Cambridge University Press, pp. 1–24; S. Morillo (2006), '*Expecting Cowardice: Medieval Battle Tactics Reconsidered*', *Journal of Medieval Military History*, 4, pp. 65–73; S. Morillo (2001), 'Cultures of Death: Ritual Suicide in Medieval Europe and Japan', *The Medieval History Journal*, 4, pp. 241–57. See also R. Abels (2008), 'Cultural Representation and the Practice of War in the Middle Ages', *Journal of Medieval Military History*, 6, pp. 1–31; J. France (2005), 'Close Order and Close Quarter: The Culture of Combat in the West', *The International History Review*, 27, pp. 498–517.
41. M. Strickland (1996), *War and Chivalry: The Conduct and Perception of War in England and Normandy, 1066–1217*, Cambridge: Cambridge University Press, pp. 98–131; J. A. Lynn (2003), 'Chivalry and Chevauchée: The Ideal, the Real, and the Perfect in Medieval European Warfare', in J. A. Lynn (ed.), *Battle: A History of Combat and Culture*, Philadelphia, PA: Westview, pp. 73–110; M. Keen (2004), *Chivalry*, New Haven, CT: Yale University Press.
42. G. de Charny (1996), *A Knight's Own Book of Chivalry*, trans. E. Kennedy, Philadelphia: University of Pennsylvania Press.
43. G. J. Brault (2010), *Song of Roland: An Analytical Edition; Introduction and Commentary*, Philadelphia: University of Pennsylvania Press.
44. A Norman French poem attributed to the jongleur Ambroise describing the course of the Third Crusade: Ambroise (2003), *The History of the Holy War*, trans. M. Ailes, Woodbridge: Boydell & Brewer.
45. This is the chivalric biography of a twelfth-century English knight, most commonly used by researchers as a window to understanding twelfth-century

nobility and tourneying in England: *The History of William the Marshal* (2016), trans. N. Bryant, Woodbridge: Boydell & Brewer; D. Crouch (2002), *William Marshal: Knighthood, War and Chivalry, 1147–1219*, London: Routledge.

46. *History of William the Marshal*, pp. 195–6.
47. P. Leverage (2010), *Reception and Memory: A Cognitive Approach to the* Chansons de Geste, Amsterdam: Editions Rodopi; S. Kay (2005), *The* Chansons de Geste *in the Age of Romance: Political Fictions*, Oxford: Oxford University Press; M. A. Newth (2005), *Heroes of the French Epic: A Selection of* Chansons de Geste, Woodbridge: Boydell & Brewer.
48. *The* Chanson d'Antioche*: An Old-French Account of the First Crusade* (2011), trans. S. Edgington and C. Sweetenham, Aldershot: Ashgate. The edition used here is *La Chanson d'Antioche* (1977), ed. S. Duparc-Quioc, Paris: Librairie Orientaliste Paul Geuthner. Two useful books on the poem are L. A. M. Sumberg (1968), *La Chanson d' Antioche, étude historique et littéraire*, Paris: Picard; R. F. Cook (1980), *'Chanson d'Antioche', chanson de geste: le cycle de la croisade est-il epique?*, Amsterdam: John Benjamins B. V.
49. Strickland, *War and Chivalry*, pp. 132–58.
50. Orderic Vitalis' *Ecclesiastical History* is our most detailed source for the battle: Orderic Vitalis (1854), *Ecclesiastical History of England and Normandy*, trans. T. Forester, London: Henry G. Bohn, III, pp. 480–6.
51. G. Duby (1990), *The Legend of Bouvines: War, Religion and Culture in the Middle Ages*, trans. C. Tihanyi, Cambridge: Cambridge University Press, pp. 110–21; for Axpoel, see p. 114.
52. Ibid., pp. 114–21; J. F. Verbruggen (1997), *The Art of Warfare in Western Europe during the Middle Ages from the Eighth Century to 1340*, Woodbridge: Boydell & Brewer, pp. 239–60.
53. Leo VI, *Taktika*, XX.36, p. 550.
54. Treadgold, 'Byzantium, the Reluctant Warrior', pp. 209–33; W. E. Kaegi (1983), 'Some Thoughts on Byzantine Military Strategy', *The Hellenic Studies Lecture*, Brookline, MA: Hellenic College Press, pp. 1–18; Haldon, *Warfare, State and Society*, pp. 34–46; G. T. Dennis (1997), 'The Byzantines in Battle', in K. Tsiknakes (ed.), *Byzantium at War (9th–12th c.)*, Athens: National Research Institute, pp. 165–78.
55. Byzantine political ideology incorporated a theory of imperial and Christian Roman *philanthropia*, which directly influenced the attitude to warfare and killing; see J. F. Shean (2010), *Soldiering for God: Christianity and the Roman Army*, Leiden: Brill; C. M. Odahl (1976), 'Constantine and the Militarization of Christianity: A Contribution to the Study of Christian Attitudes toward War and Military Service', unpublished DPhil dissertation, University of California, San Diego, especially pp. 9–59. See also the papers in the collective volume: J. Koder and I. Stouraitis (eds) (2012), *Byzantine War Ideology between Roman Imperial Concept and Christian Religion:*

Akten Des Internationalen Symposiums (Wien, 19.–21. Mai 2011), Vienna: Austrian Academy of Sciences Press. For a very interesting comparison between the warrior cultures of Japan and Western Europe, and the role religion played in shaping their attitude to killing and committing suicide, see Morillo, *Cultures of Death*, pp. 244–8.

56. J. Shepard (1985), 'Information, Disinformation and Delay in Byzantine Diplomacy', *Byzantinische Forschungen*, 10, pp. 233–93. See also the collection of papers in S. Franklin and J. Shepard (eds) (1992), *Byzantine Diplomacy*, Aldershot: Ashgate.
57. N. Oikonomides (1997), 'Το όπλο του χρήματος' [Money as a Weapon], in K. Tsiknakes (ed.), *Byzantium at War (9th–12th c.)*, pp. 261–8.
58. On the importance of stratagems in ancient Greek and Roman military tradition, which was inherited by the Byzantines, see J. E. Lendon (1999), 'The Rhetoric of Combat: Greek Military Theory and Roman Culture in Julius Caesar's Battle Descriptions', *Classical Antiquity*, 18, pp. 273–329 (mainly pp. 290–5); E. L. Wheeler (1988), 'Πολλά τα κενά του πολέμου: The History of a Greek Proverb', *Greek, Roman and Byzantine Studies*, 29, pp. 153–84.
59. *Strategikon*, VII, p. 65.
60. Leo VI, *Taktika*, XX.51, p. 554.
61. G. T. Dennis (2008), 'The Anonymous Byzantine Treatise *On Skirmishing* by the Emperor Lord Nicephoros' [hereafter *On Skirmishing*], in G. T. Dennis (ed.), *Three Byzantine Military Treatises*, Washington, DC: Dumbarton Oaks, 19, p. 214.
62. *Praecepta Militaria*, IV.195–207, p. 50.
63. Kekaumenus (1965), *Strategikon*, ed. B. Wassiliewsky and V. Jernstedt, Amsterdam: Hakkert [hereafter Kekaumenus], pp. 9–10.
64. *The History of Leo the Deacon: Byzantine Military Expansion in the Tenth Century* (2005), trans. A. M. Talbot, Washington, DC: Dumbarton Oaks, II.3, p. 73.
65. *Michaelis Pselli Scripta minora: magnam partem adhuc inedita* (1941), ed. E. Kurtz and F. Drexl, Milan: Società editrice 'Vita e pensiero', p. 181.
66. *On Strategy*, 6, p. 22.
67. Constantine Porphyrogenitus (1985), *De Administrando Imperio* [hereafter *DAI*], ed. (Greek text) G. Moravcsik, trans. R. J. H. Jenkins, Washington, DC: Dumbarton Oaks Texts, 4, pp. 50–2. Cf. *On Strategy*, 6, p. 22.
68. On Anna Comnena's projection of her father as an ideal sovereign compelled to wage war to rescue an embattled empire, see I. Stouraitis (2012), 'Conceptions of War and Peace in Anna Comnena's *Alexiad*', in J. Koder and I. Stouraitis (eds), *Byzantine War Ideology*, pp. 69–80.
69. Anna Comnena (2000), *The Alexiad*, trans. E. A. S. Dawes, Cambridge, ON: In Parentheses Publications, XII.v, p. 220.
70. Ibid., XIII.iv, p. 235.
71. Widukind of Corvey (1935), *Res Gestae Saxonicae Sive Annalium Libri Tres*, ed. P. Hirsch, Hanover: MGH Scriptores rerum Germanicarum in

usum scholarum 60, p. 148. I have not had access to the newest edition: Widukind of Corvey (2014), *Deeds of the Saxons*, trans. B. S. Bachrach and D. S. Bachrach, Washington, DC: Catholic University of America Press.

72. Amatus of Montecassino (2004), *The History of the Normans*, trans. P. Dunbar, Woodbridge: Boydell & Brewer, I.15.
73. Guillaume de Pouille (1963), *La geste de Robert Guiscard*, ed. M. Mathieu, Palermo: Bruno Lavagnini. In this study, I have used G. A. Loud's translation in English for the electronic series Medieval History Texts in Translation: *William of Apulia, The Deeds of Robert Guiscard*, p. 17, available at: www.leeds.ac.uk/arts/downloads/file/1049/the_deeds_of_robert_guiscard_by_william_of_apulia (accessed 25 June 2018).
74. *The Russian Primary Chronicle, Laurentian Text* (1953), eds S. H. Cross and O. P. Sherbowitz-Wetzor, Cambridge, MA: Medieval Academy of America, p. 88.
75. Ibn Khaldun (1969), *The Muqaddimah: An Introduction to History*, trans. by Franz Rosenthal, 3 vols, Princeton, NJ: Princeton University Press, pp. 85–6.
76. Ibid., p. 86; al-Ansari (1961), *A Muslim Manual of War*, ed. G. T. Scanlon, Cairo: American University at Cairo Press, p. 59.
77. Al-Ansari, *Muslim Manual of War*, p. 59.
78. *History of Leo the Deacon*, p. 43; T. Maniati-Kokkini (1997), 'Η επίδειξη ανδρείας στον πόλεμο κατά τους ιστορικούς του 11ου και 12ου αι.' [Demostrating Bravery in War According to the Historians of the 11th and 12th c.] in K. Tsiknakes (ed.), *Byzantium at War*, pp. 239–59; C. Whately (2016), *Battles and Generals: Combat, Culture, and Didacticism in Procopius' 'Wars'*, Leiden: Brill, pp. 169–71; M. van Creveld (2013), *Wargames: From Gladiators to Gigabytes*, Cambridge: Cambridge University Press, pp. 39–47. The most famous duel is the biblical one between David and Goliath, but they are also found in other Near Eastern civilisations. Van Creveld points out the semi-mythological status of these early encounters.
79. *The Book of Jihad*, pp. 460–1; D. G. Tor (2007), *Violent Order, Religious Warfare, Chivalry and the Ayyar Phenomenon in the Medieval Islamic World*, Würzburg: Ergon, especially pp. 241–9; H. T. Norris (2009), 'The Sacred Sword of Maslamah B. "Abd Al-Malik"', *Oriente Moderno/Studies on Islamic Legends*, 89, pp. 389–406; V. Christides (1962), 'An Arabo-Byzantine Novel, Umar b. Al-Nu'man compared with Digenes Akritas', *Byzantion*, 32, pp. 549–604.
80. Strickland, *War and Chivalry*, p. 125.
81. G. Malaterra (2005), *The Deeds of Count Roger of Calabria and Sicily and of his Brother Duke Robert Guiscard*, ed. K. B. Wolf, Ann Arbor: University of Michigan Press, *Deeds*, II.10.
82. Ibid., III.20.
83. William of Apulia, *Deeds*, p. 23.
84. The historians of the Norman expansion in the south make it perfectly clear that their enemies – the Lombards, the 'Greeks' and the Muslims – were

treacherous and cunning and should not have been trusted: 'At this time Drogo and Guaimar, the leaders of the Normans, died; the latter treacherously killed by the citizens of Salerno and his own relatives, the former murdered by the local people at Montilari, whom he trusted too much' (William of Apulia, *Deeds*, p. 18); 'The Apulian Lombards are always a most treacherous race, and they conceived a secret plot to murder all the Normans throughout Apulia on the same day' (Malaterra, *Deeds*, I.13); 'The Greeks are indeed the most treacherous of people' (Malaterra, *Deeds*, II.29). See the example of Hugh of Gercé, the commander of the eastern Sicilian city of Catania, who wished to enhance his prowess and chivalric reputation by performing feats of arms, only to be drawn into a trap after the Muslims refused to give battle and feigned a retreat. Notice that the more experienced Roger urged Hugh to avoid leaving the walls of Catania 'for he feared the latter's [Ibn al-Werd, the Muslim commander] cunning stratagems' (Malaterra, *Deeds*, III.10).

85. Strickland, *War and Chivalry*, pp. 128–31; J. Gillingham (1988), 'War and Chivalry in the History of William the Marshal', in P. R. Coss and S. D. Lloyd (eds), *Thirteenth Century England II: Proceedings of the Newcastle Upon Tyne Conference, 1987*, Woodbridge: Boydell & Brewer, pp. 251–63. On the importance of oath-taking and the keeping of prior agreements between the brothers Robert and Roger Hauteville, and the local people of the Sicilian town of Gerace, see Malaterra, *Deeds*, II.26–8. Low cunning seems not to have been dishonourable, as shown in the example of Abelard's rebellion against his uncle, Robert Guiscard, and the latter's cunning method of regaining the castle of Gargano from his nephew (Malaterra, *Deeds*, III.6).
86. Strickland, *War and Chivalry*, p. 131.
87. G. Theotokis (2010), 'The Norman Invasion of Sicily, 1061–1072: Numbers and Military Tactics', *War in History*, 17, pp. 381–402.
88. Malaterra, *Deeds*, I.16.
89. Amatus, *History of the Normans*, I.22, p. 51.
90. Malaterra, *Deeds*, II.33.
91. G. Theotokis (2015), 'Promoting the Newcomer: Myths, Stereotypes, and Reality in the Norman Expansion in Italy during the XIth Century', *Porphyra*, 24, pp. 28–38.
92. Malaterra, *Deeds*, I.36.
93. J.-C. Cheynet (2013), 'Réflexions sur le "pacifisme byzantin"', in C. Gastberger et al. (eds), *Pour l'amour de Byzance: hommage à Paolo Odorico*, Frankfurt am Main: Peter Lang, pp. 63–73. See also the second part of I. Stouraites (2012), '"Just War" and "Holy war" in the Middle Ages: Rethinking Theory through the Byzantine Case-study', *Jahrbuch der Österreichischen Byzantinistik*, 62, pp. 250–64.
94. Haldon, 'Blood and Ink', pp. 281–95.

95. C. von Clausewitz (2007), *On War*, M. Howard and P. Paret (trans.), Oxford: Oxford University Press, pp. 28–9; B. Heuser (2002), *Reading Clausewitz*, London: Pimlico, pp. 44–9; Heuser, *Evolution of Strategy*, pp. 9–15.
96. E. Chrysos (2003), 'Ο πόλεμος έσχατη λύση' [War as the Ultimate Solution], in A. Avramea, A. Laiou and E. Chrysos (eds), *Βυζάντιο: Κράτος και Κοινωνία – Μνήμη Νίκου Οικονομίδη / Byzantium: State and Society – In Memory of Nikos Oikonomides*, Athens: National Research Institute, pp. 543–63.
97. J. Koder and I. Stouraites (2012), 'Byzantine Approaches to Warfare (6th–12th Centuries): An Introduction', in J. Koder and I. Stouraites (eds), *Byzantine War Ideology*, pp. 9–15.
98. On the concept of just war in Byzantium, see Stouraites, '"Just War" and "Holy War"', pp. 227–50.

2

Byzantine and Arab Strategies and Campaigning Tactics in Cilicia and Anatolia (Eighth–Tenth Centuries)

> The general should be on the alert for news about the equipping and movement of a large army, both cavalry and infantry, especially at that time of the year when one expects large armies to be assembled, usually in August. In that month, large numbers would come from Egypt, Palestine, Phoenicia, and southern Syria to Cilicia, to the country around Antioch, and to Aleppo, and adding some Arabs to their force, they would invade Roman territory in September.[1]

This passage comes from the anonymous military treatise *On Skirmishing*, written probably around the end of the 960s under the auspices of the Emperor Nicephorus Phocas and probably by the pen of his brother Leo – the *strategos* of Cappadocia and later 'Domestic of the West'. It encapsulates the spirit of raiding and guerrilla warfare in the eastern provinces of the empire as it had developed in the last two centuries. The Muslim troops that are mentioned in the treatise were both cavalry and infantry forces made up of volunteers for the jihad, as well as regular troops from the Arab lands in the interior (*al-ᶜawāṣim*) and from the borderlands (*al-thughūr*).[2] When referring to the Byzantine scouting parties dispatched to gather intelligence, the author of the treatise mentions the number 6,000–12,000 for the invading force of Arabs.[3] Such a force would have been well within the capabilities of Sayf ad-Dawla to muster, as it is confirmed by the accounts of Yahya ibn Said of Antioch and Ibn Zafir, although it is impossible to be more precise regarding the exact numbers of different units or the ratio between infantry and cavalry forces.

Led by the emir as the leader of the jihad, such raids served both an economic and ideological function; first, their main aim was to loot and devastate the countryside, destroy the economic centres of the invading regions, disrupt commerce and everyday life, and undermine the emperor's authority.[4] They also offered an opportunity for the Muslim warriors to perform their religious and military duties against the infidel in the spirit of constant warfare for the expansion of the *Dar al-Islam*.[5] The religious-ideological element of the Byzantine–Muslim conflict of the period leading up to the middle of the tenth century, as echoed in *On Skirmishing*, is abundantly clear as the author of the treatise invokes God several times in the recommendation he makes regarding strategy and battle tactics.[6] Hence the question that emerges is the following: what is the kind of warfare that dominated in the geographical area under consideration, and what does it suggest about the strategy and strategic goals of the Arabs in the region of eastern Anatolia for the decades leading up to the middle of the tenth century?

In view of the wider debate between modern scholars such as Rogers, Gillingham and Morillo over the term 'Vegetian strategy', I will ask whether historians can characterise any strategy applied by the Arabs and the Byzantines in the operational theatres of the East before the 960s as essentially 'Vegetian'. Accordingly, a basic principle that has to be kept in mind is that the party wanting to expand and conquer – the aggressor – would often be more willing to seek a decisive battle, while the party already controlling the territories – the defender – would wish to deny their enemy this.[7] Therefore, what would seem reasonable in this case would be for the Arab invaders, being in enemy territory and far away from their supply bases, to seek a decisive battle in order to confirm their conquests. Was this the case for the period up to the 960s? We read in our treatise:

> It is your duty, General, to search very carefully for the enemy who are making a serious effort to avoid you so they can send out their raiding parties to plunder our lands. Your mind must be alert so that no plan or trick of theirs will ever get by you.[8]

Here one finds a definite idea of the 'Vegetian strategy' of the Arab raiding parties in the period preceding the middle of the tenth century.

If the Arabs, then, were to avoid a pitched battle with the Byzantines unless the odds were overwhelmingly in their favour, what exactly were their strategic aims?

> When large numbers of the enemy wander about our country ravaging, destroying, and making plans to besiege fortified places, they will indeed be on their guard to avoid being ambushed by the Roman units; in fact, they will be devising plans to ambush us.[9]

It is obvious that the main strategic aim of the Arab raiders was to loot, destroy, besiege key economic and strategic centres, take prisoners and return to their homelands laden with booty, rather than bog themselves down into any sort of permanent conquests in the Anatolian plateau. In fact, any large-scale Muslim territorial expansion in Anatolia had been abandoned since the last major siege of Constantinople in 717–18.[10] Facing the Byzantine forces in pitched battle was contemplated only as a desperate solution and was to be generally avoided – hence the relatively low number of major pitched battles between Arab and Byzantine forces before the 950s.[11] Did the Byzantine armies, though, adopt any similar kind of battle-avoiding strategy in the operational theatres of eastern Anatolia and northern Mesopotamia during the same period?

On Skirmishing makes it clear that the local general should hasten to engage the invading forces only if they consist of a small number of men – the so-called *monokoursa*.[12] This was a small-scale raid where the party was composed exclusively of cavalry for greater mobility; they were usually led by a local commander of the border areas and could have been launched at any time of the year. This, however, was not the case when the Byzantine officers had to face large-scale invasions of the type analysed at the beginning of this chapter ('περί συναθροίσεως και κινήσεως μεγάλου φοσσάτου' – 'On the Gathering and Moving of a Large Army'):

> The general must make it one of his highest priorities and concerns to launch secret and unexpected attacks upon the enemy whenever possible ... Still, instead of confronting the enemy as they are on their way to invade Romania, it is in many respects more advantageous and convenient to get them as they are returning from our country to their own. They will be worn out and much the worse for wear ... They are likely to be burdened with a lot of baggage, captives, and animals. The men and their horses will be so tired that they will fall apart in battle.[13]

The most important aspect of this frontier strategy for the Byzantines was the 'shadowing' of the enemy forces. Following and harassing the enemy by exploiting one's own knowledge of the local terrain was one aspect;[14] keeping a close watch on their column and camp in order to attempt ambushes on forage parties was another.[15] Large-scale expeditions launched in September were allowed to invade friendly territory and proceed to their targets, while being followed closely and harassed by detachments of select men who controlled the mountain and valley passes through which the invaders would return home – a strategy followed, in principle, by tacticians since antiquity.[16] The invaders' logistical difficulties would be maximised by the evacuation of the local population and the removal of livestock and crops, or even their destruction.[17]

The basic idea behind this strategy was the wearing down of the invading army while the Byzantines would have time to concentrate a significant number of reinforcements from the neighbouring themes and from the areas further in the interior of Asia Minor, forces that would converge on the area in a pincer movement to flush the invaders out of Byzantine territory.[18] Finally, local captains were encouraged to operate independently and launch raids in enemy territory in an attempt to force the enemy commander to return home and protect his people.[19]

The defensive strategy that is described in *On Skirmishing* (c. 960) certainly reflects the strategic reality of earlier periods on the eastern Byzantine borders, but what evidence modern historians possess shows centuries of trial and error in the aforementioned operational theatres which, however, had reached a high degree of sophistication and efficiency by the time the treatise was commissioned by Nicephorus Phocas shortly before his death.[20] In fact, the essence of what we read in this work – the avoidance of battle, destruction of resources, and the shadowing and harassing of the enemy – goes back as early as the third quarter of the eighth century, when Theophanes describes Leo IV's preparations to defend Anatolia against Abbasid counterattacks that came in 778 after a Byzantine expedition into Syria the previous year:

> The Emperor [Leo IV] arranged with his *strategoi* that they should not meet the Arabs in the field, but secure the fortresses and bring in men to guard them. He also sent officers to each fortress, who were to take three thousand picked men to follow the Arabs closely so that their raiding party could not disperse. Even before this they were to burn whatever fodder was to be found for the Arabs' horses. After the Arabs had been in Dorylaion [central Anatolia] for fifteen days they ran out of supplies and their animals were starving; there were heavy losses amongst them.[21]

Overall, historians should view this strategy as the empire's response to the situation in its eastern provinces – a pragmatic reaction of a political and military mechanism in the face of numerically superior invading forces. The military organisation of the *themata* that had sprung out of the late Roman system after the Byzantine defeats of the seventh century precluded any sort of linear defence on the borders of the sort that the Roman *limitanei* (frontier troops) represented.[22] Certainly, there were times when substantial forces would defend and intercept smaller invading parties on the borders or, more likely, on the mountain passes – hence the establishment of the *kleisourai*.[23] During the eighth century, however, we see the development of a defence-in-depth strategic planning, where large areas of imperial territory on the borders were left undefended, or perhaps with small forces guarding key outposts at the enemy's rear, and subject to regular raids and devastation.[24] One may question exactly how deep this buffer zone extended and what strategic areas it shielded further inland.

Byzantine Military Tactics

The ultimate target of any large-scale invasion of Byzantium, like the ones that were undertaken by the Arabs in 674 and again in 717, was Constantinople. I have already highlighted, however, that the siege of the capital in 717–18 was the last conquest expedition that would be launched against Byzantium by a Muslim army for the next seven and a half centuries. We are allowed a clear view of the zones that were protected by the thematic buffer zone, or perhaps the zones that the central government was interested in protecting from devastation, by studying the system's reaction to the large-scale raids launched by the Arabs after 718.

In the Phrygian city of Akroinon on the western edge of the Anatolian plateau, Leo III defeated an Umayyad army of some 20,000 in 740.[25] Almost four decades later, in 778, Theophanes reported on the Arab invasion of that year that was allowed to reach Doryleum in the north-western Anatolian plateau, just a few days' distance from Bithynia and the major imperial *aplekto* (military camp) of Malagina. In 838, Theophilos had to face a great Abbasid invasion army – again divided into two great bodies invading Anatolia from the south (Cilicia) and the north (Armeniakon). He was eventually defeated near the fortress-*aplekto* of Dazimon in the Armeniakon theme, a disaster which gave the Abbasid caliph, al-Mu‵tasim, the opportunity to sack Ancyra and Amorion, two key strategic cities in the central Anatolian plateau.[26] Finally, in 863, Emperor Michael III intercepted a raid by Umar, the emir of Melitene, who was riding along Anatolia and had already looted the Pontic city of Amisos before attempting to head back across the plateau. The Byzantines defeated Umar with a large force on the banks of the River Lalakaon north-east of Ankara.[27] To this list, we should add Umar's 860 raid 'deep into Anatolia' and the governor of Hims' naval raid that sacked Attaleia in the same year.[28] Both went unopposed.

It appears that Byzantine strategic planning in Asia Minor had developed into a sort of defence in depth, which included three zones:[29]

1. The frontier zone of the *themata* and the *kleisourai* of eastern Asia Minor, where the local forces would attempt to block the way and turn back any small-scale incursion by enemy forces (otherwise, this zone would have been left undefended and at the mercy of the invading forces proceeding to the next zone further inland, gradually transforming it into a 'no-man's land'[30]).
2. The second zone, which included the themes of the Anatolian plateau where the local forces would garrison key fortresses and towns on the roads leading further west, either to the Aegean coast or to

Constantinople, shadowing and harassing the invaders according to the recommendations of the treatise *On Skirmishing*. If they were successful and the danger posed by these troops to the invader army's lines of communication appeared too great, then the invaders would retreat to their country and no imperial army would be mobilised. If, however, an invading army seemed undeterred in its march through the Anatolian plateau and further west towards the Bithynian coast, an imperial army would then be mobilised to protect the third zone.
3. The third zone: the fertile coastal plains of Bithynia, western Asia Minor and, of course, the capital.

It is no coincidence that all major battles where the emperor himself took to the field were fought in the second zone of defence, in the heart of Anatolia. Byzantium developed a relationship between the capital and the provinces comparable to that of a centralised modern nation state; 'Romania' was gradually defined not by its fluctuating, shifting and porous territorial borders, but by its secure and stable centre where political and religious authority resided.[31] The emperors seem to have been more preoccupied with their personal security and preventing coups in the capital, thus focusing on what they saw as the heart of the Byzantine Empire – the imperial court;[32] the destruction of distant towns and provinces, the taking of thousands of prisoners and the decline of the local economy and agriculture were prices they were willing to pay. The intensity of their gaze towards the capital is all the more apparent if we consider the fact that a defeat on the battlefield could have had disastrous implications, not only militarily but also politically, as the outcome of any battle was viewed as God's will and a defeat could be interpreted as God's disfavour towards the emperor.[33]

Whether this defensive strategy was successful or not depends on what the strategic goal was. If we consider the aim to have been the protection of the capital and its environs, and the prevention of any permanent establishment of the Arabs in the Anatolian plateau, then certainly this policy of defence in depth produced satisfactory results. Where it largely failed, however, was in the protection of the frontier populations. While fewer and fewer Arab raids were reaching the Bithynian and Aegean coasts by the second half of the ninth century, and the empire was in no danger of being destroyed, the incursions over the Taurus and Anti-Taurus Mountains and into Cappadocia, Seleucia, Charsianon and, especially, Cilicia – where the greatest disasters occurred for the Byzantine army – went on every year for decades to come at the expense of local populations and economies.

The most important strategic development for the region of eastern Asia Minor in the ninth century was the establishment of three Muslim bastions used for razzias into the Anatolian plateau at Tarsus, Melitene and Theodosiopolis.[34] The latter was the capital of the Umayyad emirate of Kalikala (est. 700), but it was the least powerful of the three as it was just a Muslim enclave in an otherwise Christian region. It was, however, strategically important for the Byzantines because it commanded two invasion routes into Armenia, one from Theodosiopolis down the River Araxes and into Armenia, and another over the hills to Manzikert. In 931, and again in 949, Byzantine forces led by Theophilus Curcuas, grandfather of the future emperor John Tzimiskes, captured Theodosiopolis, expelling its Muslim population and resettling it with Greeks and Armenians.[35]

Melitene was conquered by the Umayyads as early as 638 and rose to prominence in the ninth century as a base for razzias into Anatolia under its semi-independent emir, Umar al-Aqta. It was significantly weakened after the emir's defeat at the Battle of Lalakaon in 863, then besieged by Basil I in 873 and again in 882, but it was only recovered for the empire during the campaigns of John Curcuas in 927–34. Melitene was also the only of the three towns in question situated west of the Anti-Taurus Mountains. Its location gave the Arabs a secure base from where to launch their raids in the plateau without having to encounter any resistance from the natural defences of Anatolia.[36]

Finally, the most strategically important city of the three was the Cilician city of Tarsus, in Abbasid hands just two years before Caliph al-Muʿtasim launched his Cappadocian campaign in 831. It passed under Tulunid control in 878 and the Byzantines besieged it unsuccessfully in the same year, and again in 882.[37] The Abbasid caliph recovered direct control of the frontier regions in 896, including Tarsus, until they passed into the hands of the Hamdanids in the middle of the tenth century. In fact, the Syrian Muslims often showed themselves disinclined to accept caliphal control, preferring the authority of their local emirs instead, who were launching annual razzias in the jihadist spirit of the *thughūr*. Contemporary sources describe extensive barrack-style accommodation and the gathering of volunteers from all corners of the Muslim world.[38]

The Relationship between the Geography of Anatolia and the Arab Invasion Routes

The value of geography to the strategist has been long recognised.[39] Vegetius noted that 'the good general should know that a large part of a victory depends on the actual place in which the battle is fought'; fourteen centuries later, Clausewitz wrote that 'in these ways the relationship between warfare and terrain determines the peculiar character of military action'.[40] This is what modern historians have identified as 'military geography', which 'assists the formulation, preparation, and execution of military plans ... [and] provides the foundation for, and the means to develop, a coherent and selective mission-oriented assessment of the environmental matrix at the tactical, operational, and strategic levels'.[41]

Therefore, as in every conflict, the outcome of a campaign relies on how well a military leader can grasp and take advantage of both the physical (the diversity of terrain features, weather patterns, etc.) and the human landscape (political structures, population distribution and settlement, road networks, etc.) that affect a military operation. For a historian to grasp the full extent of the strategic threat and danger that the Arab raids launched from the aforementioned cities posed to the economy of Anatolia, first they need to have an understanding of the geography of Asia Minor and, particularly, of the region to the east of the Cappadocian capital city of Caesarea.[42] Otherwise, it is difficult to surmise all the parameters that determined the *how*, the *why*, the *where* and the *when* war was conducted in the region and period under discussion.

The basic characteristic of the geography of the Asia Minor peninsula is the contrast between the high mountains along all its coasts and the high plateau that is formed in its heart. The Black Sea coast is dominated by a range of steep mountains that extend along the entire length of the coast, separating it from the inland Anatolian plateau. To the west, towards Bithynia, the mountains tend to be low but they rise in the easterly direction to heights greater than 3,000 metres in the Pontic Mountains – a range that runs roughly east–west, parallel and close to the coast, and extends north-east to Georgia and the Caucasus. Anatolia's Mediterranean coast is separated from the interior by steep ranges, known as the Taurus Mountains, that run along the entire length of the coast. The south-facing slopes in Lycia and Pamphylia rise steeply from the Mediterranean coastal plain, but slope very gently on the north side towards the Anatolian plateau. Stretching inland from the Aegean coastal plain, the Anatolian plateau occupies the area between the two mountain zones of the coastal ranges in the north and south. The semi-arid highlands of Anatolia are considered the heartland of the country at an elevation of 600–1,200 metres from west to east. They contain basins of fertile agricultural land named after the cities located on their edges, like Ikonion (modern Konya), Melitene (modern Malatya) and Caesarea (modern Kayseri), although the view of a self-sufficient – but not in any way rich – economy should not be distorted by today's modern methods of farming and irrigation.[43]

The most rugged country can be found in the east where the Pontic and Taurus mountain ranges converge with the Armenian Highlands to form a formidable geological barrier to north–south movement towards the heartlands of Mesopotamia and Syria.[44] The Taurus is a mountain complex that divides the Mediterranean region of Anatolia from the central Anatolian plateau, extending along an arc from Antalya and the Pisidian interior in the west to the upper reaches of the Euphrates and the Tigris Rivers in the east. More rugged and less dissected by rivers than the Pontic Mountains, the Taurus rises sharply from the coast to high elevations, reaching altitudes of over 3,700 metres north of Adana.[45] The region where the Taurus meets the Armenian Highlands and the Hakkari mountain range is known as the Anti-Taurus and can be seen as the continuation of the Taurus in a south-west to north-east arc; Lake Van is also in the area, at an elevation of some 1,500 metres, along with the headwaters of the Tigris and the Euphrates.

To the south and east of the Taurus and Anti-Taurus Mountains lies the zone of the so-called 'Fertile Crescent'. This zone stretches up the Nile through Egypt, north via Palestine and Syria, to the plains of northern Syria, borders the mountains of Anatolia and Armenia and then turns south, following the Tigris and Euphrates Rivers through Iraq and the Persian Gulf, barred from the east by the Zagros Mountains in Iran. This zone, and in particular the areas of northern Syria and northern Mesopotamia, which have a higher average of annual rainfall, along with the regions to the north of the Taurus and Anti-Taurus Mountains, became the operational theatre of war between the Byzantines and the Muslims throughout the ninth and tenth centuries. Whittow has distilled the strategic importance of the Fertile Crescent in one major point: in the pre-industrial society of the medieval Middle East, thus long before the discovery of oil in the region, any state of more than merely regional significance had to control one or more of the principal agricultural zones, and any lasting hegemony was probably impossible without the resources of either Egypt or, until the ninth century, Iraq.[46] With all of the above in mind, what were the invasion routes taken by the Arab raiding parties to lead them over the Taurus and Anti-Taurus into Anatolia?

> The road which they may plan on taking might lead from the passes in Seleukia and the theme of Anatolikon, up to the Taurus Mountains which border on Cilicia, as well as Cappadocia and Lykandos. In addition, there are the regions about Germanikeia and Adata, also Kaisum Danoutha, Melitene and Kaloudia,[47] and the region beyond the Euphrates River bordering on the country called Chanzeti, and the hostile country as far as Romanoupolis.[48]

Strategies and Campaigning Tactics

In this short paragraph, Anderson identifies the three major invasion routes of the Arabs from their bases at Tarsus, Melitene, Germanikeia (Marash) and northern Mesopotamia into imperial territories through the passes of the Taurus and Anti-Taurus mountains – the natural frontier between the Anatolian plateau, and Mesopotamia and Syria.[49] The first invasion route took the armies from Tarsus, Anazarbos and Adana either south-west to the coastal themes of Isauria and Pamphylia, targeting the port-cities of the south coast like Seleukia, Sykai, Attaleia and Myra, although these targets were more often preferred for naval raids from the Syrian ports, as we see in the case of the 860 raid mentioned by Tabari. If the target of the raiding party was Cappadocia and Charsianon, then one available route was through Anazarbos and across the Anti-Taurus through the River Saros and Kiskisos to Caesarea, or the more direct pass of the Cilician Gates and Podandos. From there, they could either march north to Develi Kara-Hisan and Caesarea or turn west through another pass called 'Maurianon' by way of Tyana and Loulon – a Byzantine fortress-*aplekto* commanding the northern approaches to the Cilician Gates – which was the regular route across the Taurus into Cilicia.

The second route had Germanikeia as its starting point and through the mountain pass of Adata and the valley of the River Pyramos the leader of the invading parties could proceed north-west to Caesarea following the old Roman road almost always taken by Byzantine armies through Arabissos (a second route is through Kokusos), the Kuru Tchai and Arasaxa.[50] The Abbasid army under Afshin, which invaded Byzantine lands in 838, had followed this road up to Tzamandos and then proceeded north to Sebasteia, while the previous year Emperor Theophilos had marched through there to sack Armosata and Sozopetra.[51]

Finally, the armies leaving Melitene had to march over the Anti-Taurus through the Tokhma Su (Melas), the Godilli Dagh and the *kleisoura* of the Lykandos-Tzamandos,[52] to Tzamandos and Caesarea; the northerly route could also take them to Sebasteia and Amaseia through the Kuru Tchai and the theme of Tephrike-Leontokome (est. 879). Chanzeti was the Byzantine name for the city of Anzitene, the military centre of which was the fortress Hanzit, one of the Greek frontier fortresses near the Euphrates, between Melitene and Samosata. According to Anderson, Romanoupolis has been identified with Palu, the *kleisoura* which lies on the road between Palu and Kharput that led to Khliat and Lake Van.[53]

We should note that the Arabs rarely succeeded in advancing further inland in the Anatolian plateau, as there were several cities blocking their advance. In addition to that, the pattern of roads and the network of communications in Asia Minor were subject to constraints, with armies

– whether large or small – having to face several difficulties when crossing or campaigning in Asia Minor, in particular the long stretches of road through arid and exposed countryside and the rugged mountainous terrain separating coastal regions from the central Anatolian plateau. It is precisely these features that the author of *On Skirmishing* advises his readers to use against an invading force:

> The general should take all his infantry and cavalry and again move in front of the enemy. He should occupy the mountain heights and also secure the road passing through. And since all the roads, as we said, leading to the enemy's country through all the themes which we have listed and which we have seen with our own eyes are difficult to travel, being in the mountains which form the frontier between both countries [Taurus Mountains], hasten to seize passes before they do and without delay launch your attack directly against them.[54]

In theory, the road network of Asia Minor could lead an invading army as far west as Nicaea and the Bithynian coasts. Hence, from Sebasteia, the Arabs could march north-west to Amaseia and then directly north to the rich Paphlagonian ports of Sinope and Amisos, or further west to Ankara following the valley of the River Halys and straight to Doryleum or Amorium into the heart of the Anatolian plateau. From Caesarea, they could have proceeded north to Charsianon and then west to Ankara or further north to the Pontic ports. The Arab raids in their majority, however, did not proceed further inland than the themes of Cappadocia, Charsianon, Lycaonia and Isauria, and no siege of a major city is reported; instead, the sources inform us of the struggles for smaller but key strategic fortresses such as Loulon, Koron, Adata (Hadath) and so on.[55]

In this chapter, my intention has been to highlight the kind of warfare that dominated in the region of the eastern borders of the empire that neighboured its Muslim enemies, and to explain the strategy and strategic goals of each opponent in the region of eastern Anatolia for the period leading up to the Byzantine expansion of the mid-tenth century. The objective was to bring to the foreground the dangers posed to the policies of the imperial government, and to the stability of the empire and its provinces in general, by the Muslim emirates that had sprung up and grown into formidable rival powers along the frontiers that were broadly defined by the mountain ranges of the Taurus and Anti-Taurus. In a sense, the underlying aim was to illustrate the way that these razzias over the Taurus and Anti-Taurus contributed to the political instability, militarisation and economic, commercial, agricultural and demographic decline of central and eastern Asia

Strategies and Campaigning Tactics

Minor, and to highlight how and why these emirates had become a 'thorn' in the side of the empire's eastern frontiers.

By doing so, I have expanded on several questions that focused on the key role played in this conflict by the cities of Tarsus, Melitene and Theodosiopolis and in what precise ways the regional geography of eastern and central Anatolia shaped the type of warfare that was waged in the region. In relation to the last point, of particular importance were the invasion routes taken by the Muslim raiding parties that led them over the Taurus and Anti-Taurus into Anatolia and how these were defined by the topography of central and eastern Anatolia with its high plateaus, river valleys and mountain ranges that could funnel invading armies into the heart of Asia Minor.

Accordingly, I emphasised the 'shadowing' of the enemy forces as the most important aspect of Byzantine strategic planning, a strategy which can be broken down into two basic approaches: (1) following and harassing the enemy by exploiting one's own knowledge of the local terrain, and (2) keeping a close watch on the enemy's column and camp in order to attempt ambushes on forage parties. Put simply, large-scale expeditions launched in September and led by the emir himself were left to invade friendly territory, while being followed closely and harassed by detachments of picked men who controlled the passes through which they would return home. The invaders' logistical difficulties would then be maximised by the clearing of the local population from the cities and villages, and the removal of livestock and crops, or even their destruction.

Two key aspects of the defence-in-depth strategy demonstrate the degree of decentralisation and autonomous command structure that distinguished the local thematic armies of the period. One distinctive feature was the pincer movement designed to clear out the enemy forces by having several smaller friendly forces converging on the area from the neighbouring themes. Another was the degree of independence of the local commanders when it came to making decisions; they were encouraged to attack the enemy when opportunity arose and organise regular raids over the border to force the enemy commander to return home and protect his people. Naturally, the Byzantine commanders were not always able to respond successfully to the Arab raids and in the following chapter I will demonstrate the difficulties faced by captains operating in the mountainous regions of northern Mesopotamia, Cilicia and Cappadocia when confronted with a classic raid of this kind. Failure to shadow the invading forces in the manner described by contemporary tacticians, along with inadequate intelligence, could lead to the defending forces being outmanoeuvred by an experienced commander, with disastrous results.

Notes

1. *On Skirmishing*, 7.4–10, p. 162.
2. For more on these Muslim administrative zones, see Asa Eger, *Islamic–Byzantine Frontier*, pp. 23–181; M. Miotto (2015), 'Ααουάσιμ και Θουγούρ, το στρατιωτικό σύνορο του Χαλιφάτου στην ανατολική Μικρά Ασία' [*Awāṣim* and *Tuġūr*: The Military Frontier of the Islamic State (Caliphate) in Eastern Anatolia], *Byzantiaka*, 32, pp. 133–56; P. Wheatley (2000), *The Places Where Men Pray Together: Cities in Islamic Lands, Seventh Through the Tenth Centuries*, Chicago, IL: University of Chicago Press; Bonner, 'Naming of the Frontier', pp. 17–24; Bonner, 'Emergence of the *Thugur*'; Bosworth, 'City of Tarsus', pp. 268–86; M. Canard (1986), 'Al-ᶜAwāṣim', in P. Bearman et al. (eds), *Encyclopedia of Islam*, I:A–B, Leiden: Brill, pp. 761–2.
3. *On Skirmishing*, 14.45–8, p. 192.
4. *On Skirmishing*, 14.4–8, p. 191; 17.4–13, p. 204; 20.4–8, p. 218. Cf. *Strategikon*, 8.I/30, p. 82; 9, p. 97.
5. Bonner, *Jihad in Islamic History*, especially chs 1 and 5.
6. *On Skirmishing*, preface, pp. 146–9; 3, pp. 154–7; 15.7–12, p. 198. I am not implying that God was invoked only when fighting against the Muslims or heretics, but rather to underline the religious overtone of the conflict in the East: 'There seemed nothing grand [in fighting] the barbarians in the West . . . but were he [the emperor Romanus III] to turn to those living in the East, he thought that he could perform nobly' (Psellus, *History*, p. 28). Stouraites has highlighted the use of God and religion in the political legitimisation of warfare as well as in the demarcation of an enemy group, whether Muslim or Christian; see Stouraites, '"Just War" and "Holy War"', pp. 227–64; I. Stouraites (2011), '*Jihād* and Crusade: Byzantine Positions towards the Notions of "Holy War"', *Byzantina Symmeikta*, 21, pp. 11–63 (specifically for Leo VI's *Taktika* 19–26). On the religious element in Byzantine wars, see J. F. Haldon (1995), '"Fighting for Peace": Justifying Warfare and Violence in the Medieval East Roman World', in R. W. Kaeuper, D. G. Tor and H. Zurndorfer (eds), *The Cambridge World History of Violence*, vol. 2: AD 500–AD 1500, Cambridge: Cambridge University Press, forthcoming; N. Oikonomides (1995), 'The Concept of "Holy War" and Two Tenth-Century Byzantine Ivories', in T. S. Miller and J. Nesbitt (eds), *Peace and War in Byzantium: Essays in Honour of G.T. Dennis*, Washington, DC: Dumbarton Oaks Research Library and Collection, pp. 62–86; A. Kolia-Dermitzaki (1991), *Ὁ βυζαντινός «ἱερός πόλεμος». Ἡ ἔννοια καί ἡ προβολή τοῦ θρησκευτικοῦ πολέμου στο Βυζάντιο* [Byzantine 'Holy War': The Concept and Evolution of Religious Warfare in Byzantium], Athens: Historikes Monografies 10.
7. The anonymous author of *On Skirmishing* clearly identifies exactly who was the aggressor and who the defender: preface, p. 147; 1.4–12, p. 151; 4.36–9, p. 158.

8. *On Skirmishing*, 14.4–8, p. 191.
9. Ibid., 20.4–8, p. 218.
10. Ahrweiler, 'L'Asie mineure', pp. 7–12. For a more detailed analysis, see J. Howard-Johnston (2010), *Witnesses to a World Crisis: Historians and Histories of the Middle East in the Seventh Century*, Oxford: Oxford University Press, pp. 495–516; R.-J. Lilie (1976), *Die Byzantinische Reaktion auf die Ausbreitung der Araber: Studien zur Strukturwandlung des byzantinischen Staates im 7. und 8. Jhd*, Munich: Institut für Byzantinistik und Neugriechische Philologie der Universität München.
11. J. F. Haldon (2001) *The Byzantine Wars: Battles and Campaigns of the Byzantine Era*, Stroud: Tempus, pp. 67–97; Runciman, *Romanus*, pp. 124–5.
12. *On Skirmishing*, 3.4–10, p. 155.
13. *On Skirmishing*, 4, pp. 156–9. Cf. Leo VI, *Taktika*, 17.60, p. 415. Kekaumenus also warns the general not to return by the same road for fear of being ambushed by the enemy while laden with booty; see Kekaumenus, pp. 25–6.
14. *On Skirmishing*, 12, pp. 186–8; 16.53–5, p. 202; 23.38–40, p. 230. Cf. Leo VI, *Taktika*, 17.62, p. 417.
15. Ibid., 9, p. 171; 10, 35–45, 80–90, pp. 175–6; 14, pp. 195–7. See also Kekaumenus, pp. 21, 23.
16. Frontinus, *Stratagems*, 1.iv–vi, pp. 27–55; Onasander (1928), *Strategikos*, in The Illinois Classical Club (ed.), *Aeneas Tacticus, Asclepiodotus, Onasander*, Loeb Classical Library, London: Heinemann, 7, pp. 403–5; 10.2, p. 415; *Vegetius: Epitome of Military Science* (2001), trans. N. P. Milner, Liverpool: Liverpool University Press, 3.10, pp. 86–9; *Strategikon*, 10.2, pp. 107–8; Leo VI, *Taktika*, 9.25 and 27–8, p. 162–4; 17.37, p. 406.
17. *On Skirmishing*, 2, p. 153; 8, p. 165.
18. Ibid., 4, pp. 156–9.
19. Ibid., 20, pp. 218–19.
20. B. Ekkebus (2009), 'Heraclius and the Evolution of Byzantine Strategy', *Constructing the Past*, 10, pp. 73–96.
21. Theophanes Confessor, *Chronicle*, pp. 624–5.
22. Haldon, *Byzantium in the Seventh Century*, pp. 208–53; J. F. Haldon (1993), 'Military Service, Military Lands, and the Status of Soldiers: Current Problems and Interpretations', *Dumbarton Oaks Papers*, 47, pp. 10–12; W. E. Kaegi (1989), 'Changes in Military Organisation and Daily Life on the Eastern Frontier', in *He kathemerine zoe sto Byzantio*, Athens: National Institute of Byzantine Research, pp. 507–21.
23. *Kleisourai* were special frontier districts which constituted independent commands and were created from subdivisions of the *themata*, the *turmai*, from which they were detached. The first *kleisourai* were established in Seleukeia, Charsianon and Armeniakon in the first half of the ninth century. This system suggests an awareness in Constantinople of the need for greater autonomy at the local level; see Haldon, *Warfare, State and Society*, pp. 79, 114. Kaegi has questioned the efficiency of this system in defending the

Anatolian hinterland: W. E. Kaegi (1967), 'Some Reconsiderations on the Themes: Seventh–Ninth Centuries', *Jahrbuch der Österreichischen Byzantinistik*, 16, pp. 39–53.

24. Haldon, *Warfare, State and Society*, pp. 77–85.
25. Although this was only one of the two divisions that had invaded Anatolia: K. Y. Blankinship (1994), *The End of the Jihad State: The Reign of Hisham Ibn Abd Al-Malik and the Collapse of the Umayyads*, New York: State University of New York Press, pp. 168–73.
26. Haldon, *Byzantine Wars*, pp. 78–82; Treadgold, *Byzantine State and Society*, pp. 437–41.
27. G. L. Huxley (1975), 'The Emperor Michael III and the Battle of Bishop's Meadow (A.D. 863)', *Greek, Roman, and Byzantine Studies*, 16, pp. 443–50; Haldon, *Byzantine Wars*, pp. 83–7.
28. Al-Tabari (1985), *History: The Crisis of the Abbasid Caliphate*, trans. G. Saliba, New York: State University of New York Press, 34, p. 167.
29. Haldon, *Warfare, State and Society*, pp. 78–9.
30. I have put the modern term 'no-man's land' in inverted commas because it is easy for such terminology to lead to misconceptions regarding the habitation patterns and socio-economic development of the Islamic–Byzantine frontier regions, associating them with images from the desolate areas between the opposing trenches during World War I. The perception that the frontier zones were somehow 'empty' has been revised by Asa Eger who, rather, puts forward three layers of interaction between the diverse societies that penetrated the vaguely demarcated frontiers: external (competition for resources), internal (relationship between the centre and the periphery) and ideological (military and religious conflict); see Asa Eger, *Islamic–Byzantine Frontier*, pp. 1–21.
31. P. Magdalino (2010), 'Byzantium = Constantinople', in L. James (ed.), *A Companion to Byzantium*, Oxford: Blackwell, pp. 43–54.
32. I underline Leo VI's deliberate policy of non-campaigning (the first emperor to adopt such a policy since Justinian), which became a trend until the advent of Nicephorus Phocas in 963. See S. Tougher (1998), 'The Imperial Thought-World of Leo VI, the Non-Campaigning Emperor of the Ninth Century', in L. Brubaker (ed.), *Byzantium in the Ninth Century: Dead or Alive?*, Aldershot: Ashgate, pp. 51–60. See also C. Holmes (2010), 'Provinces and Capital', in L. James, *A Companion to Byzantium*, Oxford: Blackwell, pp. 55–66; J. Haldon (1995), 'Strategies of Defence, Problems of Security: The Garrisons of Constantinople in the Middle Byzantine Period', in G. Dagron and C. Mango (eds), *Constantinople and its Hinterland*, Aldershot: Ashgate, pp. 143–55.
33. For example, *Strategikon*, 7.11, p. 72.
34. J.-C. Cheynet (2001), 'La conception militaire de la frontière orientale', in A. Eastmond (ed.), *Eastern Approaches to Byzantium*, Aldershot: Ashgate, pp. 57–69.

35. Yahya ibn Said al-Antaki (1932–57), *Histoire de Yahya-ibn-Sa'īd d'Antioche, continuateur de Sa'īd-ibn-Bitriq* [hereafter *Histoire de Yahya*], ed. I. Krachkovskii and A. A. Vasiliev, *Patrologia Orientalis*, Paris: Firmin-Didot, vol. 18, fasc. 5, p. 768; *Theophanes Continuatus*, p. 428; Runciman, *Romanus*, p. 139.
36. For Melitene, see B. A. Vest (2007), *Geschichte der Stadt Melitene und der umliegenden Gebiete: Vom Vorabend der arabischen bis zum Abschluss der türkischen Eroberung (um 600–1124)*, Hamburg: Armenian Research Center.
37. Whittow, *Orthodox Byzantium*, p. 314.
38. Bosworth, 'City of Tarsus', pp. 268–86.
39. R. S. Harmon, F. H. Dillon III and J. B. Garver Jr (2004), 'Perspectives on Military Geography', in D. R. Caldwell, J. Ehlen and R. S. Harmon (eds), *Studies in Military Geography and Geology*, London: Kluwer, pp. 7–20; P. Doyle and M. R. Bennett (2002), 'Terrain in Military History: An Introduction', in P. Doyle and M. R. Bennett (eds), *Fields of Battle, Terrain in Military History*, London: Kluwer, pp. 1–7; K. Cathers (2002), '"Markings on the Land" and Early Medieval Warfare in the British Isles', in Doyle and Bennett (eds), *Fields of Battle*, pp. 9–17.
40. Vegetius, *Epitome of Military Science*, 3.13, p. 92; Clausewitz, *On War*, p. 109.
41. Harmon et al., 'Perspectives on Military Geography', p. 8.
42. For more on the joint research project run by the University of Birmingham and Princeton University, using simulation techniques to study the movement and sustainability of historic armies, modelled after the campaign of Romanus IV Diogenes in Anatolia that culminated at Manzikert in 1071, see J. F. Haldon et al. (2013), 'Marching across Anatolia: Medieval Logistics and Modeling the Mantzikert Campaign', *Dumbarton Oaks Papers*, 65/66, pp. 1–27. For information on specific provinces, see the *Tabulae Imperii Byzantini* (1976–2004), Vienna: Verlag der Österreichischen Akademie der Wissenschaften, 10 vols.
43. For an analysis of the physical environment and climate of Asia Minor, including a discussion about the problems associated with trying to relate climate and environment to changes in the social, political and cultural life of medieval Anatolia, see Haldon, *The Empire That Would Not Die*, pp. 215–48; Laiou and Morrisson, *Byzantine Economy*, pp. 8–13; Whittow, *Orthodox Byzantium*, pp. 29–30. For the difficulties encountered by an army on the march whilst travelling across Anatolia, see J. F. Haldon (2006), 'The Organisation and Support of an Expeditionary Force: Manpower and Logistics in the Middle Byzantine Period', in N. Oikonomides (ed.), *Το εμπόλεμο Βυζάντιο* [Byzantium at War], Athens: National Hellenic Research Foundation, pp. 131–2; B. S. Bachrach (2006), 'Crusader Logistics: From Victory at Nicaea to Resupply at Doryleon', in J. H. Pryor (ed.), *Logistics of Warfare in the Age of the Crusades*, Aldershot: Ashgate, pp. 43–62.

44. W. M. Ramsay (1903), 'Cilicia, Tarsus, and the Great Taurus Pass', *The Geographical Journal*, 22, pp. 357–410.
45. www.ecogeodb.com/ECO_Detail.asp?P=Climate&CN=Turkey&C=TUR
46. Whittow, *Orthodox Byzantium*, p. 31. See also M. F. Hendy (1985), *Studies in the Byzantine Monetary Economy, c. 300–1450*, Cambridge: Cambridge University Press, pp. 613–18; J. L. Teall (1959), 'The Grain Supply of the Byzantine Empire, 330–1025', *Dumbarton Oaks Papers*, 13, pp. 87–139, especially pp. 117–39.
47. Hellenised form of the Arabic name for Claudias. This fortress was situated on the Euphrates near Melitene; see J. G. C. Anderson (1897), 'The Road-System of Eastern Asia Minor with the Evidence of Byzantine Campaigns', *Journal of Hellenic Studies*, 17, p. 30.
48. *On Skirmishing*, 23.7–13, p. 229.
49. On these invasion routes, see Ahrweiler, 'L'Asie mineure', pp. 8–10; Anderson, 'Road-System', pp. 22–44; W. M. Ramsay (1972), *The Historical Geography of Asia Minor*, New York: John Murray, pp. 315–17.
50. For more on these roads and for the location of *Kaisum Danoutha* mentioned by the author of *On Skirmishing*, see Anderson, 'Road-System', pp. 28–9.
51. J. B. Bury (1909), 'Mutasim's March through Cappadocia in A.D. 838', *Journal of Hellenic Studies*, 29, pp. 120–1. See also Basil I's campaign against Tephrike in 872, which followed the same route: J. G. C. Anderson (1896), 'The Campaign of Basil I against the Paulicians in 872 A.D.', *The Classical Review*, 10, pp. 136–40.
52. *DAI*, 50.154–9, p. 240.
53. Anderson, 'Road-System', pp. 26–7; *DAI*, 50.127–32, p. 238.
54. *On Skirmishing*, 24.65–77, p. 235.
55. For a guide on the relationship between information, geography, topography, road networks and fortifications in medieval Anatolia, including a very useful categorisation of the different types of strongholds and systems of fortifications developed by the Byzantines after the Muslim expansion, see J. F. Haldon (2013), 'Information and War: Some Comments on Defensive Strategy and Information in the Middle Byzantine Period (ca. A.D. 660–1025)', in A. Sarantis and N. Christie (eds), *War and Warfare in Late Antiquity*, 2 vols, Leiden: Brill, pp. 373–93.

3
The Empire's Foreign Policy in the East and the Key Role of Armenia (c. 870–965)

> If these three cities, Khliat and Arzes and Perkri, are in the possession of the Emperor, a Persian [Arab] army cannot come out against Romania, because they are between Romania and Armenia, and serve as a barrier (φραγμός) and as military halts (απλίκτα) for armies.
> —*De Administrando Imperio*, 44.125–8, p. 204

This is, perhaps, one of the most significant statements for the strategic aims of the Byzantine governments in the tenth century, written in the years 948–52 by Emperor Constantine VII in his monumental work regarding the administration of the Byzantine Empire, intended for his son and heir to the throne, Romanus. It does not simply highlight the strategic importance of Armenia to the eastern frontiers of the empire, something which becomes more apparent to the reader of the *De Administrando Imperio* if they compare the length of the so-called 'Caucasian chapters' to the rest of the work, but it also underlines the strategic importance of the fortress-towns around Lake Van and the Diyar-Bakr as 'buffer zones' between Armenia and the caliphate – the towns of Khliat, Arzes, Perkri, Manzikert, Mayyafariqin and Amida. In order to understand the strategic role of these towns to imperial policy in the East in the first half of the tenth century, and the imperial expansion into northern Mesopotamia, we should first examine Byzantine foreign policy from the wars of Basil I against the Paulicians to the imperial armies sent against the cities of Armenia and the Jazira (Upper Mesopotamia) by Romanus I Lecapenus.

This chapter examines the political reasons behind the empire's involvement in Armenia and northern Mesopotamia in the first half of the tenth century, the wars with the Muslims, its delicate diplomatic negotiations with the Armenian princes, and the emergence of a new enemy in the East. It does not claim to break new ground in the study of the Byzantine Empire's foreign policy and diplomacy in Armenia in the tenth century; Jonathan Shepard has produced two magnificent papers on the Byzantine notion of frontiers and imperial expansionism and on the empire's foreign policy in the East, focusing on Armenia. This chapter, rather, is looking to put things into perspective and explain a paradox: since Armenia was strategically far more important to the Byzantine government than Cilicia and Syria, Constantinople did not contemplate any territorial expansion in the region but rather the forging of diplomatic ties with the local *naxarars*.[1] If that was the case, then how can we explain the extensive gains of territory in Cilicia and Syria in the 950s–60s and the massive mobilisation of manpower for a war that lasted for decades and reached legendary proportions on both sides of the borders? What led to this escalation of violence between Constantine VII and Sayf ad-Dawla of Aleppo? What indications do we have about this change of policy by the empire and when exactly can we trace it in time? What were the deeper and long-term implications for Byzantium's strategic thinking, its military and political organisation?

Byzantium's Policy in the East: From Basil I to Romanus I

Byzantium's aggressive policy in the East scored its first success as early as the 870s during the reign of Basil I (867–86). Basil's reign was marked by the troublesome ongoing war with the heretical Paulicians, centred on Tephrike on the Upper Euphrates, who rebelled, offered their alliance to the Arabs and raided as far as Nicomedia, Nicaea and Ephesus, which they sacked. In the first years of his reign, Basil's attention was focused on the West and the imperial possessions in Dalmatia and Sicily, which were threatened by the Muslims of Sicily. The empire's inability to devote adequate land and naval forces in both operational theatres manifested in the period between 868–71. After the establishment of the theme of Dalmatia (868), a failed naval expedition against the Muslims of Sicily and an ill-judged anti-Arab alliance with the Frankish emperor Louis II that cost him Calabria, Basil led an army against the Paulician leader Chrysocheir in the spring of 871 that was defeated and he himself was nearly captured. A second expedition the following year, this time led by the emperor's son-in-law Christopher, Domestic of the Scholae, caught up with Chrysocheir near Dazimon and the Paulician leader was finally captured and killed.

With the Paulicians significantly weakened and ensconced in the last remaining stronghold of Tephrike, and with their Arab allies on the defensive, the Byzantine government could focus on stabilising its eastern frontiers and neutralising any Arab threats in Armenia and west of the Taurus and Anti-Taurus Mountains. Basil led an expedition against Melitene in 873 which, although failing to take the city, sacked Sozopetra and Samosata and several of the remaining Paulician strongholds in the region. By the end of the decade, the Byzantines would claim victories over the Tarsiots at Podandus and Adata, while a raiding army would reach northern Mesopotamia through Germanikeia and Adata, which they plundered.[2]

The emperor won valuable support in the East by breaking an alliance with the Armenian Ashot I. Ashot was the most powerful of the Armenian princes – belonging to the important noble family of the Bangratids – and was handed the crown by the Abbasid caliph, al-Mu^{c}tamid, in 885, pursuing the Arab policy of appointing an Armenian prince as chief client in the region along with an Arab governor (or *ostigan*) during the last century. Even though Ashot chose to remain on good terms with both Constantinople and Baghdad, his son and successor Smbat followed a clearly pro-Byzantine policy, which diverged from that of his Arab overlords and provoked a strong reaction from Yusuf, the *ostigan* of Armenia – namely, a large-scale invasion that saw Smbat replaced by Gagic Arsdrouni of Vaspourakan in 909.[3] Although this alliance would nevertheless prove invaluable for Leo's successors, for the time the situation in Armenia seemed precarious and to be leaning in favour of the Arabs. On the rest of the frontiers of the empire there had been little significant change during the last two centuries, despite the annual raids and counter-raids that penetrated the frontiers in the summer seasons. Some strategic successes, however, are noted by Constantine VII in his *De Administrando Imperio*, notably the occupation and rebuilding of the important *kleisourai* of Lykandos and Tzamandos on the Mesopotamian frontier and the annexation of the territory of Prince Manuel of Tekes, which was established as the theme of Mesopotamia by Leo VI.[4]

Until the death of Leo VI in 912, the Byzantines had not committed themselves to any definite war in a specific part of their frontiers in the East. From 915 onwards the attention of the government would be focused on Armenia – with intermediate breaks due to the wars with Symeon – and it was a period that would signal the first major counterattack by the empire on its eastern borders since the Arab expansion in the seventh century. The state of Armenia in 913 was pitiable; Smbat had surrendered himself to Yusuf and been executed, and his son Ashot had sought refuge in Constantinople. The emir, Yusuf, ruled the whole of Armenia from the city of Dvin, and almost all the Christian kingdoms south and east of the Caucasus, such as Iberia and Abasgia, were either completely within the Muslim sphere of influence or powerless to offer any help to Armenia.[5]

Constantinople may have looked like the last resort for a desperate king, but the affairs of the empire were not in a good state either. Alexander had reigned for just one year (912–13) and the next two years were dominated by the power struggle between the rival regents of the eight-year-old Constantine VII, Empress Zoe and the Patriarch Nicolas Mysticus, and the ever-present threat from Symeon. However, in spite of the political instability in the capital and the Bulgar khan who was threatening Adrianople and the environs of Constantinople, Empress Zoe answered an appeal for help by Ashot. The Byzantine government was fully aware of the strategic importance of Armenia, not just for the stability of the eastern frontiers of the empire but for the safety of the whole of Anatolia, and it seemed determined not to allow the region to fall under the control of the Abbasids. Therefore, Zoe's government launched an aggressive policy to expand its sphere of influence in the region between the Caucasus and the northern Euphrates; it was an effort to establish control and order in this strategic but volatile region, as the series of campaigns led by Curcuas in Armenia and Mesopotamia in the 920s–40s demonstrate. Later on, I return to the question of whether this was part of a conscious long-term policy of territorial expansion by the Byzantine governments of the period.

The campaigns of Byzantium on the Armenian and Mesopotamian frontiers can be divided into two periods: the first can be placed between the return of Ashot to Armenia in 915 and the end of the Bulgarian threat in 927, a peace treaty that enabled Romanus to release the empire's energies to the war in the East; the second period takes us to the coup of December 944 and the deposition of Romanus Lecapenus by his two sons. The strategic goals of the imperial armies throughout this period were the preservation of a pro-Byzantine Armenia – that was the case until 926, after which year it was only deemed necessary to prevent any aggressive movements by Subuk, emir of Azerbaijan (the successor to Yusuf),[6] and the establishment of control over the cantons of Taron and Vaspourakan, especially in the important towns of Khliat, Matzikert, Perkri and Arzes around the Lake Van,[7] and in northern Mesopotamia (Melitene, Samosata, Edessa and the regions facing the themes of Lykandos and Mesopotamia).

In 927, repeated devastations of its hinterland from armies led by Curcuas, his brother Theophilus and the *strategos* of Lykandos Mleh (Lykandos was upgraded to *thema* in 917)[8] forced the Emir of Melitene to agree to send troops to reinforce the Byzantine forces in exchange for an imperial decree of immunity from further attacks. In the same year, Curcuas' army raided the area of Samosata and managed to force himself into the Arab-held city of Dvin, before being flushed out by the population and the city's garrison.[9] Next year, Curcuas raided Vaspourakan and reduced the towns of Khliat and Bitlis, and Mleh attempted to infiltrate Melitene by performing a ruse. Although his attempt was discovered by the Melitenians, they accepted a Byzantine garrison at their citadel.[10]

The emir of Azerbaijan beat back further invasions of eastern Armenia by 929, but a significant turning point was the sacking of Theodosiopolis in 930 and the siege of Melitene in 934 – allegedly with some 50,000 troops under Curcuas and Mleh – that led to its surrender and the expulsion of its Muslim population.[11]

In 931, further attempts to besiege Samosata also brought the Byzantines into contact with the emerging emir, Nasir ad-Dawla (929–67), who was establishing his power in Mosul. It was to be a further five years until the final conquest of Samosata, but this time there was another Arab commander who would play a protagonist's role in the region's politics for the next four decades; the failure to prevent the capitulation of the city of Samosata marked the first contact between Sayf ad-Dawla, the 'Sword of the Dynasty', and the Byzantines.[12] A rebellion by the Daylamite governor of Arzan – east of Martyropolis – in the Jazira kept Sayf occupied for another year, but on 9 October 938 he inflicted a severe defeat on the domestic at Hisn-Ziyad (Harput), north-east of Armosata. According to Ibn Zafir, Sayf's *ghulam* (slave-soldier) troops attacked and dispersed a corps of 20,000 Greek 'patricians' in the centre of the Byzantine formation, thus turning the battle in favour of the Arabs.[13]

The emir managed to prevent the siege of Theodosiopolis in 939, taking by surprise the troops that were building a fortress near the city, while his march through Manzikert and Mayyafariqin went unopposed by the local Armenian toparchs. This success was followed up the following spring (940) by a campaign deep into Byzantine territory through Taron and Chaldia; Sayf marched north through the Bitlis Pass in the south-western shores of Lake Van, forcing several Arab-Armenian princes like Gagik of Vaspourakan and Ashot of Taron to submit before capturing numerous strategic fortresses in the region. He retreated to the Jazira only after the arrival of Curcuas – strange, if we consider that his brother Theophilos Curcuas was the general of Chaldia.[14] Ibn Zafir also writes about an audacious campaign by Sayf ad-Dawla against Coloneia, following an offensive letter sent to Romanus Lecapenus who had complained about his invasions at a time when Constantinople and Baghdad had been at peace since 938.[15]

Sayf's successes in Armenia against the domestic appeared, to both the Muslim and Christian worlds, to be an astonishing achievement, owing largely to the propaganda launched by poets like Mutanabbi, as I will show later. After nearly two decades of campaigning by the Curcuas brothers and Mleh, who enjoyed almost no resistance in their expansion of the Byzantine sphere of influence in Armenia and northern Mesopotamia, a new player was involved in the game. The young Muslim emir seemed determined not to leave Byzantium to dominate the strategic region of northern Mesopotamia and Armenia and to win back all the ground and cities that were lost to the Christians since 915, gradually emerging in the minds of the Muslims as the champion of the jihad against the infidels.

Byzantine campaigning in the East resumed in 942, after the Rus' attack on Constantinople had been thwarted, with Curcuas making up for the losses of the previous years by initiating a counterattack that was to last roughly three years. He raided as far south as Aleppo, in the winter (January) of 942, taking some 10,000 Arab prisoners from the town of Hamus according to one Arab source.[16] In the autumn of the same year, he moved north-west to Mesopotamia, penetrating the region of Diyar-Bakr and raiding the towns of Mayyafariqin, Arzan, Amida and Nisibis. During this time, Sayf was preoccupied with events in Baghdad and the struggle for power between several factions and dynasties, including the Buyids who eventually gained control of the Abbasid capital in 945.[17]

I should point out that in none of these cases were the Byzantines contemplating any kind of permanent territorial expansion – these remained just raids to capture prisoners that could later be exchanged for ransom money, and to enhance the emperor's influence in the region and his popularity back in the capital.[18] Melitene, one of the bastions of razzias in the region of Upper Mesopotamia, accepted a Byzantine garrison in its citadel but retained its Arab emir as the emperor's *kouratoreia*[19] until the 970s, when the first Byzantine governor is reported.[20] A more characteristic example for the non-expansionist policy of the Byzantines in this period, however, is the siege of the city of Edessa (summer 943– spring 944) and the capture of the Ιερόν Μανδήλιον (*Hieron Mandylion*, or 'Holy Shroud'), one of the most significant Christian relics (a towel on which Christ had allegedly dried His face and left His impression on it). Its return to the 'relic-hungry' city of Constantinople would constitute a major propaganda triumph for the precarious government of Romanus Lecapenus. Even though he had crowned himself emperor, made his son Christopher co-emperor, and married his daughter Helena to Constantine Porphyrogenitus in 921, many still viewed Romanus as a mere usurper of the throne. In order to secure this relic from the Muslim population of Edessa, Curcuas offered to spare the city and its population, release any prisoners and make peace; while the Edessans were awaiting their orders from Baghdad, the domestic spent his summer ravaging Mesopotamia.[21]

With a peace treaty signed (only to be broken by Sayf ad-Dawla in 950), 200 prisoners exchanged and a majestic triumph staged in Constantinople for the return of the victorious general with the Μανδήλιον, perfectly timed to coincide with the day of the Assumption of the Virgin

Mary on 15 August (944), the period of Byzantine–Arab wars in Armenia and northern Mesopotamia ended and was to be followed by six years of relative peace. These three decades saw the involvement of the empire in the politics of Armenia and the establishment of a pro-Byzantine government under the Bangratids, and the expansion of the Byzantine sphere of influence in two directions: (1) Armenia, Taron, Vaspourakan and Diyar-Bakr; (2) northern Mesopotamia, with the most strategically important conquests being those of Melitene and Edessa. For the first fifteen years, the Arabs seemed to be on the defensive, with the domestic John Curcuas launching several invasions almost annually, until the emergence of a new dynasty with two powerful brothers, who were to establish their emirates in the regions of northern Syria and Upper Mesopotamia, Nasir and Sayf ad-Dawla. The latter proved to be the nemesis of the Byzantine generals in the East for another quarter of a century.

But what, exactly, attracted Byzantium and the Hamdanids to the region of Armenia, Taron, Vaspourakan and northern Mesopotamia? For the former, the reasons can be divided into three categories: (1) political and diplomatic, (2) social and cultural, and (3) geographical. As Jenkins and Shepard have noted, chapters 43 to 46 of the *De Administrando Imperio* – the so-called 'Armenian chapters' – illustrate the empire's foreign policy in the East and offer a detailed account of the *kastra* (fortress-towns) and the local family connections in the principalities of Armenia proper, Taron, Vaspourakan (the environs of Lake Van) and Iberia: a rich and up-to-date body of material which stands out as being more coherent and more detailed with regard to current affairs in the mid-tenth century than any other section of that work.[22] The places in which Constantine was particularly interested were those on the borderlands with the caliphate, and specifically the regions of Taron and around Lake Van, such as Manzikert, Khliat, Perkri and Anzen, and the key city of Theodosiopolis. These chapters can be described best as a survey of these principalities, the main *kastra/kleisourai* of these regions and an attempt to explain the manipulation of local politics and family connections by imperial agents, such as in the case of Kritorikios of Taron who was 'honoured' with the title of *magistros* and a significant annual stipend, and was appointed as the military governor of Taron.[23] In other cases, Armenian potentates handed over their lands voluntarily when they were under pressure from other powerful magnates, as happened with Tornikios of Taron, who called on the emperor to take over his country and make him an imperial vassal in order to put an end to the depredations of his aggressive cousins.[24]

Constantine's special interest in the internal politics and family connections of the Armenian *naxarars* is certainly linked to Sayf ad-Dawla's expedition of 940. As we saw earlier, in that year Sayf managed to push deep into Byzantine territory in Chaldea after invading through the Bitlis Pass, forcing several Armenian princes in Taron and Vaspourakan to submit to him before subduing numerous strategic fortresses in the region. This aggressive policy of the Hamdanid emir in the Jazira and Lower Armenia, and the fact that he was able to penetrate so deep into territory that was deemed to be under Byzantine control after Curcuas' campaigns of the previous decade, would certainly have alarmed central government. I will examine the strategic importance of the fortress-towns around Lake Van later on, but the emperor's insistence on the control of these towns as a sort of 'buffer zone' is specifically stated at the end of chapter 44 of his *De Administrando Imperio* (written in 948–52, about a decade after Sayf ad-Dawla's expedition of 940):

> If these three cities, Khliat and Arzes and Perkri, are in the possession of the Emperor, a Persian [Arab] army cannot come out against Romania, because they are between Romania and Armenia, and serve as a barrier (φραγμός) and as military halts (απλίκτα) for armies.[25]

Imperial strategy dictated the administration and disposition of the empire's resources in men and material to the best possible effect; the entire history of Byzantium's foreign policy reflects this principle and the tenth century is no exception.[26] As Constantine VII writes in the preface to his work:

> I set a doctrine before thee, so that being sharpened thereby in experience and knowledge, thou shalt not stumble concerning the best counsels and the common good. First, in what each nation has power to advantage the Romans, and in what to hurt, and how and by what other nation each severely may be encountered in arms and subdued.[27]

What we see through the words of Constantine is the military edge of Byzantine diplomacy taking shape; if the emperor could succeed in enhancing his influence over the region of Armenia by handing out titles, money and court wives, and not by direct occupation that would have cost the government in men and money better spent elsewhere, that was the ideal foreign policy for the central government. The ground was more fertile in Armenia for the use of diplomacy than in any other region on the eastern frontiers of the empire.

The social and cultural reasons that drew Byzantium in Armenia and the region of the Caucasus pertain to the presence of Armenian migrants in Byzantium and the influence they would have exhorted over the shaping of the imperial foreign policy in the East. Armenia became the primary recruiting pool for the Byzantine army after the loss of the Balkans to the Avaro-Slavs in the sixth century, with the Armenians quickly turning into the most prominent group in the ranks of the imperial forces. In fact, Charanis believes that in the ninth and tenth centuries, at the height of the Middle Byzantine phase, the Armenians may have formed something like twenty-five per cent of the armed forces of the empire.[28] All of these soldiers would also have accepted the Chalcedonian rite of Orthodox Christianity without much difficulty, thus improving the religious ties between Constantinople and Armenia. As Whittow has pointed out, the Armenian *naxarars* (especially the Bangratids) wanted ties with Byzantium and were prepared to compromise in the field of ecclesiastical controversy, prompting attempts for the unification of the churches in the early tenth century.[29]

The policy of the Byzantine governments to recruit Armenians into the imperial army was also combined with the forced transfer of populations that has been recorded as early as the reign of Tiberius, in 578, when as many as 10,000 Armenians were removed from their homes and settled on the island of Cyprus.[30] Inherited from the pagan Roman Empire, this practice was frequently resorted to throughout the history of the Byzantine Empire and it can be viewed as a measure to defend large regions from foreign invasion. Political instability at home, especially after Armenia's conquest by the Arabs, and the threat of external enemies from the south and north also prompted Armenian (and Iberian–Georgian) populations to migrate voluntarily to imperial lands; until the mid-tenth century these populations would settle primarily in Lesser Armenia, the Pontic frontier and the themes of Cappadocia and Armeniakon.[31]

It was these troubled conditions at home and the attractions of Constantinople to the minds of several young aspiring Armenians that was to see a great number of individuals of Armenian origin rising through the levels of the aristocracy, acquiring titles and fame that would lead them to establish families which would dominate Byzantine politics in the tenth century. In a state as centralised as the Byzantine one, where military leadership could lead to political and social leadership, and even to the throne, the imperial court became a magnet to anyone with ambitions of power and glory. At the end of the ninth century, it was Leo VI who defined in straightforward terms the place of *eugeneia* (nobility) in his choice of generals and their subordinates and, in a way, he seems to have put merit on an equal footing with birth when it came to war.[32] In the tenth century, it was based on this quality of *eugeneia* that the generals exercising command in Asia Minor were appointed.

Therefore, impoverished individuals of Armenian origin such as Basil[33] and Romanus[34] – if one focuses on the ninth and tenth centuries – came to establish two imperial dynasties, one of which was among the most glorious in the history of the Byzantine Empire, the Macedonian or Basilid[35] (867–1056) and the Lecapenid (920–45). Others arrived in the capital to serve in the army and it was their military and diplomatic skills, their loyalty to the reigning dynasty and imperial favour that earned them high offices, titles, lands and money. In fact, from the middle of the eighth century, the Armenian element in the list of *strategoi* in Asia Minor during the reign of Leo IV, provided by Theophanes, was unambiguous.[36]

As there are numerous exhaustive studies that trace the origins and history of these great eastern families of Byzantium, it would suffice here to mention the names of the Scleroi, the Phocades, the Curcues, the Maleinoi, the Dalassenoi, the Musele and the Tornikioi.[37] These families would come to form a landed aristocracy in the provinces of eastern Anatolia, which would monopolise military offices in a hereditary manner and through an extensive network of clients and kinsmen on both sides of the borders, while they would also tend to exercise semi-feudal control over the local thematic troops, largely Armenian in composition, who were settled in their expanding estates, as the epic poem of *Digenes Akritas* and the *Strategikon* of Kekaumenus vividly illustrate.[38]

Finally, geographic, topographic and climatic factors dictated the importance of Armenia and the regions around Lake Van for the defence of Anatolia and for any potential offensive campaigns in Mesopotamia. One of the reasons was simply that the distance between Armenia and the centres of Muslim power in Mesopotamia, such as Baghdad or Mosul, was far shorter than that between them and the cities and ports of Cilicia. Other topographic factors include the terrain of Mesopotamia, the Taurus and Pontic Mountains, and the difficulties they presented to a campaigning army in terms of traversable passes and en-route supplies. As I have pointed out earlier, the Pontic Mountains are less rugged and more dissected by rivers than the Taurus and Anti-Taurus. This meant that any large army marching through the Cilician Gates or the other passes over the Taurus would have been more exposed to threats from smaller Byzantine detachments, which could easily have barred their invasion routes to the north, than if they were marching through Koloneia and Theodosiopolis.[39] Thus any army from Iraq or Syria could have bypassed the Taurus by marching north following the Euphrates and entering Anatolia through Taron and Vaspourakan.

As Kaegi has underlined in his article comparing the Mesopotamian campaigns of Julian (363), Heraclius (627–8) and John Tzimiskes (974), any successful invasion of Mesopotamia required the cooperation of the Armenian princes.[40] As the campaign routes through the Syrian Desert present great logistical difficulties for large armies, long-lasting Roman traditions advocated for a march through the alluvial plains of the Jazira and the banks of the Euphrates or the Tigris.[41] Along the Euphrates, supplies were available only through the local population, but its cooperation with an invading force could not be guaranteed. The Tigris route – following the east bank of the river – was less arduous and easier to supply, even though it was more vulnerable to surprise attacks. To follow this route, however, would have required a march through either the Bitlis Pass, immediately south-west of Lake Van, or the Rawanduz Pass, south-west of Lake Urmia, as in the cases of Heraclius' Persian campaigns of 625 and 627–8 respectively.[42]

From what I have pointed out so far, it is easy to understand that Armenia was strategically more important to the Byzantine government than Cilicia or Syria, and the central government did not contemplate any territorial expansion in the region but rather the forging of diplomatic ties with the local *naxarars* and the neutralisation of several key fortress-towns around Lake Van and Upper Mesopotamia. If that was the case, however, how can we explain the paradox of the extensive territorial gains on the other side of the empire's eastern frontiers – in Cilicia – in the third quarter of the century and the massive mobilisation of manpower for a war that lasted for decades? It all comes down to the personal and, as these were interconnected, political image of the Byzantine emperor as a sovereign chosen by God to rule by divine providence and protect God's people. In this case, it was the personal image of Constantine Porphyrogenitus and those of his predecessors, going back to the founder of the 'Macedonian' dynasty, Basil.

Constantine Porphyrogenitus was the fourth member of an imperial dynasty that was established some eight decades before he was crowned emperor at the age of forty in January 945, helped by a clique of palace officers faithful to his house. Since 920, he had been pushed aside by the co-emperor Romanus Lecapenus, the former *droungarios* (commander) of the imperial fleet, leaving him cut off from all power and patronage. Added to that was the fact that he was also the illegitimate, but only, son of Leo VI, born out of wedlock. Desperate to produce an heir to his throne, Leo VI had married his mistress Zoe Karbonopsina on 9 January 906, but only after she had given birth to the future Constantine VII at the end of 905. This, however, constituted his fourth marriage and was, therefore, uncanonical in the eyes of the Church. In order to help with his legitimation, his mother gave birth to him in the 'purple room' of the imperial palace, hence his nickname *Porphyrogenitus*.

The origins of Constantine's father (Leo 'the Wise') are also obscure and it is highly likely that he was the child of the deposed and murdered Emperor Michael III.[43] Whatever the case, Leo was crowned co-emperor in 870 at the age of four, but remained second in the line of succession to Basil's oldest son, Constantine, until the latter's death in 879. The way that the founder of the Macedonian dynasty, Basil, achieved power in Constantinople must have seemed repellent even to his contemporaries. His case is typical of a young man of obscure origins who drew the attention of Emperor Michael III in Constantinople, gradually being promoted and becoming a leading member of the imperial court (the emperor's bodyguards – the *excubitai*) and crowned co-emperor on 26 May 866. He finally murdered his patron on 24 September of the same year, after Michael had begun to favour another courtier – Basiliscus.

Coming from a family with such an obscure past, while himself acquiring full and unequivocal power in the capital only after a quarter of a century, it could have been expected that Constantine would seek demonstrative evidence for the aptness of his rule in the repertory of imperial ideology. Victory celebrations presented the perfect propaganda tool for any political person in Byzantium who wished to demonstrate to the real source of imperial power – the imperial court and the people of the capital – his *felicitas*, his divine favour and his political power. A remnant of the empire's past, it was since the days of the Roman Republic that even a mediocre achievement on the battlefield would be used to justify the celebration of a triumph; the Byzantine period was no exception, especially during the period of the 'Macedonians'.[44] In order to safeguard his position, Basil had exploited the victory celebrations in conjunction with the propagation of victorious omens, which significantly helped to enhance his authority as God's viceregent on earth[45] (frequently also described as *theophylaktos*, 'protected by God'), regardless of his humble origins.[46] In total, he had celebrated three triumphal victories in the capital, the first on the very next day after his rise to the throne and upon the arrival of news of the deliverance of Christian prisoners of war;[47] most important, however, was his second triumph after the conclusion of his 873 campaign in the East. Although his expedition against the towns of Samosata and Zapetra was a success, his failure outside Melitene forced him to divert his army to Paulician territory for a quick invasion campaign to avoid concluding the campaigning season with a defeat.[48]

The years between 945 and 949, however, did not offer many opportunities to celebrate a victory in the East, and the earliest recorded triumph of

The Empire's Foreign Policy in the East

Constantine came only after ten years in power, in 956. This was a period of relative peace between the Byzantines and Sayf ad-Dawla until Rabi^c II 336/October–November 947, when the latter confirmed his possessions in Syria and concluded a peace treaty with the Ikhshidids of Egypt – until then the nominal suzerains of the region. After that period, in the summer of 948, a Byzantine incursion into Syria was intercepted by Sayf, who managed to win back the booty and prisoners, while the only Byzantine success was the taking and destruction of Adata (Hadath) in 949.[49] The taking of Theodosiopolis is recorded by Yahya in Rabi^c I 338/September–October 949,[50] and it would have come too late in terms of planning for an expedition that would leave its mark on Constantine's reign as it had done on his father's three and a half decades ago.

The first years of the reign of Constantine VII are linked to the failed expedition to conquer Crete from the Umayyad Arabs in 949. Since its capture in 837, the island had increased pirate activity in the southern Aegean Sea and the eastern Mediterranean to a degree that significantly affected commercial navigation in the region and towards the capital; hence, military action was deemed necessary. The emperor himself noted the depredations of pirates in the islands of the Cyclades and the Aegean,[51] but perhaps the most vivid description of the danger posed by the Arabs comes from Leo the Deacon:

> Now this Emperor Romanos [II] decided to eradicate, with the support of the Almighty, the tyranny of the Arabs of Crete, who were arrogant and had murderous intentions against the Romans. For they exalted immeasurably over the recent disaster suffered by the Romans, and were plundering the shores of the Roman Empire on a large scale.[52]

This description should be coupled with the alleged speech of Nicephorus Phocas to his officers after the capture of Chandax in 961:

> I think that none of you is unaware of the cruelty and ferocity of the descendants of the maidservant, and the raids and enslavement that they have murderously perpetrated against Romans (and this when they were living on an island that was subordinate to [the Romans], although it had come to the Agarenes through the wickedness of fortune) . . . Therefore Providence has by no means tolerated that these liars, these most evil beasts, these lazy gluttons feed forever off the Christian people, but with the help of the Almighty, it has brought us here to repay them sevenfold the evil fortunes they have mercilessly brought upon us.[53]

But why was Crete such a thorn in the side of the empire and what was its strategic importance?

From a navigational point of view, Crete was the key to the eastern Mediterranean. Its geographic importance in the east can be compared to the one of the Balearics for the western Mediterranean and Sicily for the central. From its capital at Chandax, on the north coast of the island, the Muslim corsairs could attack merchant ships sailing from the Ionian Sea, Italy, Sicily, Dalmatia and, of course, the empire's commercial powerhouse, Venice. A ship coming from the aforementioned regions could navigate along the east coast of the Ionian Islands, stopping at Zante and Modon (in the southern Peloponnese) before having to sail along either the north or the south coast of Crete depending on its final destination. If a ship wished to continue towards the coasts of Syria and Palestine, it also had to sail through the islands of Rhodes, Karpathos and then along the Pamphylian and Lycian coasts of Anatolia to Cyprus and Cilicia. Since the Muslims controlled the last two of these bases and were within striking distance of the first two, their fleets virtually dominated the European (as opposed to the North African) sailing routes to and from the Middle East. However, even if a ship wished to continue north to the Aegean and the capital, along either the Greek coast or the Asiatic one, once again it would find itself in dangerous waters as neighbouring islands such as Naxos, Paros and Cythera – in between Crete and the Peloponnese – were held by the Cretan fleets as advanced naval bases. It was these bases that formed a naval stepping stone for operations further north into the Aegean Sea, as in the case of the sack of Thessaloniki by Leo of Tripoli in 904.[54]

In terms of overall strategy, the struggle was waged for the control of the islands and the mainland bases which dominated these sailing routes in the eastern Mediterranean, and Crete presented an obvious target for the Byzantine government.[55] Christides has dismissed outright the concept presented by contemporary and later Byzantine sources and some modern historians of the period that Crete was a 'corsair's nest' or that the island was supported solely by plundering, piracy and the slave trade, thus crippling commercial activity in that part of the Mediterranean. Rather, Crete was a Muslim frontier state, a bastion of jihad of the sort one could find in Sicily or the naval equivalent of Tarsus and the rest of the frontier cities of the Byzantine–Arab *thughūr*. While the attacks by other Muslim fleets, such as those from North Africa or Sicily, were in general simply local razzias, those of the Muslims of Crete very often aimed at the establishment of permanent bases further up the Aegean Sea in the spirit of the war of attrition against the empire. More or less the same strategic objectives were set by the fleets of Tarsus and Tripoli, as seen in 904, where one can also distinguish a direct correlation between land and naval warfare;[56] fortunately for Byzantium, however, there was never a direct coordination of activities between Crete and its Muslim counterparts in Cilicia and Syria.[57]

What is clear from the evidence that we have from the primary sources is that the island of Crete had been a strategic target for the Byzantines since the first year of Constantine's sole emperorship in 945. This is because, according to Ibn Idhari, the emperor had sent ambassadors to the Umayyad caliph in Cordoba as early as AH 334/August 945–August 946 to secure the neutrality of the Spanish Muslims, to whom the Cretan Muslims were related, in the case of an imperial expedition against Crete or even an alliance against a common enemy, the Shi°a Fatimids of North Africa.[58] Constantine would certainly have been aware of his father's failed expedition against Crete in 911, as his *De Ceremoniis* is the most detailed account we have on the equipping of the army that sailed from the capital that year.[59] What is more difficult to prove is whether Constantine was deliberately following his father's strategy, hoping that a victory would redeem his imperial house from the stigma of the failure of 911. What is certain, however, is that Constantine wished to highlight his blood relation with the founder of the dynasty, Basil, and the continuity of his reign in the imperial house of the 'Macedonians', a notion that stood out in his father as well, regardless of the enmity that existed between him and Basil.[60]

In the opening years of his rule as senior emperor, Constantine staked his prestige on recovering Crete, thus putting himself in the honourable if unsuccessful tradition of his father's policy to recapture the island from the Arabs. However, as the Cretan campaign of 949 ended in disaster, it was humiliating and politically damaging for the Constantine's prestige and competence as a sovereign appointed by God and it made a great impression upon the nobility and the people in the capital.[61] What followed was an equally disastrous period of incessant raids conducted by Sayf ad-Dawla, the arch-enemy of the Byzantines in Cilicia throughout the decade, which resulted in some of the most spectacular and humiliating defeats of Byzantine arms for many decades. Since the Byzantine strategy of the period was clearly defensive, as we have already seen, and did not involve any kind of territorial expansion, then – to return to the question that I posed at the beginning of this chapter – how can we explain the extensive gains of territory in Cilicia and Syria in the following decades? The answer lies in the failed expedition in Crete and the propaganda war against an emerging enemy of the empire in the East, whom the Byzantines had already encountered in the 940s, the 'Sword of the Dynasty' Sayf ad-Dawla.

In order to understand why this conflict began as a propaganda war that 'got out of control', we should examine the Byzantine notion of frontiers and its link to the Byzantine strategic planning of the period. As Shepard has explained in his study on the expansionism of the Byzantine emperors, one of the major cultural changes that took place in the empire in the mid-seventh century was the lack of protracted literary attention to, or substantive discussion of, frontiers as physical barriers or limits, or as dividing lines between polities in Byzantine writings.[62] It is true that the notion of linear frontiers as they existed in the Roman period, with the most characteristic examples being those of the Rivers Danube and Rhine, Euphrates and Tigris, and Hadrian's Wall in Britain, had long ago disappeared from the tenth-century reality of warfare. As I have pointed out before, the basic idea behind the establishment of the *themata* in the seventh century was defence in depth, which on the one hand allowed the invaders to march deep into imperial territory, thus extending their supply lines and becoming bogged down with booty and prisoners, and on the other hand allowed for precious time for friendly armies to gather and harass the invading forces in an attempt to flush them out to their own country. As a result, a sort of 'no-man's land' had developed in the frontier regions between Byzantium and its neighbours, with the associated repercussions for the local communities that I already examined in Chapter 2.

The reason 'frontiers' and 'frontier affairs' did not receive the required attention by the Byzantine chroniclers can be distilled down to two main points: first, it must have been demeaning, to the point of being insulting, to the emperor as a God-appointed sovereign of the Byzantine Empire and his ability to rule by divine providence to point out how much the Roman Empire had been diminished to the East and the West and 'mutilated'. Hence, the ideology of 'limited *Oikumene*' that was promoted by the Macedonian dynasty emphasised an emperor of a genuine Eastern Roman Empire that had its roots in the East and originated with the 'Eastern' emperors Diocletian and Constantine.[63] The second and most important point, however, relates to the imperial foreign policy and how this was shaped by diplomacy and diplomatic relations with Byzantium's neighbours. I have already pointed out the significance of Armenia in the *De Administrando Imperio* and the shaping of the military edge of imperial diplomacy, but to make my point I think it is necessary to repeat an extract from the aforementioned work:

> I set a doctrine before thee, so that being sharpened thereby in experience and knowledge, thou shalt not stumble concerning the best counsels and the common good. First, in what each nation has power to advantage the Romans, and in what to hurt, and how and by what other nation each severely may be encountered in arms and subdued.[64]

The Empire's Foreign Policy in the East

Thus the second point has to do with the personal and diplomatic ties between the emperor and local dignitaries, lords, chieftains or kings, which were much more important for the central government – for reasons that I have explained above – than tracts of land, which would probably have been destroyed anyway from years of raiding, and lines of forts difficult to maintain in the spirit of the old Roman *limes* (frontiers).

Warfare was seen primarily as a matter of subjecting or sacking cities, and breaking the power of troublesome border emirs, rather than any territorial expansion per se.[65] Geographic borders still existed in the Byzantine literature, like the Rivers Danube and Euphrates and 'the mountains' of Anatolia such as the Taurus and Anti-Taurus, while Theophanes Continuatus refers to the mounting pressure on the 'Ρωμαϊκῶν ὁρίων ἐσχατιαί' (the farthest point of the Roman extremities/borders) by the Tarsiots.[66] It is clear, however, that the principal aim was the sacking of cities and the defeat (and humiliation) of the empire's enemies. Below are some extracts from contemporary sources of the tenth and eleventh centuries that support this point (emphasis added):

> The so-called John the magister, Curcuas, excellent in the affairs of war he managed to *collect many trophies* [from the field of battle] and to *expand the limits of Roman territory* and to *conquer numerous cities* from the Agarenoi.[67]
>
> And annihilated the Agarenoi ... and those who witnessed the victorious Nicephoros [Phocas] were filled with surprise (ἔκπληξιν) and admiration (θάμβος) when he *cut through and defeated the armies of the godless Hamdan* ... and the *cities*, and *towns* and *countries* [of the enemy] he *burned with fire* and he *brought back prisoners* [with him] and *made them come to terms with the Romans*.[68]
>
> In that year [961] Nicephor Phocas the magister, who had already been promoted to domestic of the Scholai of the East by Constantine [VII], he won *many victories* over the eastern Agarenoi, and the Emir of Tarsus called Karamonin and the [emir] of Chalep [Aleppo] Hamdan and the [emir] of Tripoli Izeth he *utterly humiliated*, he sent a multitude of elite soldiers and a well-equipped fleet against the Saracens of Crete.[69]
>
> With such victories and stratagems, the general *prevailed* over numerous hosts of barbarians and *destroyed* them, *breaking the insolent arrogance* of Hamdan and *reducing him to ignoble and unmanly cowardice and flight*.[70]
>
> For the enemy, it is a matter of great importance, and they will make use of every device to assail you when you do not expect it, so that they may overwhelm you, to *the harm and destruction* of the people of Christ, the *dishonour* of the mighty Romans, and *the exultation and swollen pride* of the arrogant sons of Hagar, who deny Christ our God.[71]

But the most vivid example found in Byzantine literature about the propaganda war against the Arab emir, which does not involve the notion of territorial expansion but rather a 'higher' and more 'honourable' struggle for Christianity and the reputation of the empire and its glorious soldiers, is the speech that was allegedly read out – probably in late 950 – to the soldiers returning from the eastern campaign of that year during the review and before being disbanded for the winter.[72] We read in the first oration:

> With confidence in this hope [in Christ], and after entrusting your souls to it, you have set up such trophies as these against the enemy, you have striven for such victories as these, which have reached every corner of the world, and have made you famous not only in your native lands but also in every city. Now your wondrous deeds are on every tongue, and every ear is roused to hear them.[73]

These are comments that reveal something more important than simple clichés of imperial propaganda; they bear witness to the changes in Byzantine military policy in the middle of the tenth century and they shed light on the question of morale and motivation in the armies of the time. Hence, the victory of Leo Phocas over the Hamdanids that year (950) seems to have been exploited for propaganda purposes rather than for its real strategic value, thus restoring some much-needed prestige to the regime of Constantine Porphyrogenitus after the humiliation of the Cretan expedition. It is made perfectly clear in the aforementioned extract that the war against the Emir of Aleppo was not a territorial war for the acquisition of lands in Cilicia and northern Syria, but rather an ideological war for the humiliation and destruction of a dangerous (and defiant) enemy of the empire and, of course, God. In fact, the author of the oration does not identify any of the regions that would later become the operational theatres of war in the East; the clear target of the propaganda is one and only:

> In truth, the Hamdanid has no power. Do not believe in his skills and wiles – he is afraid, he is devious, and without a reliable force, in mortal fear of your onslaught and driven back headlong by it, he is trying to put fear in your minds with ruses and deceptions.[74]

Between the composition of the aforementioned oration in late autumn 950 and the next example of Byzantine propaganda against Sayf ad-Dawla, by which I mean the victory celebrations in the capital in 956 and the second oration of 958, this was a period of very damaging raids conducted by the Arab emir that culminated in the battle at Hadath in 954

– damaging not only for the provinces that had to bear the brunt of the Arab attacks but, more importantly, for Constantine's prestige among his military commanders and courtiers in the capital. In line with the strategy of containment manifested in *On Skirmishing*, Constantine made further attempts to pacify that section of the eastern borders by making overtures – which included an exchange of prisoners – to Sayf. These were defiantly rebuffed by the emir and, instead, were used by Mutanabbi to enhance his patron's stance in the Muslim world as champion of the jihad.[75] Why would Sayf, however, have taken such a defiant stance against the Byzantine emperor? A brief inroad into his past should be able to give an answer to this question.

Sayf ad-Dawla was born in 916 into the Hamdanid family, a branch of the Banu Taghlib, an Arab tribe residing in the Jazira since pre-Islamic times.[76] The Taghlib traditionally controlled a large area around Mosul until the ninth century, their power having diminished by the centralised policy of the government in Baghdad. After 895, Sayf's uncle, Husein ibn Hamdan, raised troops among the Taghlib in exchange for tax remissions from the Abbasid government and established a commanding influence in the Jazira by acting as a mediator between the authorities in Baghdad and the Arab and Kurdish populations of the region. Husein was a successful chieftain who had distinguished himself in wars against the heretic Kharijites and the Tulunids. His younger brother Ibrahim was governor of the Diyar-Rabi°a (roughly in northern Iraq) in 919 and Sayf ad-Dawla's father, Abdallah, served as emir of Mosul in 905/6–913/4, and again in 925/6. As power in Mosul was relegated in 935 to Abdallah's eldest son, al-Hasan (the future Nasir ad-Dawla), Sayf ad-Dawla originally served under his elder brother in the latter's attempts to establish his control over the weak Abbasid government in Baghdad during the 940s.

After the failure of these endeavours in 943, and the brothers' expulsion from Baghdad, Sayf turned to Syria where he was confronted by the Ikhshidids of Egypt, vassals of Baghdad and nominal masters of the region since 935. In Syria, Sayf faced a double threat, the first and obvious one being the Ikhshidids and their own ambitions in the region, and the second the destructive raids of the Arab-Bedouin tribes. After two wars with the Egyptians in 945 and 947, he managed to keep northern Syria and Aleppo, but conceded southern Syria and Damascus to the Ikhshidids, thus turning it into some sort of buffer zone between the Egyptian lands in Palestine and the Byzantine Empire to the north in Cilicia.

Northern Syria at this time was controlled by a number of Arab tribes, which had been resident in the area since the Umayyad period, and in many cases even before that. The region from the Orontes to beyond the Euphrates was controlled by the still largely nomadic tribes of the Banu Numayr, Banu Ka°b and the Kushayr, as well as the Banu Kilab around Aleppo. The main problem with these Bedouin tribes was their regular raids, in accordance with their nomadic lifestyle, against the more settled Arab communities that resulted in regular outbreaks of rebellion between 950 and 954, which had to be put down repeatedly by Sayf ad-Dawla.[77] The final settlement came in 955 after the brutal suppression of a tribal rebellion, where Sayf forced several tribes into the desert, offering them a stark choice – submit or die.

The motives behind Sayf's decision to rebuff Constantine's overtures in the early 950s and his incessant raids in Cilicia and Mesopotamia can be traced back to a political situation not dissimilar to the one the Byzantine emperor had found himself in after 945, and especially in 949 after the failed expedition in Crete. Sayf was a newcomer to the region trying to establish his power against the odds and against many enemies on different fronts, both Muslim and Christian, and with barely any support from members of his own family-clan (his brother Nasir was occupied in Mosul where he was attempting to establish his own emirate). His wars with the Ikhshidids had cost him southern Syria and Damascus, although a settlement was easily reached in 947, since the Egyptians did not view any territorial expansion in northern Syria and Cilicia as strategically viable. Sayf's biggest problem was internal: the Arab-Bedouin tribes of the Syrian Desert and the Jazira, and their raids against the sedentary populations of the region. Although neutralising some might have been possible, instead the Emir of Aleppo considered using them as mercenary (light) troops in his wars in the north and east – the famous Αραβίται (*Arabitai*), who were granted special attention in the *Praecepta Militaria* of Nicephoros Phocas.[78] Therefore, winning victories for Islam, for himself and for his booty-hungry troops was essential if Sayf was to remain in power and legitimise his rule.

In order to reinforce his fame as a champion of jihad, Sayf used Mutanabbi's poetry as the ideal tool for his propaganda; in return for the invaluable publicity that Mutanabbi provided, the poet was well paid in money, clothing, horses, land and status. Mutanabbi's poems belong to a large category of lyric poetry that originated in pre-Islamic Arabia known as *qasida* (Arabic for 'intention'), which I will discuss in more detail in a following chapter. Here is a sample of this epic propaganda poetry that was so skilfully used by the emir of Aleppo against the Byzantines between 948 and 956:

The Empire's Foreign Policy in the East

Extracts from the *Panegyric to Saif al-Daula, commemorating the building of Marash in 341 (952 AD)*[79]

22. May the people of the frontiers enjoy your judgement concerning them, and that you, God's 'party', have become a 'party' to them,
24. so on one day with horsemen you drive the Byzantines from them, and on another day with bounty you drive away poverty and dearth.
25. Your expeditions are continuous, and the Domesticus in flight, his companions slain and his properties plundered;
26. he came to Mar'ash, deeming the distant near as he advanced, and when you advanced he retreated, deeming the near distant.
30. But he turned his back, when the thrusting waxed furious – when his soul remembered the sharpness, he felt his flank,
31. and he abandoned the virgins, the patriarchs and the townships, the dishevelled Christians, the courtiers, and the crosses.
40. For a good reason the Caliph has made him ready against the enemy and named him the Sharp Sword [Sayf ad-Dawla], to the exclusion of all others.

Perhaps the most famous of Mutanabbi's odes was composed to glorify Sayf ad-Dawla's recapture of the strategic frontier fortress-town of Hadath (Adata) in October 954. After a victorious battle, where not only did the army of Bardas Phocas, the Domestic of the Scholae, melt away after suffering heavy casualties, but many members of the aristocracy were captured and held to ransom:[80]

9. [Sayf] built her [Hadath] and upraised [her], as shaft beat against shaft, while the waves of doom clashed all around her.
10. She suffered [a derangement] the like of madness, but came to dawn with amulets upon her – to wit, the corpses of the slain.
11. The [predatory] Nights put everything they have taken beyond [the] reach [of the owners thereof], yet they are bound to repay the debt of what they take from you.
12. How could the Byzantines and Russians ever hope to destroy her, when such thrusting [of your army's weapons] afforded her foundations and pillars?
13. They came against you hauling [such a mass of] iron [armour that it was] as if they crawled on coursers with no legs [to hold them Up].
22. You stood [your ground] when death was not in doubt for anyone who did so, it was as if you were in the very eyelid of Destruction as he slept!
25. You pressed both their wings upon the heart in a grip, dealing death to the secondaries and primaries beneath it,
26. with blows that fell on skulls while victory was not yet won, and which went down to lower throats as victory came.
29. You scattered them all over Uhaydib, as dirhams are scattered over a bride.
30. You were leading you up your horse's eagle nests in the mountains, around which you let a rich pasture,

31. when they slipped, you made them walk on their bellies, like snakes.
32. Dare he [Domesticus] always attack you when his neck was always reproaching his face?

As was the case with Constantine's speech to his soldiers in 950, so Mutanabbi's poetry does not involve any notion of territorial expansion in Cilicia and in regions beyond the Taurus and Anti-Taurus Mountains. The main objective of the emir is the defeat and humiliation of his enemies: 'Your expeditions are continuous, and the Domesticus in flight, his companions slain and his properties plundered'; and 'he [Domesticus] abandoned the virgins, the patriarchs and the townships, the dishevelled Christians, the courtiers, and the crosses', while underlining the important association of his nickname: 'For a good reason the Caliph has made him ready against the enemy and named him the Sharp Sword [Sayf ad-Dawla], to the exclusion of all others.'

The build-up of Sayf's image as the leader of jihad continues with a question: 'How could the Byzantines and Russians ever hope to destroy her [Hadath], when such thrusting [of your army's weapons] afforded her foundations and pillars?' Mutanabbi gives an answer in the following verses comparing the imperial mail-clad army comprising many nationalities as a 'scene of ludicrous turmoil',[81] opposite which Sayf and his men stood against all odds 'as if you were in the very eyelid of Destruction as he slept!' And, of course, they prevailed, with the starkest contrast in the battle scenes being that between Sayf and Bardas; the emir is portrayed as a bold and daring leader ('You stood [your ground] when death was not in doubt for anyone who did so'), while Bardas is clearly depicted as a coward ('Dare he [Domesticus] always attack you when his neck was always reproaching his face?').

It is from this period, the middle of the 950s, that we can witness the beginning of a new policy for Constantine VII, to 'raise the stakes' in his conflict with Sayf ad-Dawla. This policy would lead to the subjugation of the Emirate of Aleppo in 962, after which year the emir was unable to draw sufficient resources in money and manpower to intercept the Byzantines advancing south into Cilicia and northern Syria, thus leaving the road wide open for Nicephorus' armies to move against their main targets – the cities of Cilicia, Tarsus and Mopsuestia. We have several indications about a change of policy from the central government dating from about this time and, if we wish to be more ambitious in our analysis – although this always bears the risk of over-confidence in our primary sources – at about the year 955. The proliferation of military treatises during this period, such as the *Syntaxis Armatorum Quadrata* of the mid-950s,[82] on which the author of the later *Praecepta Militaria* (c. 969) heavily relied to compose his work, and the innovative and constantly revised tactics which they incorporate – implying that they had only recently come into effective use – are certainly indications of a new and aggressive strategy from the Byzantine government in Cilicia and Syria.

Arabic sources of the period mention for the first time in detail the composition of the invading Byzantine armies. Contemporaries like Mutanabbi, Abu Firas and the late twelfth-century Ibn Zafir wrote about Armenian, Rus', Bulgarian and Slav troops that accompanied Bardas Phocas on his expeditions. If we couple this with the exaggerated numbers of troops reported for the Byzantine armies invading Cilicia and Syria, then we gain a picture of a central government mustering sizeable numbers of soldiers of different nationalities for its wars in the East. However, as I will show in much more detail in a following chapter, Muslim sources were providing inflated figures for Byzantine armies long before the Battle of Hadath; thus, this does not constitute definitive proof of a Byzantine change of policy. Neither this nor the fact that the main Muslim sources – Ibn Zafir may have drawn some of his information for this period from a source that had access to Abu Firas' work (see Chapter 4) – provide us for the first time with the composition of the foreign contingents of the Byzantine army just for the Battle of Hadath, and nowhere else, should necessarily lead us to assume that the Byzantine government had introduced large numbers of mercenary Russian, Armenian and Slav troops into its ranks only for its campaign against Sayf ad-Dawla.[83]

In fact, Rus' and Bulgarian soldiers – of unknown numbers – can be identified in Mesopotamia since 947, while the treaty that immediately followed the Russian siege of Constantinople in 941 would have provided ample opportunities for mercenaries from the north to enter Byzantine service. If we combine this with the fact that this is also the first mention of the deployment of *kataphraktoi*[84] by the Byzantine armies, something which had led many historians to assume that this heavily armed cavalry unit was introduced into the ranks of the Byzantine army during the reign of the Emperor Nicephorus II Phocas after four centuries of absence, then our suspicion should become greater. As the Battle of Hadath acquired legendary proportions both for contemporary and later Muslim chroniclers, more details about the battle would have been included for posterity to magnify the glory of Sayf ad-Dawla as the champion of jihad; indeed, I have shown how Mutanabbi's portrayal of the Byzantine mail-clad army comprising many nationalities was described as a 'scene of ludicrous turmoil' by an authority in the field; thus, Sayf's victory would have seemed even more glorious.[85]

Constantine's military oration of 958 also refers to these mercenary troops (emphasis added): 'When several contingents of these foreign people *recently* joined you [and the eastern armies] on campaign.'[86] This information, however, does not necessarily imply that these mercenary soldiers were employed only some years before the writing of this oration just for the wars against the Hamdanid emir. Rather, I believe it is due to the introduction of a new and aggressive policy and new battle tactics that required both light and heavy, but disciplined, cavalry and infantry forces operating together in the field of battle under a new command structure that these corps of soldiers – the *kataphraktoi* and the Armenian and Rus' infantry – were simply coming into prominence. Further, the much warmer climate of Cilicia and Syria – compared with Anatolia there is a significant difference of some seven degrees Celsius[87] – had a major impact on the extension of the campaigning period between October and April, contrary to the three summer months of the Arab period, as noted in *On Skirmishing*, which required great numbers of professional troops to be maintained on the field for longer periods, as opposed to the part-time thematic soldiers.

Another indication about the change of policy of the Byzantine government was the replacement of the ageing Domestic of the Scholae, Bardas Phocas, in 955. If we believe Skylitzes' comments on the military ingenuity of the senior of the Phocades: 'Whenever served under another, he showed himself to be a fine commander; but once authority over the entire land forces depended on his own judgement, he brought little or no benefit to the Roman realm.'[88] The dismissal of a high-ranking official of the imperial army as the Domestic of the Scholae was more of a political, rather than military, decision and should not be related directly to the failures of Bardas in the field of battle against Sayf ad-Dawla. Ascending to the high office of *domesticus* or that of *strategos* depended on family ties, on the relationship with the emperor and other variables, as well as on military ability.[89] After all, it was the brilliant John Curcuas who had also been dismissed in December 944 by Romanus Lecapenus' sons simply because he was loyal to their father; thus, it is reasonable to assume that the office of the second-in-command of the army should never have been held by a political enemy of the emperor. However, the fact that Bardas was replaced by his son Nicephorus, the *strategos* of Anatolikon – which meant that the office remained with the family of the Phocades, a family whose support and high visibility in military affairs were essential to the restoration of Constantine in 945 – and that no political persecution of its members took place leads us to assume that this important change of command signifies a change in imperial policy as well.

The Empire's Foreign Policy in the East

Finally, a propaganda event dating in 956 constitutes another significant indication of a change of policy against the Hamdanids – the victory procession in the capital with the ritual trampling of Abu'l Asair. This is another example of how minor strategic victories in the East can be used to increase the popularity of a government that badly needed some sign of military success. In 956, Bardas' second son, Leo, *strategos* of Cappadocia, captured a Hamdanid party led by Sayf ad-Dawla's cousin, Abu'l Asair, whose mission was to rebuild the fortress of Arandas near Duluk (Doliche, in between the Aleppo–Germanikeia road).[90] This success was exploited to maximum effect and we should pay special attention to two processional innovations. Our first-hand source for this ritual, the *De Ceremoniis* – and in particular a specific section of the second book that was probably compiled between 957 and 959[91] – talks about the revival of the *calcatio*, a Roman ritual that had not been used in processions since the crushing of Thomas the Slav's rebellion in 823. This involved the ritual trampling of the enemy leader, with the *protostrator* pushing the emperor's lance in the captive's neck while the *psaltes* (ecclesiastics) were singing: 'What God is great like our God? You are the God who works wonders.'[92] This rather theatrical humiliation of the Hamdanid leader's cousin was intensified by the fact that it took place in the Forum of Constantine, very important in itself because it reinforces the links between Constantine VII and his grandfather Basil, who had also staged his triumphs there, but also because the site was tailor-made for an emperor who had spent his entire life in the capital and had never led a victorious army back through the Golden Gate.

An unavoidable conclusion from the detailed description of the victory procession of 956 is the emperor's wish not just to humiliate the Hamdanid dynasty, but also to involve as much as possible[93] (1) the people of the capital, by staging his triumph in one of the busiest sites in Constantinople – also a religious centre for the merchants, containing a shrine to the Virgin Mary which had played a major role in the triumphs of Basil I – and (2) the army and the family of the Phocades, as this was the first time since the early Byzantine period that any theme commanders had participated in a victory parade entering the capital. The fact that this event took place in 956, some eleven years after Constantine's rise to power as sole emperor and not earlier is certainly an indication of a change in his policy against the Hamdanids.

In his military oration addressed to his generals of the East in 958, Constantine set out his strategy against the Arabs of Cilicia and Syria. He clearly referred to his troops as 'champions and defenders of the Byzantines' praying to God that the 'Christ-loving tagmata and themata will intimidate their adversary', not missing out on highlighting twice the military revolution that had been taking place for the last few years.[94] The supplication of divine intercession is more evident in the latter than in the previous oration of 950, as this time Constantine used the combination of the *stauros nikopoios* (victory-bearing cross), holy water and the prayers of monks from Mount Athos to raise the morale of his armies.[95] He also intended to lead them to battle himself, a wish which was much better defined and stated in 958 than eight years before. The target of this propaganda – and of the upcoming campaigns – is clear in the second half of this lengthy oration, where Constantine refers to the 'impious Hamdanid and the Christ-hating Tarsiots', and the rumour that 'they are brave and have acquired a host invincible in war, wherefore out of terror and weakness you avoided engaging them in combat'.[96]

In order to answer this rumour, the emperor brought to the attention of his generals the military accomplishment of Basil Hexamilites, the *strategos* of the theme of Cibbyraeots, who defeated an Arab fleet and raided Tarsus in the autumn of 956,[97] 'arming his host with the utmost zeal and inspiring speeches, the kind of campaign he conducted and the number of officers and the huge host of Tarsiots he took prisoners'. However, this is not the only victory given as an example by the emperor, who also writes about the military accomplishments of the armies of the West against the Arabs in Longobardia in 956, 'when they won victories against the enemy – take our word for it that they mastered and subdued those who opposed Our Majesty'.[98] These wars were fought for the defence of the Christian realm and not for any kind of aggressive acquisition of territories, something which is further underlined in the final paragraph of the oration:

> Sturdy and invincible champions of the Byzantine people . . . we [the people of the capital] will embrace you as victors appearing as triumphant conquerors against the enemy . . . we will kiss your bodies wounded for the sake of Christ in veneration as the limps of martyrs.[99]

There should be little doubt that the strategy of Constantine and his son in this period was one of containment and an attempt to create a sort of demilitarised zone rather than the acquisition of territory beyond the Taurus Mountains, although the control of the trade routes passing through Aleppo and Tarsus would have seemed quite lucrative.[100] The decision for an all-out war against Sayf ad-Dawla targeting Aleppo, the capital and seat of his power, had been taken if not by 955 then perhaps a few years later. However, whether this was the main strategic objective of the Byzantine government is another matter. The Byzantine strategy in Cilicia and northern Syria between the years 959 and 965 has been examined in a 2008 study by Garrood, and it suffices for me to summarise his main points.[101] The primary aim of the Byzantine push in the eastern marches of the Anatolian provinces was Cilicia and the cities of Tarsus and Mopsuestia; other aims were secondary, including the offensive against the city of Aleppo, which served to subdue the power of Sayf ad-Dawla and clear the way for the success of the primary objective. These two objectives were interconnected: Cilicia could not be subdued without the neutralisation of the Hamdanid forces and the power of the Emir of Aleppo would be strong as long as the Cilician cities remained in his sphere of influence.[102]

The main points made in this chapter are related to the political and strategic importance of Armenia proper – and more specifically the cantons of Taron and Vaspourakan – as the 'back door' of any enemy invasion routes into Anatolia. The empire applied a sort of soft diplomacy that enabled negotiation, flattery and/or intimidation to win over the local *naxarars*. In this context, and bearing in mind that the empire never contemplated any kind of permanent territorial expansion in the East in the 950s–60s, the nature of warfare with the Hamdanid dynasty of Aleppo seems quite peculiar. If we look at the political background of both protagonists – Constantine VII and Sayf ad-Dawla – and their place within their respective courts, including the dire state of their internal political situations and their desperate need for a military success, then it all seems to fall into place. By the end of the 950s, this war had already escalated into an all-out conflict between the emperor and the Aleppan emir, where no one could afford (politically) to succumb. In the end, it would be the vast resources that Byzantium poured into the wars in the East that turned the tide in their favour by 962.

Notes

1. *Naxarar* was a hereditary title given to houses of the ancient and medieval Armenian nobility.
2. Treadgold, *Byzantine State and Society*, pp. 455–61; Whittow, *Orthodox Byzantium*, pp. 314–15.
3. Runciman, *Romanus*, pp. 120–31.
4. *DAI*, 50.115–31, 152–66, pp. 238–40. See also Runciman, *Romanus*, pp. 120–2.
5. Runciman, *Romanus*, pp. 128–9; M. Canard (1948), 'Les Hamdanides et l'Armenie', *Annales de l'institut d'études orientales d'Alger*, 7, pp. 82–3.

6. Runciman, *Romanus*, p. 136.
7. *DAI*, 44.112–15, 125–8.
8. *DAI*, 50.161–3, p. 240.
9. Canard, *Hamdanides*, p. 739.
10. Canard, *Hamdanides*, p. 733; A. A. Vasiliev (ed.) and M. Canard (trans.) (1935), *Byzance et les Arabes, 867–959*, 2 vols, Madison, WI: Institut de philologie et d'histoire orientales, II.1, pp. 267–8; Treadgold, *Byzantine State and Society*, p. 480.
11. *Theophanes Continuates*, p. 415; Canard, *Hamdanides*, p. 735.
12. Canard, *Hamdanides*, p. 742; Canard, 'Les Hamdanides et l'Armenie', pp. 84–94; Treadgold, *Byzantine State and Society*, p. 483.
13. Ibn Zafir, pp. 121–2; Abu Firas, p. 358; Canard, *Hamdanides*, p. 743.
14. Vasiliev, *Byzance et les Arabes*, 2.I, pp. 285–8; Whittow, *Orthodox Byzantium*, pp. 319–20; Canard, *Hamdanides*, p. 745; Canard, 'Les Hamdanides et l'Armenie', 85–95; Treadgold, *Byzantine State and Society*, p. 484.
15. Ibn Zafir, p. 123; Canard, *Hamdanides*, p. 746; Runciman, *Romanus*, p. 143. This expedition is also mentioned by Abu Firas, p. 358. For the Byzantine-Abbasid peace treaty of 938, see *Histoire de Yahya*, 18, p. 710.
16. Al-Gauzi in Vasiliev, *Byzance et les Arabes*, 2.II, p. 174.
17. For these events in Iraq and Sayf ad-Dawla's battle with a Buyid army, see *Histoire de Yahya*, 18, pp. 728–42.
18. Shepard, 'Constantine VII', pp. 20, 27–33.
19. *Kouratoreia* is a term used to designate an imperial demesne. For more on the central government's policy to increase the number of *kouratoreia* in eastern Asia Minor in the second half of the tenth century and during the reign of Basil II (976–1025), see J. Howard-Johnston (1995), 'Crown Lands and the Defence of Imperial Authority in the Tenth and Eleventh Centuries', *Byzantinische Forschungen*, 21, pp. 75–100.
20. Canard, *Hamdanides*, p. 804.
21. *Histoire de Yahya*, 18, pp. 730–1; *Theophanes Continuates*, pp. 429–31; Symeon Magister, *Chronicon*, p. 338; John Skylitzes, *Synopsis of Byzantine History*, pp. 223–4.
22. *DAI*, p. 12; Shepard, 'Constantine VII', p. 24; J. Shepard (1995), 'Imperial Information and Ignorance: A Discrepancy', *Byzantinoslavica*, 56, pp. 110–11.
23. *DAI*, 43.61–71, p. 190.
24. *DAI*, 43.163–85, p. 196.
25. *DAI*, 44.125–8, p. 204.
26. Haldon, *Warfare, State and Society*, pp. 38–9. See also the following two chapters in E. Jeffreys, J. Haldon and R. Cormack (eds) (2008), *The Oxford Handbook of Byzantine Studies*, Oxford: Oxford University Press: Haldon, 'Army', pp. 554–61; Brandes and Haldon, 'Revenues and Expenditure', pp. 562–9.

27. *DAI*, preface, p. 44.
28. P. Charanis (1963), *The Armenians in the Byzantine Empire*, Lisbon: Fundação Calouste Gulbenkian, pp. 16–21, 32–3; P. Charanis (1959), 'Ethnic Changes in the Byzantine Empire in the Seventh Century', *Dumbarton Oaks Papers*, 13, pp. 28–43; C. Toumanoff (1971), 'Caucasia and Byzantium', *Traditio*, 27, p. 131.
29. Whittow, *Orthodox Byzantium*, p. 207; Runciman, *Romanus*, p. 156; Charanis, 'Cultural Diversity', pp. 18–20. For Toumanoff's view of an imposition of the Chalcedonian rite over the Armenians, who would 'rather have the [Muslim] turban', see 'Caucasia and Byzantium', pp. 146–7.
30. For Tiberius' transfer of Armenians, see *The History of Theophylact Simocatta* (1986), ed. and trans. by M. Whitby and M. Whitby, Oxford: Oxford University Press, III.15, p. 143. For the integration of the Armenians into the Byzantine Empire, see J.-C. Cheynet (2014), 'Les Arméniens dans l'armée byzantine au Xe siècle', *Travaux et mémoires*, 18, pp. 175–92; ibid. (1996), 'Les Arméniens de l'empire en orient de Constantin X à Alexis Comnène', in N. G. Garsoïan (ed.), *L'Armenie et Byzance: histoire et culture*, Paris: Publications de la Sorbonne, pp. 67–78; N. G. Garsoïan (1998), 'The Problem of the Armenian Integration into the Byzantine Empire', in H. Ahrweiler and A. E. Laiou (eds), *Studies on the Internal Diaspora of the Byzantine Empire*, Washington, DC: Dumbarton Oaks Research Library, pp. 53–124; G. Dédéyan (2002), 'Reconquête territoriale et immigration arménienne dans l'aire cilicienne sous les empereurs macédoniens (de 867 à 1028)', in M. Balard and A. Ducellier (eds), *Migrations et diasporas méditerranéennes: Xe–XVIe siècles: actes du colloque de Conques, Octobre 1999*, Paris: Publications de la Sorbonne, pp. 11–32; G. Dédéyan (1993), 'Les Arméniens sur la frontière sud-orientale de Byzance, fin IXe–fin XIe siècles', *La frontière, travaux de la maison de l'Orient*, 21, pp. 67–85; G. Dédéyan (1975), 'Immigration arménienne en Cappadoce au XIe siecle', *Byzantion*, 45, pp. 41–117. On the transfer of populations in the Byzantine Empire, see J.-C. Cheynet (2003), 'Les transferts de population sous la contrainte à Byzance', in L. Feller (ed.), *Travaux et recherches de l'UMLV. Littératures. Sciences humaines. Les déplacements contraints de population*, Marne-la-Vallée: Université de Marne La Vallée, pp. 45–70; P. Charanis (1961), 'The Transfer of Population as a Policy in the Byzantine Empire', *Comparative Studies in Society and History*, 3, pp. 140–54.
31. Toumanoff, 'Caucasia and Byzantium', p. 132; Charanis, 'Ethnic Changes', p. 29.
32. Leo VI, *Taktika*, 4.3, pp. 46–8.
33. Basil was born to peasant parents in late 811 at Charioupolis in the Byzantine theme of Macedonia (an administrative division corresponding to the area of Adrianople in Thrace) and was, most likely, of Armenian stock. He appeared for the first time at the Byzantine court as *protostrator* (literally,

'first equerry' or the imperial stable master) for Theophilos, a relative of Michael III; see John Skylitzes, *Synopsis of Byzantine History*, pp. 121–2; Treadgold, *Byzantine State and Society*, p. 455. Tobias, however, has argued that it is impossible to be sure of Basil's Armenian or Slavo-Armenian origins: N. Tobias (2007), *Basil I, Founder of the Macedonian Dynasty: A Study of the Political and Military History of the Byzantine Empire in the Ninth Century*, New York: Edwin Mellen Press, p. 264. For more on the author of the *Life of Basil*'s efforts to trace Basil's genealogy back to Tiridates, an Arsacid king of Armenia, see W. Treadgold (2013), *The Middle Byzantine Historians*, Basingstoke: Palgrave Macmillan, pp. 167–70.

34. Romanus Lecapenus was born in the Cappadocian town of Lecape and possibly descended from the impoverished branch of the princes of Gabeliank. He was the son of an Armenian peasant who rescued the Emperor Basil I from the enemy in a battle at Tephrike and had been rewarded with a place in the imperial guard; see *Oxford Dictionary of Byzantium*, s.v. 'Lekapenos', pp. 1203–4; Garsoïan, 'Armenian Integration', p. 66; Runciman, *Romanus*, p. 63.

35. I agree with Toumanoff that the geographical misnomer 'Macedonian' should be replaced by the more appropriate 'Basilid'; see Toumanoff, 'Caucasia and Byzantium', p. 196 n. 99.

36. 'The emperor Leo mobilised the Roman army: 100,000 men, invaded Syria under the command of Michael Lachanodrakon of the Thrakesians, the Armenian Artabasdos of the Anatolics, Tatzates of the Bucellarii, Karisterotzes of the Armeniacs, and Gregory, son of Mousoulakios, of the Opsikians' (Theophanes Confessor, *Chronicle*, p. 623).

37. J.-C. Cheynet (2008), *La société byzantine. L'apport des sceaux*, 2 vols, Paris: Association des amis du Centre d'histoire et civilisation de Byzance (for the Bourtzes, Brahamioi, Dalassenoi, Phocades, Maleïnoi, Argyroi, Diogenoi and Krateroi); J.-C. Cheynet (1996), *Pouvoir et contestations à Byzance (963–1210)*, Paris: Publications de la Sorbonne; J.-C. Cheynet (1986), *Études prosopographiques*, Paris: Publications de la Sorbonne (for the Bourtzes, Brahamioi and Dalassenoi families); J.-C. Cheynet (1986), 'Les Phocas', in G. Dagron and H. Mihaescu (eds), *Le traité sur la guérilla (De velitatione) de l'Empereur Nicéphore Phocas (963–969)*, Paris: Editions du Centre national de la recherche scientifique, pp. 289–315; J.-C. Cheynet (1991), 'Fortune et puissance des grandes familles (Xe–XIIe siècle)', in V. Kravari, J. Lefort and C. Morrisson (eds), *Hommes et richesses*, II, pp. 199–213; V. Blysidu (2001), *Αριστοκρατικές οικογένειες και εξουσία (9ος–10ος αι.). Έρευνες πάνω στα διαδοχικά στάδια αντιμετώπισης της Αρμενο-Παφλαγονικής και Καππαδοκικής αριστοκρατίας* [Aristocratic Families and Power (9th–10th c.): Investigations on the Successive Stages of Dealing with the Armenian-Paphlagonian and Cappadocian Aristocracy], Thessaloniki: Vanias; Whittow, *Orthodox Byzantium*, pp. 335–57.

38. I. Ševčenko (1979–80), 'Constantinople Viewed from the Eastern Provinces in the Middle Byzantine Period', *Harvard Ukrainian Studies*, 3/4, pp. 726–46; A. Pertusi (1974), 'Tra storia e leggenda: Akritai e Ghazi sulla frontiera orientale di Bisanzio', in M. Berza and E. Stănescu (eds), *Actes du XIVe Congrès international des études byzantines: Bucarest, 6–12 septembre 1971*, Bucharest: Editura Academiei Republicii Socialiste România, I, pp. 285–382; N. Oikonomides (1979), 'L'épopée de Digénes et la frontière orientale de Byzance aux Xe et XIe siècles', *Travaux et mémoires*, 7, pp. 375–97; P. Magdalino (1989), 'Honour among *Romaioi*: The Framework of Social Values in the World of Digenes Akrites and Kekaumenos', *Byzantine and Modern Greek Studies*, 13, pp. 183–218.
39. Howard-Johnston, 'Heraclius' Persian Campaigns', pp. 32–3.
40. W. E. Kaegi (1991), 'Challenges to the Late Roman and Byzantine Military Operations in Iraq (4th–9th Centuries)', *Klio*, 73, pp. 586–94.
41. W. E. Kaegi (1981), 'Constantine and Julian's Strategies of Strategic Surprise against the Persians', *Athenaum*, 69, pp. 209–13.
42. Kaegi, *Heraclius*, pp. 122–55; Whittow, *Orthodox Byzantium*, pp. 199–201.
43. C. Mango (1973), 'Eudocia Ingerina, the Normans, and the Macedonian Dynasty', *Zbornik Radova Vizantološkog Instituta*, 14–15, pp. 17–27. Tougher disagrees over the certainty surrounding Michael III's paternity of Leo; see S. Tougher (1997), *The Reign of Leo VI (886–912): Politics and People*, Leiden: Brill, pp. 42–67.
44. M. McCormick (1986), *Eternal Victory: Triumphal Rulership in Late Antiquity: Byzantium, and the Early Medieval West*, Cambridge: Cambridge University Press, pp. 11–34, 131–78.
45. I am referring here to the well-established concept of a God-appointed emperor, which had originated and developed from the Roman pagan concept of the god-emperor, through the intersection of Christianity, with the sovereign's power preserving its divine character. The most detailed documentary source that displays this sense of divine and human order is the *De Ceremoniis*; see A. Cameron (1987), 'The Construction of Court Ritual: The Byzantine Book of Ceremonies', in D. Cannadine and S. Price (eds), *Rituals of Royalty, Power and Ceremonial in Traditional Societies*, Cambridge: Cambridge University Press, pp. 106–36. Other works on the divine character of the Byzantine emperor include Haldon, *The Empire That Would Not Die*, especially pp. 1–25; R.-J. Lilie (1984), 'Des Kaisers Macht und Ohnmacht. Zum Zerfall der Zentralgewalt in Byzanz vor dem vierten Kreuzzug', in R.-J. Lilie and P. Speck (eds), *Varia* I (Poikila Byzantina 4), Bonn: R. Habelt, pp. 9–120; H. Ahrweiler (1975), *L'ideologie politique de l'empire byzantine*, Paris: Presses Universitaires de France, pp. 129–47; J.-C. Cheynet (2002), 'Les limites du pouvoir à Byzance: une forme de tolérance?', in A. Nikolaou (ed.), *Ανοχή και καταστολή στους μέσους χρόνους: Μνήμη Λένου Μαυρομάτη* [Tolerance and Suppression in the Middle Ages: In Memory of Lenos Mavromatis], International Symposia 10,

Athens: National Research Foundation, Institute of Byzantine Research, pp. 15–28; S. Runciman (1977), *The Byzantine Theocracy*, Cambridge: Cambridge University Press, pp. 5–50; G. Ostrogorsky (1956), 'The Byzantine Emperor and the Hierarchical World Order', *Slavonic and East European Review*, 35, pp. 1–14. See also a very interesting article by Canard, who compares the notion of the divine character of the sovereign in Constantinople, Baghdad and Cairo: M. Canard (1951), 'Le cérémonial fatimite et le cérémonial byzantin, essai de comparaison', *Byzantion*, 21, pp. 378–420. Although the term 'political theocracy' is an inadequate definition of the empire's political system, in recent years there has been a tendency to modify substantially the view of the role of the emperor as a despot as an extension of the Hellenistic paradigm: 'To be sure, the Byzantine emperor was expected to do God's work, but it is no coincidence that the will of the Byzantine God was that the emperor work hard to benefit the republic' (A. Kaldellis [2015], *The Byzantine Republic: People and Power in New Rome*, Cambridge, MA: Harvard University Press, pp. 32–61 [extract quoted from p. 53]). Although Kaldellis ends up by dismissing the idea that the emperor was appointed to rule by God and had a duty to imitate Christ (pp. 48–9), I would adopt Haldon's more conciliatory tone of a 'republican monarchy', where the ruler's authority was moderated by the people and where the role of Christianity in the shaping of political ideology has been overrated (*The Empire That Would Not Die*, p. 17).

46. 'Since he [Michael III] had shortly before adopted Basil and was aware that Basil stood out from the rest both in prowess and in wisdom and was capable of making up for his own shortcoming in steering the ship of universal state; since, furthermore, divine Providence herself urged him to this end . . . the imperial crown was placed upon Basil's head . . . it was placed there by the hand of Michael who was reigning at that time, but through the decree and choice of Christ who reigns forever'. *Vita Basilii (Chronographiae Quae Theophanis Continuati Nomine Fertur Liber Quo Vita Basilii Imperatoris Amplectitur)* (2011), Corpus Fontium Historiae Byzantinae, vol. 42, ed. and trans. I. Ševčenko, Berlin: Walter de Gruyter, 18.28–41, pp. 72–4.
47. *Vita Basilii*, 29, pp. 112–13.
48. *Vita Basilii*, 40, pp. 146–7; McCormick, *Eternal Victory*, pp. 154–5.
49. Canard, *Hamdanides*, pp. 761–2; Vasiliev, *Byzance et les Arabes*, 2.I, pp. 317–19.
50. *Histoire de Yahya*, 18, pp. 767–8.
51. *DAI*, 22.44–8, p. 96.
52. *History of Leo the Deacon*, pp. 58–9. See also John Skylitzes, *Synopsis of Byzantine History*, p. 236; I. Kaminiatis (2000), Για την άλωση της Θεσσαλονίκης [For the Sack of Thessalonike], Athens: Kanaki, 12, p. 86. See also the very detailed footnotes in Vasiliev, *Byzance et les Arabes*, 2.I, pp. 320–1, n. 1 and p. 322, nn. 2–4.
53. *History of Leo the Deacon*, I.6, p. 65.

54. *Vita Basilii*, 60–1, pp. 214–20; Kaminiatis, *Εις την άλωσιν της Θεσσαλονίκης*. On Byzantium's naval strategy in the eastern Mediterranean and the 'necessity' of possessing both Cyprus and Crete, see Lounghis, *Byzantium in the Eastern Mediterranean*, pp. 14–22, 28–31 and 87–101; V. Christides (1984), *The Conquest of Crete by the Arabs (ca. 824): A Turning Point in the Struggle between Byzantium and Islam*, Athens: Academy of Athens, pp. 81–96, 157–68; V. Christides (1981), 'The Raids of the Moslems of Crete in the Aegean Sea: Piracy and Conquest', *Byzantion*, 51, pp. 76–111, especially pp. 79–99. See also D. Tsougarakes (1988), *Byzantine Crete: From the 5th Century to the Venetian Conquest*, Athens: St. D. Basilopoulos; K. M. Setton (1954), 'On the Raids of the Moslems in the Aegean in the Ninth and Tenth Centuries and their Alleged Occupation of Athens', *American Journal of Archaeology*, 58, pp. 311–19; G. C. Miles (1964), 'Byzantium and the Arabs: Relations in Crete and the Aegean Area', *Dumbarton Oaks Papers*, 18, pp. 1–32.
55. J. H. Pryor (2000), *Geography, Technology, and War: Studies in the Maritime History of the Mediterranean, 649–1571*, Cambridge: Cambridge University Press, pp. 103–11.
56. Runciman, *Romanus*, p. 123.
57. Christides, *Conquest of Crete*, pp. 37–40, 157–68; Christides, 'Raids of the Moslems of Crete', pp. 79–99.
58. Ibn Idhari in Vasiliev, *Byzance et les Arabes*, 2.II, pp. 218–19. The *De Ceremoniis* also reports the reception of a Muslim embassy from Cordoba: Constantine Porphyrogenitus (1829–30), *De Ceremoniis Aulae Byzantinae* [hereafter *De Ceremoniis*], Corpus Scriptorum Historiae Byzantinae, vols 5–6, ed. I. Reiski, Bonn: Webber, col. 1273.
59. *De Ceremoniis*, cols 1212–24.
60. P. A. Agapitos (1989), 'Η εικόνα του Αυτοκράτορα Βασιλείου Ἀ στη φιλομακεδονική γραμματεία 867–959' [The Image of Emperor Basil I in the pro-Macedonian Literature 867–959], *Ελληνικά*, 40, pp. 285–322. Constantine's wish for continuity is clear if we consider the following points:

   1. His commissioning of the *Life of Basil*, a choice of work that may strike us as strange if we bear in mind that Constantine never knew Basil and that he must have suspected he was not his real grandfather. The preface of the work, however, underlines Constantine's wish for the *Life of Basil* to work as a history of all Basil's noteworthy acts that could serve as examples for future emperors. See Treadgold, *Middle Byzantine Historians*, pp. 165–80.
   2. A laudatory poem on the death of Leo VI commissioned right after his death and found in the Skylitzes manuscript in Madrid: I. Ševčenko (1969–70), 'Poems on the Deaths of Leo VI and Constantine VII in the Madrid Manuscript of Scylitzes', *Dumbarton Oaks Papers*, 23/24, pp. 201–10. In verses 40–4, the author attempts to legitimise Constantine's claim to the imperial throne despite the Church's objection to Leo's

fourth marriage. Then follows the confirmation of the blood relation between Constantine and Basil in verses 59–62:

> (53) Forthwith the heir to Leo's throne,
> Of lineage imperial,
> The purple-clad sun doth arise:
> His name is Alexander.
> A star is rising side by side
> With Master Alexander:
> 'Tis Constantine, child issued from
> The loins of Emp'ror Leo.
> O City, sing, intone the praise
> Of Basil's noble offspring.

3. Numismatic evidence from the reigns of Constantine and his predecessors. At first, both Leo VI and his son Constantine used the issuing of coins to promote their individuality as sovereigns. However, whenever the issue of family succession was in doubt, and when both emperors felt it was the right time to crown their sons as co-emperors, we return to the traditional formula of two long-bearded standing figures holding a cross between them. See A. R. Bellinger (1956), 'The Coins and Byzantine Imperial Policy', *Speculum*, 31, pp. 70–81.

61. *Vita St. Pauli iunioris*, ed. H. Delehaye, reprinted in T. Wiegand (ed.) (1913), *Der Latmos, Milet: Ergebnisse der Ausgrabungen und Untersuchungen seit dem Jahre 1899*, Berlin: G. Reimer, III/1, p. 122. Skylitzes also recalls a century later the shameful (αἰσχίστως) retreat of the Romans from their military camp in Chandax: *Synopsis of Byzantine History*, pp. 236–7.
62. Shepard, 'Emperors and Expansionism', pp. 55–82.
63. T. Lounghis (1993), *Η ιδεολογία της Βυζαντινής ιστοριογραφίας* [The Ideology of Byzantine Historiography], Athens: Herodotos, pp. 48–9, 75, 77, 129; T. Lounghis (1999), 'La théorie de l'oecumène limité et la revision du Constitutum Constantini' [The Theory of the Limited *Oecumene* and the Revision of the *Constitutum Constantini*], in A. Dzhurova and G. Bakalov (eds), *Obshchoto i spetsifichnoto v balkanskite kulturi do kraya na XIX vek. Sbornik v chest na prof. Vasilka Tupkova-Zaimova*, Sofia: Ivanovo State University Press, pp. 119–22. For Basil I, the 'Western' imperial boundaries of the *Oikumene* were limited to southern Italy and Arab-dominated Sicily, a foreign policy that renounced a venerable tradition going back to the times of Justinian I. See Lounghis, *Byzantium in the Eastern Mediterranean*, pp. 85–7. When Anna Comnena was recalling her father's struggle to recover all provinces and restore the Byzantine boundaries existing since the Macedonian dynasty, she makes no mention of the former Italian provinces: 'Now, however, the Emperor Alexius by striking with both hands . . . enlarged the circle of his rule, for on the west he made the Adriatic Sea his frontier, and on the east the Euphrates and Tigris' (Anna Comnena, *The Alexiad*, 6.11, p. 113).

64. *DAI*, preface, 12–25, pp. 44–6.
65. Shepard, 'Emperors and Expansionism', p. 71.
66. *Vita Basilii*, 50, pp. 178–9.
67. Symeon Magister, *Chronicon*, 44, p. 337.
68. *Theophanes Continuates*, pp. 459–60.
69. John Skylitzes, *Synopsis of Byzantine History*, p. 249.
70. *History of Leo the Deacon*, 2.5, p. 75.
71. *On Skirmishing*, 15, p. 198.
72. E. McGeer (2003), 'Two Military Orations of Constantine VII', in J. W. Nesbitt (ed.), *Byzantine Authors: Literary Activities and Preoccupations*, Leiden: Brill, pp. 117–20. See also A. Markopoulos (2012), 'The Ideology of War in the Military Harangues of Constantine VII Porphyrogennetos', in J. Koder and I. Stouraites (eds), *Byzantine War Ideology between Roman Imperial Concept and Christian Religion, Akten des Internationalen Symposiums (Wien, 19.–21. Mai 2011)*, Vienna: Austrian Academy of Sciences Press, pp. 47–57. A similar scene can be found in Leo the Deacon's description of an army review before the walls of Antioch in 968: *History of Leo the Deacon*, 3.11, pp. 123–5.
73. McGeer, 'Two Military Orations', 1, pp. 117–18.
74. McGeer, 'Two Military Orations', 3, p. 119.
75. Mutanabbi in Vasiliev, *Byzance et les Arabes*, 2.II, pp. 322, 328–31.
76. Canard, *Hamdanides*, pp. 287–306. See also H. Kennedy (2004), *The Prophet and the Age of the Caliphates*, London: Longman, pp. 267–84.
77. Canard, *Hamdanides*, pp. 608–20; A. J. Cappel (1994), 'The Byzantine Response to the Arab (10th–11th century)', *Byzantinische Forschungen*, 20, pp. 116–18.
78. Canard, *Hamdanides*, pp. 608–20.
79. A. J. Arberry (1967), *Poems of Al-Mutanabbi*, Cambridge: Cambridge University Press, pp. 62–8.
80. M. Larkin (2008), *Al-Mutanabbi: Voice of the 'Abbasid Poetic Ideal*, Oxford: Oxford University Press, pp. 54–6; J. D. Latham (1979), 'Towards a Better Understanding of al-Mutanabbī's Poem on the Battle of al-Hadath', *Journal of Arabic Literature*, 10, pp. 1–22.
81. Larkin, *Al-Mutanabbi*, p. 55.
82. E. McGeer (1992), 'The Syntaxis Armatorum Quadrata: A Tenth-Century Tactical Blueprint', *Revue des études byzantines*, 50, pp. 219–29.
83. Abu Firas on the Battle of Rabat in 958: 'Il [Constantine VII] conclut la paix avec le roi des Bulgares, des Russes, des Turks, des Franks et autres peoples et leur demanda des secours' (in Vasiliev, *Byzance et les Arabes*, 2.II, p. 368). This comment implies some sort of diplomatic agreements with neighbouring rulers for the provision of contingents of troops and mercenaries. For more details on this, see G. Theotokis (2012), 'Rus, Varangian and Frankish Mercenaries in the Service of the Byzantine Emperors (9th–11th c.): Numbers, Organisation and Battle Tactics in the Operational Theatres of Asia Minor and the Balkans', *Byzantina Symmeikta*, 22, pp. 129–33.

84. The cavalry clad in full armour, mail – from the Greek κατά-φρακτος.
85. Larkin, *Al-Mutanabbi*, p. 55.
86. McGeer, 'Two Military Orations', 6, p. 131.
87. Naval Intelligence Division (1942), *Geographical Handbook Series: Turkey*, London, pp. 405–7; Naval Intelligence Division (1942), *Geographic Handbook Series: Syria*, London, p. 401.
88. John Skylitzes, *Synopsis of Byzantine History*, p. 232.
89. Very similar to the appointment of the Abbasid *qaʾid*: Kennedy, *Armies of the Caliphs*, pp. 99–100.
90. *Histoire de Yahya*, 18, p. 773; Vasiliev, *Byzance et les Arabes*, 2.I, p. 358; Canard, *Hamdanides*, pp. 793–4.
91. McCormick, *Eternal Victory*, p. 161.
92. *De Ceremoniis*, 2.19, pp. 607–12.
93. McCormick, *Eternal Victory*, pp. 164–5.
94. McGeer, 'Two Military Orations', 1, 2, 4, pp. 128–30.
95. Ibid., 3, 8, pp. 129, 133.
96. Ibid., 5, pp. 130–1.
97. Canard, *Hamdanides*, pp. 792–3; Vasiliev, *Byzance et les Arabes*, 2.I, p. 360.
98. McGeer, 'Two Military Orations', 7, p. 132.
99. McGeer, 'Two Military Orations', 8, p. 132.
100. P. von Sivers (1982), 'Taxes and Trade in the Abbāsid *Thughūr*, 750–962/133–351', *Journal of the Economic and Social History of the Orient*, 25, pp. 71–99.
101. W. Garrood (2008), 'The Byzantine Conquest of Cilicia and the Hamdanids of Aleppo, 959–965', *Anatolian Studies*, 58, pp. 127–40.
102. Garrood, 'Byzantine Conquest of Cilicia', p. 132.

4

The Byzantine View of their Enemies on the Battlefield: The Arabs

> These stratagems [hit-and-run tactics] are practised by the Persians and the Turks and the Arabs and by most of the nations. Thus, learning from them the Romans also practice them; but they also invent counter-stratagems, learning from their experience and their defeats.
> —*Sylloge Taktikorum*, 24.1

The Byzantines encountered many different nations on the battlefield during their long history. The surveys of foreign peoples in the military treatises amply illustrate their readiness not only to scrutinise and evaluate the tactics and characteristics of their enemies, but also to learn from them when necessary and adapt their tactics to the requirements of each operational theatre. This, of course, added to the long tradition of military science inherited from classical antiquity. The Byzantines may have revered the deep knowledge of the Greeks and the Romans in military matters, but the manuals compiled in the sixth and tenth centuries AD were a conscious adaptation to the geopolitical realities of their day, with the authors willing to enrich the content of their work rather than simply pass on obsolete battle tactics. Indeed, through these manuals, the Byzantines learned to understand war and its basic principles, such as order, discipline and the creation of an adequate command structure – an invaluable lesson for every civilisation.

In the middle of the tenth century, this renewal of military science came largely as a response to the increasing danger from the Arabs, whom the Byzantines had come to consider their most formidable enemy in the East. Although none of these changes appeared overnight, by the time Emperor Nicephorus Phocas launched his ambitious campaigns in Cilicia and northern Syria in the 960s, these changes had matured and the empire was ready to reap the rewards of their fruition.

The main objective of this chapter is to point out the perception of the 'other' in Byzantine sources, where 'other' refers to the 'military other', and the competence and skills of the empire's enemies as warriors. More specifically, I will focus on the following set of questions: What kinds of questions were the Byzantines asking about their enemies? What specific characteristics drew their attention and which aspects did they end up emulating? Are we able to uncover any social stereotypes on the part of the Byzantines and, if so, how did this influence their understanding of Islamic institutions and doctrines, and how did these underpin the Arabs' war efforts?

The first of the military treatises that offers a view of foreign warriors from the eyes of a contemporary of the Germanic invasions was Publius Flavius Vegetius Renatus' *Epitoma Rei Militaris*. Dated AD 383–450, it was the only work of its kind in Western Europe until Machiavelli's *Dell' Arte Della Guerra* (1521), presenting an idealised version of the organisation, battle tactics, armament and training of the Roman army of the fourth and fifth centuries AD. The author considers the careful selection of recruits as one of the main reasons for the Romans's success in conquering 'all peoples'; thus, he begins his work with a book on the recruitment and training of the Roman soldiers.[1]

Vegetius' division of Roman army recruits is geographical rather than sociological, following the same principle developed by Hippocrates on the climatic theory of human nature,[2] with the general selecting his recruits based on geographical criteria. Hence, the peoples near the sun are deemed more intelligent, but lacking in steadiness and confidence to fight at close quarters, while the peoples of the north are less intelligent but readiest for war. This rather crude division of soldiery based on geographical factors is supplemented by the usual sociological stereotypes that become common in the *strategika* of the following centuries: the Germans and the Spaniards are taller and stronger, the Africans are cunning and easily corrupted, the Greeks are intelligent and have a love for arts – something which would effeminise them in the eyes of many Western chroniclers of the tenth century onwards – and the Gauls base their power in their multitude of warriors. For Vegetius, the world is divided by the peoples that constitute recruiting grounds for the Roman army, rather than by neighbouring nations which threaten the empire.[3]

Two centuries after Vegetius compiled his work in Latin, the *Strategikon* attributed to Emperor Maurice was the first to devote an entire book to the foreign peoples neighbouring the empire.[4] What makes this chapter of the *Strategikon* so innovative and of such great value to modern historians is that, during those dangerous times of foreign invasions when the empire was threatened on all fronts, it provided its reader not just with a summary of the battle tactics of 'foreign peoples' but rather with a detailed analysis of the political, social and military organisation of all the nations that either bordered with or had threatened the empire in the past. As the author states in the introductory paragraph:

> The purpose of this chapter is to enable those who intend to wage war against these peoples to prepare themselves properly. For all nations do not fight in a single formation or in the same manner, and one cannot deal with them all in the same way.[5]

The predominant criterion of ethnicity in the *Strategikon* is the notion of *Romanitas*, which still holds the meaning of Roman citizenship, while the epicentre is, of course, Byzantium, with the foreign nations divided into four categories based on geography: the Persians, the Turks (Avar and Hunnish tribes), the Franks and Lombards, and the Slavs of the Balkans.[6]

The author's discussion of the characteristics and tactics of various peoples follows the same structure throughout Book XI, beginning with an introduction to the polity and the social and ethnic divisions of these people, their geographical and ethnological backgrounds, and the usual stereotypical characteristics that we saw in Vegetius' *Epitome of Military Science*. The author tries to present a distorted image of each nation's civilisation and achievements, compared with the Byzantines, by repeating negative contrasts.[7]

> The Persian nation is wicked, dissembling, and servile, but at the same time patriotic and obedient. The Persians obey their rulers out of fear, and the result is that they are steadfast in enduring hard work and warfare on behalf of their fatherland.[8]

The author is very much concerned about the morale of each nation exhibited in battle, and their κράσις – the physical strength and endurance of warriors in adverse conditions. He specifically stresses the fact that the Persians are 'extremely skilful in concealing their injuries and coping bravely with adverse circumstances, even turning them to their own advantage', and 'since they have been brought up in a hot climate, they easily bear the hardships of heat, thirst, and lack of food'.[9] The main part of each discussion, however, focuses on the description of the offensive and defensive equipment of the foreign warriors, their camps and battle formations, followed by the identification of weaknesses that each nation eventually displays on the battlefield, and advice on how to take full advantage of them. According to the author, 'they [Persians] wear body armour and mail, and are armed with bows and swords',[10] while he also brings the reader's attention to their main difference on the battlefield compared with other nations: 'They are more practised in rapid, although not powerful archery, than all other nations.' The description of their camps is equally short, but it summarises some key points that a general should be aware of: 'They surround themselves with a ditch and a sharpened palisade. They do not leave the baggage train within, but make a ditch for the purpose of refuge in case of a reversal in battle.'

The purpose of this style of writing – they do not attack like X (i.e. like the Romans or others do), but instead they attack like Y (a unique or peculiar tactic, which the author brings to the reader's attention) – is to highlight to the reader the differences between Persian and Roman battle tactics:

> They draw up for battle in three equal bodies, centre, right, left, with the centre having up to four hundred additional picked troops. The depth of the formation is not uniform, but they try to draw up the cavalrymen in each company in the first and second line or phalanx and keep the front of the formation even and dense.[11]

Finally, in the last part of each discussion, the author identifies the weaknesses of each nation on campaign. For example, 'they [Persians] are really bothered by cold weather, rain, and the south wind',[12] as they have been brought up in a warm climate, a key piece of information provided earlier in the discussion. Climatic factors also affect their equipment, as he specifically states 'all of which loosen their bow strings'. The author further identifies five more 'disturbances' that could disrupt the Persian armies: (1) a tightly packed infantry formation, which could withstand their *kataphraktoi* attack; (2) a flat and open battlefield, which would allow their enemies to use heavy cavalry; (3) engaging them in close-quarter combat (contrary to the 'Scythian' way of fighting) due to their lack of lancers and heavy infantry, and their dependence on foot archers instead; (4) encircling tactics; and (5) night attacks could also be very effective against the Persians.

When Leo VI commissioned his *Taktika* around the year 900, the empire was in a much different state of affairs in every aspect of its political, social and economic life than the reality in which the anonymous author of the *Strategikon* was writing three centuries before. Different times had brought different enemies to the beleaguered empire, but old ones still retained pride of place in the first of the military treatises of the tenth century. Thus, the geographical division of the *Strategikon* is kept but adjusted to the geopolitical reality of the time; the sections of the *Taktika* on the Franks and the Slavs are largely derived from the *Strategikon*, although the author does insert some important up-to-date information, such as the Christianisation of the Slavs that had taken place as recently as the second half of the ninth century.[13] The Scythian and Turkish nations of the sixth century have now been replaced by the Magyars and the rest of the steppe nations of the northern Black Sea coast but, once more, these sections of the *Taktika* are derived from the *Strategikon* with only some minor changes such as the following: 'These characteristics of the Turks are different from those of the Bulgarians only inasmuch as the latter have embraced the faith of the Christians and gradually taken the Roman characteristics.'[14]

The most significant change in the *Taktika* is the adjustment of the main criterion for the division of nations from *Romanitas* to *Christianitas*.[15] I have already examined the major cultural changes that took place in the empire in the mid-seventh century and the lack of protracted literary attention to, or substantive discussion of, 'frontiers' as physical barriers or limits, or as dividing lines between polities in Byzantine writings. At the beginning of the tenth century, the Byzantine Empire was still the epicentre of the treatise's discussion on foreign people, only by this time Constantinople considered itself as the capital of a Christian nation and defender of a universal Christian faith, classifying its enemies according to religion rather than ethnicity:

> Since the Bulgarians, however, embraced the peace of Christ [contrary to the Turks] and share the same faith in him as the Romans, after what they went through as a result of breaking their oath, we do not think of taking up arms against them.[16]

Christianity's main role for the empire was to 'Hellenise' and culturally assimilate its neighbours, as numerous passages from the *Taktika* clearly imply:

> These characteristics of the Turks are different from those of the Bulgarians only inasmuch as the latter have embraced the faith of the Christians and gradually taken the Roman characteristics. At that time, they threw off their savage and nomadic way of life along with their faithlessness.[17]
>
> He [Basil I] liberated them [Slavs] from slavery to their own rulers and trained them to take part in warfare against those nations warring against the Romans . . . As a result, he enabled the Romans to feel relaxed after the frequent uprisings by the Slavs in the past and the many disturbances and wars they had suffered from them in ancient times.[18]

By the turn of the tenth century, the main enemy of the empire was replaced by another, equally determined and ferocious as the Persians had been at the beginning of the seventh century. 'Permit us now to call to mind as best we can the nation of the Saracens that is presently troubling our Roman commonwealth', with the author proceeding to point out what he will examine in his discussion of this nation: 'What are they really like? What weapons do they make use of in military campaigns? What are their practices? How does one arm himself and campaign against them and thus carry out operations against them?' What we have here is, in essence, the same structural analysis of the characteristics and tactics of foreign people seen in the *Strategikon*.[19]

Leo begins with a very brief survey of the origins of the Saracens, calling them Arabs by race, having originated in the region of Arabia Felix (Arabian Peninsula). What is important to note here is the derogatory view he takes of the first Arab conquests, as he attributes the loss of Mesopotamia, Palestine and Egypt to 'the devastation of the Roman land by the Persians [that] allowed them [the Saracens] to occupy those lands' – although the Persians were a significant factor behind the collapse of Byzantine rule in the region, they were certainly not the only one.[20] The author also pays attention to the Saracens' κράσις as warriors with a hot temperament. Borrowing elements from the *Strategikon*, he identifies the main weakness of the Arabs as adverse weather conditions: 'This people is hurt by cold, by winter, and by heavy rain.' The reason they are reluctant to fight battles in this weather is because 'their bowstrings become slack when it is wet and because of the cold their whole body will become sluggish'.[21] Indeed, weather conditions in Anatolia were among the main reasons why the Arabs would not launch any expeditions between the months of September and November – a tactic which Nicephorus Phocas took full advantage of after 962, because 'they flourish in good weather and in the warmer seasons, mustering their forces, especially in summer, when they join up with the inhabitants of Tarsus in Cilicia and set out on campaign'.[22] The author also underlines the importance of logistics and transport restraints for the Arab raids over the Taurus Mountains and, contrary to the *Strategikon*, he highlights the difference between the Roman use of wagons and pack animals, and the Arab preference for camels, asses and mules for the faster and more practical movement of troops through rough and mountainous terrain.[23]

The author of the *Taktika* underlines the influence of the Romans on the armour and battle formations of the Arabs: 'They make use of armament, and their cavalry uses bows, swords, lances, shields, and axes. They wear full armour, including body armour, cuirasses, helmets, shin guards, gauntlets, and all the rest in the Roman manner.'[24] They also deploy their troops in their native formations which 'are both square and oblong', but 'it is as though they have been trained by experience in the other models of battle formations, so the very things they suffered from the Romans they are now busily putting into practice against them'. The *Taktika*'s emphasis on the fact that the Arabs were copying – or rather 'adapting to' – the Roman tactics is palpable, but the author immediately follows up by underlining that 'in their battle formations they [the Saracens] are *inventive* and *steadfast* and are *not frightened* by the rapid onslaught of their attackers nor do they become too *relaxed* by simulated delays'.[25] How, then, are the Arabs viewed as warriors by the author of the *Taktika*, our most detailed source on this issue?[26]

> Neither when they are pursuing nor are being pursued do they break their formation. But if it should happen that they do so, they lose their cohesion and are unable to return, only racing on to save themselves.[27]
>
> They are *bold* at the expectation of victory but very *cowardly* when victory is denied them.[28]
>
> They stand *steadfast* in their formation, bearing up valiantly under the missiles fired by the forces boldly attacking them. When they observe that their adversaries' energies are dropping, then they rise up and *fight strenuously* ... After those who had been shooting against them have discharged [their arrows], which they endure by forming a wall of shields, they quickly come together and in a body rise up and start fighting hand-to-hand. In attacking these people it is always necessary to be ready for anything.[29]
>
> They are more notable than all other peoples in *relying on good counsel* and *firm adherence* to methods of warfare.[30]

The number of adjectives in the *Taktika* concerning the Arab troops – cavalry, infantry, volunteers for jihad, or professional mercenaries – is impressive: brave, daring, audacious, steadfast, patient, strenuous and inventive are found in just three pages of the treatise.[31] The Arabs, however, may be steadfast in their battle formation, but if they do break it they lose their cohesion and flee; they may be bold when they expect victory, but they lose heart when it is denied to them. The reasons why they might lose their cohesion, their will to fight and instead choose to retreat from the battlefield is, first, their 'superstition' and, second, their motives for launching the campaign in the first place.

Calling Islam a 'superstition', the author discusses the Arabs' religious beliefs and habits by using the same negative contrasts that we come across in the *Strategikon*: 'They *appear* to show proper reverence, *but* their apparent reverence must be recognised as blasphemy.'[32] This is followed by a vivid contrast between the 'Saracen God', Allah, and the true Christian God:

> They cannot bear to call Christ God, [although He is indeed] true God and saviour of the world. They argue that God is the cause of every evil deed and they claim that God rejoices in war and scatters abroad the peoples that want to fight.[33]

What is repeatedly criticised in the *Taktika* is the Muslim warriors' belief of a vengeful God:

> They say that everything comes from God, even if it should be evil. If it happens that they suffer a setback, they do not resist since it has been decreed by God. Overthrown by the onslaught, they are completely undone.[34]

Finally, although the author of the *Taktika* (c. AD 900) does not use the term 'holy/just war' anywhere in his work,[35] he is aware that his Muslim foes are offered spiritual rewards, a recompense given by God for the moral quality of their efforts if they die in battle, which he identifies as compensation (noun μισθός – *misthos*: to denote a spiritual rather than monetary reward). Moreover, and exceptionally, the Christian emperor, while offering the usual condemnation of the 'barbarous and impious race', recommends that the Byzantines emulate the infidel as, for them, warfare is a collective effort, whereby all members of society share in the expenses (verb χορηγῶ – *chorego*: to supply the costs for any purposes, including war).[36] At the same time, he is also fully aware of their desire to obtain material goods: 'Because of the booty they have reason to expect, and because they do not fear the perils of war, this nation is easily gathered together in large numbers from inner Syria and all of Palestine.'[37] The collection of booty had always been a significant incentive for campaigning armies and it is clear that every author places importance on its control.[38] As Haldon notes, it does appear to have been an inducement to Byzantine troops, although it is never mentioned as a motive for recruitment, as in the case of the Arabs.[39]

This stereotype of the opportunistic soldier prone to the looting and destruction of rural societies is repeated by two other key sources of the tenth century, Leo the Deacon and Ioannes Kaminiates. The latter was another well-educated cleric from Thessaloniki, who held the relatively low-ranking clerical grade of *anagnostes* and served as a chamberlain in the bishop's palace. He wrote his account while imprisoned (or shortly afterwards) in Tarsus in 905–6 after experiencing first-hand the sack of the city of Thessaloniki by Arab raiders in September 904.[40] We read in Leo the Deacon's *History* (emphasis added):

> Now this Emperor Romanus decided to eradicate, with the support of the Almighty, the tyranny of the Arabs of Crete, who were *arrogant* and had *murderous intentions* against the Romans. For they *exalted immeasurably* over the recent disaster suffered by the Romans, and were *plundering* the shores of the Roman Empire on a large scale.[41]

Referring to the Cretan expedition of 960, Leo presents the Arabs as 'arrogant' and full of 'murderous intentions', with a beleaguered empire on the defensive against an inferior adversary, who took advantage of its recent military setbacks in Syria and Cilicia – or rather the change of focus of the Byzantine government – to plunder its Anatolian shores. According to Leo, these raiders were a cunning and uncivilised enemy hungry for death, destruction, money and glory. This notion of the Arabs as parasites is further enhanced where Nicephorus delivers his alleged speech to his officers during the siege of Chandax (emphasis added):

> I think that none of you is unaware of the cruelty and ferocity of the descendants of the maidservant, and the raids and enslavement that they have murderously perpetrated against Romans ... Therefore, Providence has by no means tolerated that these liars, these most evil beasts, these lazy gluttons *feed forever off the Christian people*.[42]

Finally, Leo targets Sayf ad-Dawla's motives for his campaigns against the empire, accusing him of opportunism and profiteering over the local populations (emphasis added): 'He [Sayf ad-Dawla] decided that this was an opportunity for him *to raid* all the Roman territory in the east with impunity, *to plunder it without bloodshed, amass enormous wealth, and gain eternal glory.*'[43]

The effects of the Arab naval campaigns in Byzantine territories and the terror that these spread in the local populations are vividly described in the account of the *Sack of Thessaloniki* by Ioannes Kaminiates. Raw emotions and intense accounts of individual tales of heroism and misery dominate his work. Such material could have been gathered only orally through conversations with his fellow prisoners.[44] Hence, Kaminiates writes about the slaughter of the population of the city and the wish of the Arabs to prolong the pain and suffering of their victims for as long as possible in some sort of sickening torturous game, as if they were gaining pleasure from the spectacle of condemning the inhabitants to 'slow death'. The author describes the sheer terror experienced by the inhabitants of the city when they heard rumours that the Arab fleet was approaching:

> Nothing could withstand their raids and their repeated attacks ... Their ships were fifty-four in number and each one of them was equal to an entire town due to their size, guns and armour ... All of them were thirsty for blood, real beasts, practised in the art of murdering and addicted to slaughtering and stealing.[45]

Ioannes' vivid but rather exaggerated comments are coupled with the image he wishes to paint of these Arab raiders as materialistic bandits prone to looting and destruction; he dedicates a large part of his narrative to the description of the bribery of his pursuers in order for him and his relatives to avoid execution. He portrays the Arabs as being easily manipulated by money and gold. He also quotes an order allegedly issued by Leo the Tripolites himself:

> If any prisoner has money hidden away, he should be taken apart from the rest and use it to buy off his own life. Those who do not [have money hidden away], they would be decapitated and written off from the list of the living.[46]

Hence, a sharp contrast emerges from the careful examination of these sources. On the one hand, we have the image of the Arab soldiers and volunteers for jihad, hopeful both for spiritual ('they do not fear the perils of war') but mainly for material rewards, opportunistic and greedy, willing to leave everything behind in order to embark on an adventurous campaign of destruction and pillage that may lead them to glory and money. On the other, these ephemeral and immoral values of the *Hagarene* are sharply contrasted with Christian values. They, instead, were fighting to defend their God and the rest of their fellow Christians from humiliation and death at the hands of the aggressors – the Arabs:

> For the enemy it is a matter of great importance, and they will make use of every device to assail you when you do not expect it, so that they may overwhelm you, to the harm and destruction of the people of Christ, the dishonour of the mighty Romans, and the exultation and swollen pride of the arrogant sons of Hagar, who deny Christ *our* God.[47]

This was a war between good and evil, between the true faith and superstition, between a 'real' (or 'ideal') soldier and a 'false' one, the 'moral' and 'immoral', where the spiritual and material motives of the state and the individual are contrasted with the ideals of 'just war' and jihad: 'You know how virtuous it is to fight on behalf of Christians, and how much glory the man who does so achieves for himself. This is more profitable than all wealth, more praiseworthy than all other honour.'[48]

According to the Byzantine treatises, if a Byzantine army is well equipped with men and supplies, well led and – most importantly – has God's help, it will prevail over these Muslims, who are considered to be far inferior to the imperial war machine:

> If this is how everything goes, the army of the Romans, well and properly armed, will greatly increase, especially with a large number of men chosen for their courage and nobility, and lacking nothing of what is needed, it will easily, with God's help, be crowned with victory over the barbarian Saracens. If, in our weaponry, especially our great supply of bows and arrows, our numbers and courage, and our requisite stratagems and machines, we Romans are far superior to the barbarians, and if we have the divinity as our ally in everything, we will easily achieve victory over those peoples.[49]

Heavenly support for the imperial armies is constantly repeated in every military treatise we read, making it clear that successful warfare without God's help would be impossible.[50] It is exactly this point that Constantine VII wishes to make in his propagandistic oration of 950:

> How you [soldiers] were embroiled in combat not as if against men but as if triumphing over feeble women, succeeding not as in battle or in war, but rather dealing with men as though it were child's play, even though they were mounted on horses whose speed made them impossible to overtake, even though they were protected by equipment unmatched in strength and in craftsmanship ... But since they were without the one paramount advantage, by which I mean hope in Christ, all of their advantages were reduced to nothing and were in vain.[51]

This image of the Arabs as effeminate warriors painted by Constantine in the early winter of 950, which served to promote himself and his reign in the eyes of not only his soldiers but the political establishment in Constantinople by translating a minor military victory over the Hamdanids into a great military success, is sharply contrasted by his writings in the *De Administrando Imperio*. A manual of kingcraft for the admonition and training of a young heir to the throne, and the longest and most comprehensive work on the history, politics, culture, social organisation and foreign policy of the nations that neighboured the empire, the *De Administrando Imperio* follows the same criterion for the division of the ἔθνη (*ethne*, 'nations') – religion.[52] There is no doubt that this was a confidential document, which contained a lot of information about the principles of imperial foreign policy and diplomacy. Commissioned by Constantine in 948–52, this work was intended for the eyes of the heir to the throne and, perhaps, a handful of high-ranking officials at the imperial court, thus including some rather more secret and up-to-date views of the empire's enemies. The well-educated emperor had this to say about the Fatimids, at the time only a peripheral dynasty in North Africa (Libya and Tunisia), which would become one of the empire's most dangerous competitors for regional power in the eastern Mediterranean basin by the end of the tenth century:

> They are an Arab nation, carefully trained to wars and battles; for with the aid of this tribe Mahomet went to war, and took many cities and subdued many countries. For they are brave men and warriors, so that if they be found to the number of a thousand in an army, that army cannot be defeated (αήττητον) or worsted (ακαταμάχητον). They ride not horses but camels, and in time of war they do not put on corselets or coats of mail but pink-coloured cloaks,[53] and have long spears and shields as tall as a man and enormous wooden bows which few can bend.[54]

Kaminiates also portrays the Arab raiders as brave warriors rather than the effeminate and feeble men that we read about in the oration of 950. They may be materialistic and prone to looting and killing but they possess many qualities as soldiers that bring both terror and admiration to the reader. According to the chronicler, the 'barbarians' are intelligent (ευφυείς) in devising stratagems and several other surprise attacks with which to win over a city during a siege, especially during the night. They do not show any consideration for the dangers of their undertaking, even when it seems impossible to succeed, and they think that dying while attempting to execute a daring plan of attack wins them glory and fame.[55] Kaminiates vividly portrays their rage in battle and their resilience and high morale in the following passage:

> Hearing them raging against us was truly horrifying. They were, indeed, exhibiting signs of extreme fury every time they were letting out screams and their mouths were foaming, things which revealed their demonic nature. They did not even wish to have something to eat that day, instead they carried on fighting in this hellish heat without feeling their bodies becoming tired or burned under the summer sun. They had no thoughts other than how to conquer the city and satisfy their wrath against us and, if they could not achieve that, they would rather lose their lives and die with their weapons in hand.[56]

Contrary to Kaminiates, who witnessed the siege of Thessaloniki first hand and was captured and held prisoner in Tarsus for about a year,[57] Leo the Deacon was not an eyewitness to the events that took place in the East; in fact, the only military campaign he ever had the chance to participate in was Basil II's Bulgarian expedition of 986. Did he ever experience a siege by an Arab army or the ferocious attack of Muslim elite horsemen on the battlefield? The answer is no. His comments on the fighting abilities of the Arabs are most likely the outcome of the stereotypical attitudes held by many Constantinopolitan courtiers with whom Leo would have had the chance to converse.

For the Cretan expedition of 960 we read that 'the barbarians were astonished at this strange and novel sight [the disembarkation of the Byzantine cavalry and their horses from their transport ships using ramps (κλίμακες, *klimakes*)]'.[58] This was either a gross mistake on the part of Leo, bearing in mind that he was not a military man himself, or a clear and deliberate misrepresentation of Arab naval warfare technology. As there are no depictions of Muslim merchant ships and warships before the fourteenth century, we can speculate with a reasonable degree of certainty that the Arabs would have taken over the ships and ship-building techniques of conquered peoples around the Mediterranean coast and, although there are undoubtedly some differences, Muslim ships were not greatly dissimilar or inferior to the Byzantine ones.[59] We know of the use of *chelandia* as horse transport units of the Byzantine fleet, which were equipped with a *klimaka* (κλίμακα) since the early tenth century, a ramp for the loading and unloading of the horses from the ship's gunwales, either from the stern or usually from the bow.[60] Muslim nations had a similar type of ship, known to the Byzantines as *tarida* or *tarita*, which had originally developed from a reed canoe used on the Red Sea;[61] its crucial design included a square stern with two stem posts and a fitted door or ramp which could be lowered to the coast or dock to unload men and horses, an action powered by oars and reminiscent of modern landing crafts.[62]

Other examples of Leo's propensity to praise Nicephorus and humiliate the Arabs include: 'And so in a short time the entire host of 40,000 barbarians from youth upwards *was easily killed*, victims of the Romans' swords.'[63] The massacre of 40,000 men after being surprised in a night attack in the vicinity of Chandax during the siege of the city is certainly an exaggeration, especially when no details of the battle are given. Other examples include:

> I think that none of you is unaware of the cruelty and ferocity ... and the raids and enslavement that they [Cretan Saracens] have murderously perpetrated

against Romans (and this when they were living on an island that was subordinate to [the Romans], although it had come to the Hagarenes through the wickedness of fortune).[64]

Once again, the chronicler repeats the same stereotype of the wicked and cruel Arabs, who were motivated by their individual materialistic incentives. According to his interpretation of history, the island fell under their control not because of their military prowess or the failings of the central government in Constantinople, but due to misfortune – exactly the same reason attributed to the Arab conquests of the seventh century by Leo VI.[65]

Skylitzes is careful not to use any derogatory comments to describe the Muslims as warriors or any negative contrasts in the fashion that we see in the *Taktika*, although his tendency to praise the military achievements of the Byzantine generals by exaggerating the magnitude of the victory over the Muslims does not always do justice to the fighting abilities of their enemies. Thus, for the Byzantine raids in Cilicia in 959–60 that probably intended to stamp the authority of two key lieutenants of Nicephorus on the East, Tzimiskes and his brother Leo Phocas, we read in the *Synopsis*:

> In that year [960] Nicephorus Phocas the magister, who had already been promoted to domestic of the Scholae of the East by Constantine [VII], he won many victories over the Agarenoi of the East, and the Emir of Tarsus called Karamonin and the [Emir] of Chalep [Aleppo] Hamdan and the [Emir] of Tripoli Izeth he utterly humiliated, he sent a multitude of elite soldiers and a well-equipped fleet against the Saracens of Crete.[66]

In his short account on the Byzantine interception of a Hamdanid raid in Cilicia in 960 that culminated in the Battle of Adrassos, Skylitzes comments on the outcome of the battle, which was indeed a devastating defeat for the Arabs as Phocas' troops recovered all the booty and prisoners:

> Hamdan the emir of Chalep, warlike and very active in other things ... He [Leo Phocas] met with him in a location called Adrassos and he forced him to an all-out retreat and he completely annihilated him, the numbers [of the Saracens] who fell in battle is incalculable.[67]

What stands in sharp contrast to the view of the Arab warriors shared by all the sources, lay and ecclesiastical, that have been examined thus far is the description we get from the military treatises of the period, especially the works commissioned by Nicephorus Phocas in the middle of the tenth century – *On Skirmishing* and the *Praecepta Militaria*. Both works were concerned with the empire's wars in the East, although the fact that the author of the *Praecepta* does not clearly identify the enemy in that operational theatre is a sign of the treatise's place in the military literature of the period (end of the 960s). This clearly contrasts with the earlier treatise *On Skirmishing* that deals with past events leading up to the period of the commissioning of the work, when the times and conditions it portrays were already passing into history and legend, and everyone would have been familiar with the historical and geographical context of the work. What is striking in both treatises, however, is the lack of any stereotypical descriptions of the fighting abilities of the Arabs, bearing in mind that the author of *On Skirmishing* had no sympathy for the Armenians who were serving in the imperial army, and he is keen to highlight the latter's unruliness and unreliability on the field of battle and as sentries in military outposts.[68]

By reading the treatises of the mid-tenth century we can piece together some interesting information regarding the way in which the Byzantine officers would have viewed their opponents in the East, judging by the precautionary measures they describe taking against them in their works:

> Otherwise, if the enemy find out that the public road is securely held by a large number of troops, they will advance along one of those off to the side. If this should not be well and securely guarded, the enemy will use that to find a way through and will appear to the sides or the rear of our formation, injecting confusion and fear. But if both sides are tightly guarded, then the enemy will either charge into battle and, with God's help, will be put to shame, or, struck with terror, they will take another road a number of days distant.[69]

In this passage from *On Skirmishing*, we are presented with the view that the Arabs are a smart and ingenious enemy who are always thinking of ways to outmanoeuvre their enemy in war, fully capable of injecting fear and confusion and likely to stand and fight rather than strike camp and retreat, regarding death in battle as an honour. This view is further supported by the following passages: 'They [the enemy] might make their stand in that very place, unloading the pack animals and throwing up a sort of rampart of all the things lying around, and form up for battle against us. This would cause great difficulty';[70] and:

> Our men must likewise persevere in fighting with no thought of flight until the hand of God intervenes and the enemy recoils. If it should happen that the enemy hits our cavalry units hard and repels them – God forbid – they must retire inside our heavy infantry units for protection.[71]

The Arabs were also capable of deploying large numbers of well-equipped cavalrymen that could break into enemy infantry formations, even the tightly packed and disciplined Byzantine infantry squares of the period, as we read in the *Praecepta*:

> In the likely event that the enemy gets word of these formations and in turn chooses to react with equal force and outfit heavy cavalrymen, to keep both themselves and their horses safe by means of armour, so that the spears of the infantrymen will be smashed to pieces by these men, and by using these horsemen the enemy will shatter the infantry units.[72]

In a characteristic 'Vegetian' attitude to warfare, the author of the *Praecepta* recommends to his audience to

> avoid not only an enemy force of superior strength but also one of equal strength, until the might and power of God restore and fortify the oppressed hearts and souls of our host and their resolve with His mighty hand and power.[73]

The contrast between the attitude towards the Arab soldiers shown by the authors of the mid-tenth century *strategika* and the rest of the lay and ecclesiastical historians of the period is unambiguous. For example, we read in Skylitzes' *Synopsis*:

> So outstanding was he [Leo Argyros] among his contemporaries during the reign of the emperor Michael [III] that he alone, together with his household, dared oppose the Manichaeans of Tephrike and the Hagarenes of Melitene in battle – and easily defeated them. The mere mention of his name infused terror in every adversary.[74]

This chapter has offered a detailed and comprehensive overview of the perception of the 'other' in Byzantine sources, where the 'other' refers to the main enemy of the Byzantines on the battlefields of the tenth century – the Arabs. What I noticed by going through the numerous lay and ecclesiastical sources of the period was the constant motif of the conflict between a 'real' (or 'ideal') soldier and a 'false' one, between the 'moral' and the 'immoral', where the spiritual and material motives of the state and the individual are contrasted with the ideals of 'holy/just war' and jihad, which are repeatedly appraised as the ultimate duty of a pious soldier for his God. For the Byzantine sources, the dominant perception regarding God's role in war is that He aids the righteous warriors, who struggle to protect or restore the territories of the divinely protected Byzantine Empire.

Leo the Deacon, an educated ecclesiastic in the imperial court, offers the most stereotypical view of the Arabs as warriors of jihad: they were unsophisticated in warfare and technologically inferior; they were a cunning enemy focused on money, plunder and booty, and were inherently unchivalrous and uncivilised. This type of opportunistic soldier can also be found in Ioannes Kaminiates' history of the siege and conquest of Thessaloniki in 904. He greatly despised them. There is no doubt about that, but he was also highly impressed by their fighting abilities: they were intelligent, furiously resilient, with high morale and ready to die for their cause.

Leo VI's *Taktika* even reveals a feeling of respect for the empire's Muslim adversaries on the battlefield; he describes them as formidable enemies who surpassed all foreign nations in intelligence and who had adopted Roman weapons and often copied Roman tactics. We understand that he admired the spiritual motives of the Arab warriors and he actively encouraged Byzantine officers to instil the same motivation in their soldiers. How could they not be portrayed as worthy adversaries when they had dominated the Mediterranean at the turn of the tenth century? However, Leo separates the Arabs from all the other enemies of the empire because of their religious beliefs and he twists the virtues of the Muslims into vices. There is a strict distinction between Christian and non-Christian enemies in the *Taktika*, where religion qualifies as the central and fundamental criterion for the division of the empire's neighbours, a point of reference which is also apparent in all the sources for this period.

The most striking difference of opinion, however, can be found in the works of the Emperor Constantine VII. In his oration to his troops that was, allegedly, delivered in the autumn of 950, he portrays the Arabs of Aleppo as feeble women, while in his magisterial manual on kingcraft, commissioned in 948–52, he highlights their prowess in battle and the splendour of their weapons and armour. What could be the reason behind this apparent contradiction? Can this deliberate attempt to appear objective and impartial when describing their enemies be interpreted as a sign of the level of self-confidence that the Byzantine military leadership had achieved by the 950s? Or could this have been part of the imperial propaganda to enhance the military achievements of the Byzantine nobility against a worthy opponent?

Each text was written with a different purpose in mind. Constantine VII's oration, read to his returning troops in the autumn of 950, is a good example of Byzantine propaganda literature: a small victory against the Hamdanids was exploited for propaganda purposes rather than for its real strategic value; the purpose was restoring some much-needed prestige to the regime of Constantine Porphyrogenitus after the humiliation of the Cretan expedition the year before. His real thoughts about the Arabs as warriors may be seen in his *De Administrando Imperio*, and they were much more pragmatic. Indeed, it is clear that the author of the latter work deliberately avoids any derogatory comments or negative contrasts when describing the qualities of the Arab warriors.

This is also the case for the military treatises of the period and for Skylitzes. They paint a clearer and more refined picture of their enemies as ingenious and brave soldiers, capable of injecting fear and confusion into their enemies and likely to stand their ground and even fight a losing battle rather than strike camp and retreat. It is, nevertheless, the Byzantines who have the moral high ground; it is they who will achieve eternal glory with the help of God. Perhaps the authors of the military treatises were deliberately trying to appear objective and impartial when describing their enemies, a sign of the level of self-confidence that the emerging military nobility would have attained in the two decades leading up to the climax of the conflict with the Hamdanids in the middle of the century.

Notes

1. Vegetius, *Epitome of Military Science*, I.1–3, pp. 2–4.
2. Hippocrates, *On Airs, Waters and Places*, 24.
3. G. Dagron (1987), '"Ceux d'en face": les peoples étrangers dans les traits militaires byzantins', *TravMém*, 10, pp. 207–32, p. 208.
4. The chapter is entitled 'The Tactics and Characteristics of Each Race which may Cause Trouble to Our State', *Strategikon*, 11, pp. 113–26.
5. *Strategikon*, XI, p. 113.
6. Dagron, 'Les peoples étrangers', pp. 209–16. Wiita gives an excellent commentary on Book XI: J. E. Wiita (1977), 'The Ethnika in Byzantine Military Treatises', PhD dissertation, University of Minnesota.
7. Dagron, 'Les peoples étrangers', p. 212.
8. *Strategikon*, XI.1, p. 113.
9. Ibid., XI.1, p. 113.
10. Ibid., XI.1, p. 114.
11. Ibid., XI.1, p. 114.
12. Ibid., XI.1, p. 114.
13. Leo VI, *Taktika*, XVIII.40, p. 452; XVIII.93–5, p. 470.
14. Ibid., XVIII.59, p. 458.

15. Dagron, 'Les peoples étrangers', pp. 217–18.
16. Leo VI, *Taktika*, XVIII.42, pp. 452–4. For the Franks: XVIII.74, p. 462. For the Slavs: XVIII.93, p. 470.
17. Ibid., XVIII.59, p. 458.
18. Ibid., XVIII.95, p. 470. Cf. the writings of Nicolas Mysticus, who addressed his letters to Symeon of Bulgaria at about the same time as the compilation of the *Taktika*: 'What has become of the fair hopes, which we unhesitatingly entertained, that the evil demon would no longer have strength to introduce offense between Romans and Bulgarians, or would disturb their pure love?' (p. 50); 'For since He [Christ], who willed to award peace to the world through His emptying out of his Father's bosom, made the race of the Bulgarians His own, and joined them fraternally to the Roman race, and granted to them the same honour as to us, that they should be called Christians' (p. 54); Nicolas I, *Letters*, CFHB 2, pp. 50–1 and 54. See also Kaminiates, *Για την Άλωση της Θεσσαλονίκης*, 9, pp. 81–3; John Skylitzes, *Synopsis of Byzantine History*, pp. 107–8, 213.
19. Leo VI, *Taktika*, XVIII.104, p. 474. Cf. the questions Constantine VII asks in: *DAI*, preface, pp. 45–7; 13, p. 76.
20. Leo VI, *Taktika*, XVIII.104, p. 474. Cf. the view of Ibn Hawqal, a propagandist for the Fatimids of North Africa, on the Byzantine victories of the tenth century: 'L'empire byzantin apparaît à beaucoup de musulmans cultivés et d'auteurs d'ouvrages comme très different de ce qu'il est en réalité. En effet, il est dans une situation précaire; sa puissance est insignifiante, ses revenus sont médiocres, ses populations d'humble condition, la richesse y est rare, ses finances sont mauvaises et ses ressources sont maigres ... l'empire byzantin n'approche pas l'importance du Maghreb ni sa puissance' (Ibn Hawqal [1964], *Configuration de la terre (Kitāb Surat al-arḍ)*, trans. J. H. Kramers and G. Wiet, Paris: G. P. Maisonneuve and Larose, I, p. 195.
21. Leo VI, *Taktika*, XVIII.118, p. 480.
22. Ibid., XVIII.119, p. 480.
23. Ibid., XVIII.106, p. 476.
24. Ibid., XVIII.110, pp. 476–8.
25. Ibid., XVIII.115, p. 480.
26. See also the very interesting article by T. G. Kolias (1984), 'The *Taktika* of Leo VI and the Arabs', *Graeco-Arabica*, 3, pp. 129–35.
27. Leo VI, *Taktika*, XVIII.111, p. 478.
28. Ibid., XVIII.112, p. 478.
29. Ibid., XVIII.116, p. 480.
30. Ibid., XVIII.117, p. 480.
31. A. Toynbee (1973), *Constantine Porphyrogennitus and His World*, London: Oxford University Press, pp. 382–3.
32. Leo VI, *Taktika*, XVIII.105, p. 476.
33. Ibid.

34. Ibid., XVIII.112, p. 478. Cf. *Strategikon*, VII.11, p. 72. For Kaminiates' description of the destructions that befell several Byzantine cities in the Balkans due to 'God's Will' and the sins of their citizens, see Kaminiates, *Για την Άλωση της Θεσσαλονίκης*, 12, pp. 85–7 (Thessaloniki); 44, pp. 126–7, 14, pp. 88–9 (Demetriada).
35. Δικαίαν δει τήν ἀρχήν του πολέμου γίνεσθαι: *Strategikon*, VIII.2.9, p. 84. On the theoretical distinction between 'holy war' and 'just war' in Byzantine thought, see Stouraitis, '"Just War" and "Holy War"', pp. 227–64; Stouraitis, '*Jihād* and Crusade', pp. 11–63 (specifically for Leo VI's *Taktika*, see pp. 19–26). For a comparative approach to the role of religion in warfare and an introduction into the analytical term 'war ideology' in the research on Byzantine conceptions of war and peace, see I. Stouraitis (2009), *Krieg und Frieden in der politischen und ideologischen Wahrnehmung in Byzanz*, Vienna: Fassbaender. See also the following collection of papers: Koder and Stouraitis, *Byzantine War Ideology*; J. Kelsay and J. T. Johnson (eds) (1991), *Just War and Jihad: Historical and Theoretical Perspectives on War and Peace in Western and Islamic Traditions*, Westport, CT: Greenwood; P. Stephenson (2007), 'Imperial Christianity and Sacred War in Byzantium', in J. K. Wellman (ed.), *Belief and Bloodshed: Religion and Violence across Time and Tradition*, Plymouth: Rowman & Littlefield, pp. 88–9.
36. Leo VI, *Taktika*, XVIII.122, pp. 482–3; G. Dagron (1983), 'Byzance et le modèle islamique au Xe siècle, à propos des *Constitutions tactiques* de l'empereur Léon VI', *Comptes rendus des séances de l'année de l'Académie des inscriptions et belles-lettres*, pp. 219–43. Most of the raids in the *thughūr* were conducted by regular soldiers. The volunteers for jihad (*muttawiʾa*) comprised a small but not negligible number in these armies. See Kennedy, *Armies of the Caliphs*, pp. 98–9, 106–7. For the *muttawiʾa* in Islam, see Tor, *Violent Order*, pp. 37–65.
37. Leo VI, *Taktika*, XVIII.126, pp. 484–5; see also, ibid., XVIII.130, pp. 486–7.
38. *Strategikon*, VII.14, p. 68; Leo VI, *Taktika*, XII.101; XIII.15; XVI.4, pp. 23–5; XVII.25, p. 37.
39. J. Haldon (2014), *A Critical Commentary on the 'Taktika' of Leo VI*, Washington, DC: Dumbarton Oaks Research Library and Collection, p. 374.
40. Kazhdan directly challenged the authenticity of Kaminiates' work. See A. P. Kazhdan (1978), 'Some Questions Addressed to the Scholars, who Believe in the Authenticity of Kaminiates' Capture of Thessalonika', *Byzantinische Zeitschrift*, 71, pp. 301–14. Christides more or less agreed with Kazhdan's conclusions: V. Christides (1981), 'Once Again Kaminiates' Capture of Thessalonica', *Byzantinische Zeitschriftt*, 74, pp. 7–10. Tsaras has supported the view that Gregory of Cappadocia was not a real person but rather a literary device: J. Tsaras (1988), 'Η αυθεντικότητα του Χρονικού του Ιωάννου Καμινιάτη' [The Authenticity of the Chronicle of Ioannes Kaminiates], *Βυζαντιακά*, 8, pp. 43–58. On the issue of the

'abundance of naked Ethiopians' among the Arab raiders, a point considered dubious by several modern Byzantinists, see K. Okwess-O'Bweng (1988), 'Le portrait du soldat noir chez les Arabes et les Byzantins d'après l'anonyme "Foutouh al-Bahnasâ" et "De Expugnatione Thessalonicae" de Jean Caminiatès', *Βυζαντινός Δόμος*, 2, pp. 41–7. Frendo objected to both the literary and historical arguments raised by Kazhdan: D. Frendo (2000), *John Kaminiates: The Capture of Thessaloniki*, Perth: Australian Association of Byzantine Studies, pp. xxxvii–xl. Karpozilos and Odorico have defended the authenticity of Kaminiates' work: A. Karpozilos (2002), *Βυζαντινοί Ιστορικοί και Χρονογράφοι (8ος–10ος αιώνας)* [Byzantine Historians and Chroniclers (8th–10th Centuries)], vol. 2, Athens: Kanaki, pp. 253–62; P. Odorico (2005), *Jean Caminiatès, Eustathe de Thessalonique, Jean Anagnostès – Thessalonique: chroniques d'une ville prise*, Toulouse: Anacharsis, pp. 11–24 (I have used the Greek translation of this work: Π. Οντορίκο (2010), *Ιωάννης Καμινιάτης, Ευστάθιος Θεσσαλονίκης, Ιωάννης Αναγνώστης: Χρονικά των αλώσεων της Θεσσαλονίκης*, trans. Χ. Μεσσής, Athens: AGRA). Although the case for a fifteenth-century forgery by John Anagnostes (the 'Reader'), who was the actual composer of the 1430 work, is inconclusive, we should at least examine Kaminiates' work with due caution; see Treadgold, *Middle Byzantine Historians*, pp. 121–3. I believe, however, that it is ludicrous to purport that Kaminiates' choice of χειροποιήτῳ βροντῇ (man-made thunder) pertains to the use of a crude form of canon during the siege. Frendo wrote an extensive footnote explaining Kaminiates' use of this 'puzzling' phrase by drawing parallels with the terminology in *The Miracles of St. Demetrius*. See Frendo, *John Kaminiates*, pp. 167–9.
41. *History of Leo the Deacon*, I.2, pp. 58–9.
42. Ibid., I.6, p. 65.
43. Ibid., II.1, p. 70.
44. S. Patoura (1994), *Οι Αιχμάλωτοι ως Παράγοντες Επικοινωνίας και Πληροφόρησης* [Prisoners as Agents of Communication and Information], Athens: National Research Institute, pp. 83–93; Frendo, *John Kaminiates*, pp. xxxii–xxxiii.
45. Kaminiates, *Για την Άλωση της Θεσσαλονίκης*, 18, p. 94.
46. Kaminiates, *Για την Άλωση της Θεσσαλονίκης*, 58, p. 142.
47. *On Skirmishing*, 15, p. 196, 19 (emphasis added). Cf. the letter of the Domesticus of the Scholae to the emir of Melitene: John Skylitzes, *Synopsis of Byzantine History*, pp. 140–1. See also the request for divine intersection by St Demetrius from the citizens of Thessaloniki during the Muslim siege of the city: Kaminiates, *Για την Άλωση της Θεσσαλονίκης*, 22, pp. 98–9. See McGeer, 'Two Military Orations', 2, p. 118, for Constantine describing his soldiers as 'avengers and champions not only of Christians but of Christ himself, Whom they wickedly deny'.
48. McGeer, 'Two Military Orations', 3, p. 119.

49. Leo VI, *Taktika*, XVIII.124 and 125, pp. 482–4.
50. *Strategikon*, VIII.1.38, p. 82; VIII.2.1, p. 83; *On Skirmishing*, 24, p. 236; *Praecepta Militaria*, II.44–5, p. 24; II.93–8, p. 28; II.123–4, p. 28; IV.148–50, p. 46. A few examples taken from lay and ecclesiastical sources: *DAI*, proemium, pp. 45–7; *History of Leo the Deacon*, pp. 65, 71, 127, 155–6; Theodosius Diaconus, *Expugnatio Cretae*, p. 113, col. 1026 (vers. 165–6); John Skylitzes, *Synopsis of Byzantine History*, p. 285 (Dorystolo); Michael Psellus (1899), *Chronographia*, ed. C. Sathas, London: Methuen, III.9, pp. 68–9 (Syrian expedition of 1028); Kekaumenus, pp. 10, 23.
51. McGeer, 'Two Military Orations', 1, p. 117. See also 3, p. 119.
52. Cf. the άπιστα (faithless) and άτιμα (dishonourable) nations of the north (Patzinaks etc.) with the βαπτισμένα (baptised) (Croats, Serbs) and other αβάπτιστα (unbaptised) (Zachlumi) nations of the Balkans: *DAI*, 13, p. 70; 31, pp. 148–52; 33, p. 162.
53. Probably the armour would not have been visible under the cloaks.
54. *DAI*, 15, pp. 78–9.
55. Kaminiates, *Για την Άλωση της Θεσσαλονίκης*, 28, pp. 106–7.
56. Ibid., 29, p. 108.
57. Sources report an exchange of prisoners brokered by Leo Choirosphaktes in 906: Kaminiates, *Για την Άλωση της Θεσσαλονίκης*, p. 18.
58. *History of Leo the Deacon*, I.3, p. 61.
59. Christides, *Conquest of Crete*, pp. 42–67; Pryor, *Geography, Technology and War*, pp. 28–9, 62.
60. This term is mentioned in *De Ceremoniis* for the Cretan expeditions of 911, 949 and 960–1; see *De Ceremoniis*, pp. 658–9.
61. D. A. Agius (2008), *Classic Ships of Islam: From Mesopotamia to the Indian Ocean*, Leiden: Brill. The sophistication of Arab naval technology and ship design is supported by the study of a tenth-century Arab source which Christides has dated at around the same period as the *Naumachica* of Leo VI: V. Christides (1982), 'Two Parallel Naval Guides of the Tenth Century: Qudâma's Document and Leo VI's *Naumachica*: A Study on Byzantine and Moslem Naval Preparedness', *Graeco-Arabica*, 1, pp. 52–103.
62. Christides, *Conquest of Crete*, p. 46; Pryor, *Geography, Technology and War*, p. 28; J. Pryor (1984), 'Transportation of Horses during the Era of the Crusades, Eighth Century to 1285, Part I: to c. 1285', *Mariner's Mirror*, 70, pp. 9–27; M. Bennett (2006), 'Amphibious Operations from the Norman Conquest to the Crusades of St. Louis, c. 1050–c. 1250', in D. J. B. Trim and M. C. Fissel (eds), *Amphibious Warfare 1000–1700*, Leiden: Brill, p. 54.
63. *History of Leo the Deacon*, I.7, p. 67.
64. Ibid., I.6, p. 65.
65. See also, Leo Grammaticus (1842), *Chronographia*, Corpus Scriptorum Historiae Byzantinae, vol. 34, ed. I. Bekker, Bonn: Webber, p. 274. The fall of Taormina to the Arabs in 902 'due to negligence and treason from Eustathius the drungarius of the ploimon'.

66. John Skylitzes, *Synopsis of Byzantine History*, p. 240.
67. Ibid., p. 250.
68. *On Skirmishing*, 2.11–23, p. 152.
69. Ibid., 3, p. 154; see also 9, p. 172. Cf. Kaminiates, *Για την Άλωση της Θεσσαλονίκης*, 28, pp. 106–7; 29, p. 108.
70. *On Skirmishing*, 10, p. 176.
71. *Praecepta Militaria*, II.93–8, p. 28.
72. Ibid., I.98–103, p. 18.
73. Ibid., IV.195–207, p. 50.
74. John Skylitzes, *Synopsis of Byzantine History*, p. 183.

5

Methods of Transmission of (Military) Knowledge (I): Reconnaissance, Intelligence

Military intelligence is a well-defined area of intelligence collection, processing, exploitation and reporting using a specific category of technical or human resources. There are seven major disciplines; human intelligence, imagery intelligence, measurement and signature intelligence, signals intelligence, open-source intelligence, technical intelligence, and counterintelligence. Intelligence preparation of the battlefield is the systematic, continuous process of analysing the threat and environment in a specific geographic area and it is designed to support the staff estimate and military decision-making process.[1]

The definition above provides a clear idea of the importance and complexity of intelligence-gathering in a modern-day battlefield environment. This information-gathering and analysis approach, used to provide guidance and direction to commanders in support of their strategic and tactical decisions, can be divided into three levels based on the hierarchy of intelligence collection and decision-making by the state's political and military leaders. First, strategic intelligence is concerned with broad issues, such as economics, political assessments, military capabilities and the intentions of foreign nations. Such intelligence can be scientific, technical, sociological, economic or diplomatic, and is analysed in combination with known facts about the region in question related to geography, topography, industrial capacity and known demographics. Second, operational intelligence provides support to the army commander and is attached to the formation headquarters. Third, tactical intelligence is focused on supporting the operations on a tactical level and is, thus, attached to the battle group; patrols gather intelligence information on current threats and collection priorities and then transmit it for further assessment via the reporting chain higher up the levels of command.[2]

Intelligence has been a fundamental aspect of warfare from ancient Greece and Rome to Byzantium, and from the Napoleonic Wars to the conflicts in the Middle East in the twenty-first century. By the term, according to Clausewitz, 'we mean every sort of information about the enemy and his country – the basis, in short, of our own plans and operations'.[3] Correct and accurate intelligence-gathering was, and still is, paramount for any military operation, a fact that has been acknowledged by military writers as early as the fifth and fourth centuries BC when Sun-Tzu and Aeneas Tacticus were advising their readers on the advantages of accurate intelligence in war:

> Now the reason the enlightened prince and the wise general conquer the enemy whenever they move and their achievements surpass those of ordinary men is foreknowledge ... What is called 'foreknowledge' cannot be elicited from spirits, nor from gods, nor by analogy with past events, nor from calculations. It must be obtained from men who know the enemy situation.[4]

The chapter on the 'Employment of Secret Agents' concludes with the following dictum: 'An army without secret agents [bringing intelligence] is exactly like a man without eyes or ears.'[5]

> You must always, in making your attacks upon the enemy, strive to profit from your acquaintance with the terrain; for you will have a great advantage from previous knowledge of the country and by leading the enemy into such places as you may wish, which are known to you and suitable, whether for defence, or pursuit, or flight, or withdrawal into the city either secretly or openly. Moreover, you will also know in advance what part of the country will supply you with provisions, whereas the enemy will be unacquainted, ignorant, and embarrassed in all these particulars.[6]

The aim of intelligence-gathering was to give the local commanders assigned to the protection of a specific region against enemy action as detailed a report as possible regarding several specific questions. A short compilation now called *On Imperial Expeditions*, addressed to Constantine VII's son Romanus, draws on earlier works and archives to describe what emperors should do whilst campaigning and celebrating triumphs.[7] This work devises the questions that should have been asked by a commander before the launching of an expedition; there were questions about routes towards enemy lands, about the geography and topography of the neighbouring regions and the detailed description of plains, valleys, rivers, bridges, roads and mountain ranges that could impede or facilitate the friendly or enemy movement of troops and their supply of food, water and armament, about cities, castles and other strongholds that could be besieged or avoided altogether, and about the political situation and socio-religious demography of the targeted region. In order to underline how accurate intelligence-gathering was a significant aspect of pre-war preparation for Muslim armies as well, I will give a more detailed quote describing the range of different information gathered and relayed back to the military leaders, written by Nizam al-Mulk (1018–92), the vizier of the Seljuk sultans Alp-Arslan (1063–72) and Malik-Shah (1072–92):

> It should be realised that when kings send ambassadors to one another their purpose is not merely the message or the letter which they communicate openly, but secretly they have a hundred other points and objects in view. In fact they want to know about the state of the roads, mountain-passes and rivers, to see whether an army can pass or not; where fodder is available and where not; who are the officers in every place; what is the size of the king's army and how well it is armed and equipped; what is the standard of his table and his company; what is the organisation and etiquette of his court and audience-hall; does he play polo and hunt; what are his qualities and manners, his designs and intentions, his appearance and bearing; is he cruel or just, old or young; is his country flourishing or decaying; are his troops contented or complaining; are the peasants rich or poor; is he avaricious or generous; is he negligent in affairs; is his vizier competent, religious and righteous or the reverse; are his generals experienced and battle-tried or not; are his boon-companions polite and worthy; what are his likes and dislikes; in his cups is he jovial and good-natured or not; is he strict in religious matters and does he show magnanimity and mercy or is he careless and slack; does he incline more to jesting or to gravity; and does he prefer boys or women. So that, if at any time, they want to win over that king, or oppose his designs or criticize his faults, being informed of all his affairs they can think out their plan of campaign, and knowing what to do in all circumstances, they can take effective action.[8]

This quotation from the *Book of Government* details what kind of information the ambassador to an enemy court was expected to relay back after an audience with the state's sovereign and/or high officials. The Khorasanian vizier, however, did not simply include questions regarding the geography of the enemy country, information about roads, bridges and the countryside, points which are raised in the treatise *On Military Expeditions* as well; what also seemed to him very important in terms of strategic intelligence was to focus on the character of the enemy sovereign. Knowing as much as possible about the enemy is paramount for the successful outcome of a battle or a war and, as the author of the *Strategikon* highlights in his treatise:

> Our commander ought to adapt his stratagems to the disposition of the enemy general. If the latter is inclined to rashness, he may be enticed into premature and reckless action; if he is on the timid side, he may be struck down by continual surprise attacks.[9]

Methods of Transmission: Reconnaissance, Intelligence

Therefore, questions would be asked about the judgement, appearance and conduct of the sovereign, whether he was generous, merciful, religious or negligent in state affairs, whether he was surrounded by good and trusted officials and courtiers who could affect his judgement and decisions to a significant degree, whether his appointed generals were experienced and, finally, if he played polo and hunted.[10] This would certainly not have constituted a detailed psychological profile in the manner of twentieth-century psychoanalysis, such as, for example, the report on Adolf Hitler's personality commissioned by the OSS (the forerunner to the CIA) in 1943.[11] Nevertheless, any kind of information on the character and personality of the enemy leader might reveal strengths and weaknesses in his leadership and administration that could be taken advantage of. Other crucial questions would have included the state of the sovereign's army, its numbers, armament and morale/loyalty, what rank of officers would have been posted in each fortress and town in the border areas – obviously hinting at their battle-worthiness and loyalty to the regime[12] – and what would have been the state of the country's economy, thus considering whether an invasion army could be logistically supported, if there were rich pickings to be won or if there was any public feeling of discontent with the central government to be exploited.

Intelligence about the enemy could be procured in two ways: by reconnaissance (or tactical intelligence), when a commander openly sent out scouts (either light infantry, cavalry or swift scouting ships[13]) to observe the enemy army and collect information about its numbers, composition and their general's intentions, or by espionage, when disguised or hidden agents operated in secret in enemy territory collecting information about the enemy.[14] It is not an easy task to determine where exactly espionage ends and reconnaissance begins;[15] a key point in this is, in my view, the declaration of war, as information could then have been gathered openly rather than in secrecy, although agents could still be risking their lives behind enemy lines. In his brilliant book on *Intelligence in War*, where he examines the effect that intelligence-gathering has had on conflicts around the world, from the Napoleonic Wars to the 'War on Terror', Keegan points out this exact moment in the culmination of hostilities between two nations as the key point in time:

> Espionage, usually but not necessarily a state activity, is a continuous process, of very great antiquity ... States seem always to have sought to know the secrets of each other's policy, particularly foreign but also mercantile and military policy ... Operational intelligence [I use the term reconnaissance in this study], by contrast, is specifically an activity of wartime and, at high tempo, is limited to comparatively brief periods of hostilities.[16]

The Byzantine military manuals do not fail to address the subject of intelligence-gathering but, as Koutrakou has pointed out, they do not always provide us with a clear-cut picture of it, and both the technical and lay terms used in Byzantine sources to refer to spies, special agents, commandos or, more generally, troops dispatched to conduct reconnaissance on the enemy 'offer interesting possibilities for discussion'.[17] Therefore, it is the purpose of this chapter and the one that follows to determine the existing channels used by governments across the Byzantine–Arab borders to obtain information about the enemy, as well as which professions, groups of people and places were considered ideal for the procuring of intelligence. I will also ask if there was any distinction between professional and amateur spies, and if they used oral or written ways to transmit their information. Finally, I will examine what was the specific set of skills required by a spy to conduct their task more effectively.

Constantinople was the centre of a highly efficient system for transmitting intelligence from the provinces to the capital, a product of several hundred years of military experience that had been institutionalised and professionalised by generals and emperors like Julius Caesar and Justinian.[18] We read in the treatise *On Military Expeditions*: 'Be aware of a surprise invasion of an enemy force, and for that [purpose] you should constantly write to and receive reports from the border themes, and spy on the neighbouring enemies, and learn about and report.'[19] These types of intelligence reports involved two kinds of information channels, those coming from agents and informants operating in enemy territory – which I will examine below – and information sent by sentries and scouting parties stationed in the border areas and the 'no-man's land' that formed the unofficial frontier between the Byzantine and Muslim lands.

The sentries posted all along the frontiers played an essential role in the relay of information to the local officers and, eventually, to the authorities in the capital. These men are encountered under different names in our sources: Leo VI and the author of *On Skirmishing* identify them as *viglatores* (βιγλάτορες, Lat. *vigilator*) and *caminoviglatores* (καμινοβιγλάτορες, Lat. *caminus*, meaning 'the path', and βιγλάτορες), while the author of the late ninth-century treatise *On Strategy* calls them simply *fylakes* (φύλακες, from the verb φυλάττω: to keep watch and ward, keep guard) and *profylakes* (προφύλακες, from the verb προφυλάσσω: to keep guard over a person or place). Their main duty was to spot any enemy activity 'so that when the enemy begin to move, the sentries will learn of it from the posts along the road [and] the general will have advance knowledge that the enemy are moving out and what road they plan on taking'.[20] They operated in small teams of around ten men or even fewer – 'their numbers being carefully noted down in registers' (τυπωθέντες βιγλάτορες – εν ματρικίοις απογράφεσθαι). They were responsible for patrolling a specific section of the border road network, smaller pathways and flat areas with water supplies where an enemy unit was likely to pitch camp, and were relieved of their duties every fifteen days. Constant movement and vigilance were paramount, lest they were discovered by the enemy and taken prisoner. Once an enemy incursion was detected, they would relay the message to the nearest station, some three to four miles away, until this reached the cavalry stations further inland, from where it would have been delivered to the provincial authorities.[21]

A similar kind of advance warning was also served by the corps of the signalmen posted on a configuration of warning beacons in elevated places, starting from the frontier regions and ending up in the capital itself. Both the Byzantines and Muslims used to alert their naval centres for an imminent campaign by land or sea, and used more or less the same techniques.[22] Fire signals have been known since Homeric times and it appears that the messages were conveyed by torches, fire beacons and smoke signals, all of which were still in use in Roman times.[23] In Byzantine sources, the signalmen are mentioned as those responsible for 'τους τη φροντίδα των πυρσών' (or φανόν) (the upkeep of the torches); by manipulating the flame and smoke of the signals these men could indicate to the next fire-signalling post the direction, composition and even the numerical strength (in thousands) of the invading army.[24] Essential qualities for the soldiers serving in these sensitive posts were native knowledge of the topography of the region, a courageous disposition, intelligence and agility, and great physical strength, while the author of *On Strategy* also highlights the fact that these troops should be accompanied by their families, probably to prevent them from abandoning their posts in the face of danger, and that they should receive lavish rewards to keep their morale high.[25]

Once an invading force was in hostile territory, it was expected that special units would precede the main army and baggage train to reconnoitre the ground and local road network, and locate a suitable site for the army's encampment. The *skoulkatores* (σκουλκάτορες) responsible for this task were watchmen and scouts, who formed the vanguard of the main army and were dispatched to 'learn what is going on with the enemy and inform us'. This unit was also acting, according to Leo VI, as marching-camp patrols against a surprise night attack by the enemy.[26] Other scouts, also operating in the capacity of guides, were the *minsouratores* (μινσουράτορες) or *doukatores* (δουκάτορες), trustworthy and experienced men who 'were sent to spy/reconnoitre the flat regions [ahead of the army] and determine if they have adequate supply of water'[27] and 'in addition to knowing the roads, [were] able to conduct the army through the mountain passes, [could] plan ahead, and [knew] the proper distances for the campsites'.[28]

Finally, the *prokoursatores* (προκουρσάτορες, Lat. *procursor*) was a reconnaissance unit of lightly armed cavalry numbering 500 men – 110–20 of whom would have been horse archers, while the rest were lancers – marching immediately ahead of the main body of the army and whose tactical aim was to:

> seek contact with the enemy and set ambushes if they can, so that if the enemy is advancing in disorder without proper reconnaissance they can intercept them and strike against their prokoursatores to cause them panic to overcome their main force.[29]

The latter were more lightly armed than the heavy cavalry of the *themata* or the *tagmata*, wearing only a *klibanion* – a waist-length cuirass of lamellar construction, made out of iron or leather – and carrying only a small round shield for better mobility and manoeuvring, while the *minsouratores* and *skoulkatores*, whose operational aim was to spy on and reconnoitre the enemy, would have been even more lightly armed, probably not even wearing breastplates or coats of mail. According to al-Ansari, this would have been the case so as not to be detected by enemy forces while performing their duties.[30]

The information provided by the aforementioned sources, namely both the sentries and watchmen posted along the borders, and the different units of horsemen and foot soldiers marching ahead of the main body of an imperial army on campaign, can be classified as – to use a modern term – 'real-time intelligence', essentially meaning 'who knows what' in sufficient time to make effective use of the news.[31] Questions like 'Where was the enemy yesterday?', 'What are his plans of invasion?', 'What is the composition and direction of his columns?', and 'Where he might be expected today?' formed the basis of the tactical thinking of generals from Alexander the Great to Napoleon and Ulysses Grant. In the age before the invention of the wireless telegraph, however, armies had to operate under a 'peculiar constraint': the very slow speed of communication between the field of operations – be it a battlefield, the siege of a city or simply a point of invasion – and the military leadership. Realistic estimates put the daily distance covered by a messenger (on horseback or on foot) to less than fifty miles per day, not including the time and distance required to relay back the orders assigned to him once he arrived at the headquarters.[32] As we can imagine, by the time the orders reached the front line they would have already been out of date, as amply illustrated by more recent examples from the First World War, where the confusing orders issued by frustrated general staff officers from their headquarters could not keep up with the rapid developments on the field of battle. Thus the above explain the enormous importance attached by the commanders of the pre-telegraph age to strategic intelligence – questions regarding the character of the enemy, the disposition and size of their forces and all the points that Nizam al-Mulk thought significant enough to include in his *Book of Governments*.

Surely, however, tactical intelligence collected in a 'real-time' theatre of operations could also bring a number of advantages to a commanding general, regardless of the slow speed of reporting back. The relatively fast relaying of information about any impending invasion, whether a small-scale incursion or large-scale invasion, could save lives and livestock, as it would have provided precious warning to frontier societies to seek shelter and protection inside walled towns or other fortifications.[33] The author of the *Strategikon* is keen to point out that tactical – real-time – intelligence could also indicate the numerical strength of the adversary, thus greatly influencing the decision on whether to join the enemy in a pitched battle or retreat altogether. The author is also careful to stress that only experienced scouts should be sent to perform this task:

> The arrangement of cavalry and infantry formations and the disposition of other units cause great differences in their apparent strength. An inexperienced person casually looking at them may be very far off in his estimates. Assume a cavalry formation of six hundred men across and five hundred deep . . . If they march in scattered groups, we must admit that they will occupy a much greater space and to the observer will appear more numerous than if they were in regular formation . . . Most people are incapable of forming a good estimate if an army numbers more than twenty or thirty thousand . . . Hence, if a commander wants to make his army appear more formidable, he can form it in a very thin line, extend it a long distance, or leave gaps in the line.[34]

A good example of this is the marching of the Crusaders towards Ascalon in 1099 when, on the vicinity of the city, they came across numerous herds of sheep, goats, camels and oxen belonging to the Fatimid army. Taking the animals with them, the Crusaders, thus, marched towards the city with the animals

> marching in a straight line on the left and on the right of the battle lines, although herded by no one, so that many of the heathens from a distance, seeing them marching with our soldiers, thinking all were Frankish army.[35]

If the general decided to stay and fight a pitched battle with his enemy, then good reconnaissance would also play a crucial role in his choice of battleground and battle formation:

> That general is wise who before entering into war carefully studies the enemy, and can guard against his strong points and take advantage of his weaknesses. For example, the enemy is superior in cavalry; he should destroy his forage. He is superior in number of troops; cut off his supplies. His army is composed of diverse peoples; corrupt them with gifts, favours, promises ... The foe is superior in infantry; entice him into the open, not too close, but from a safe distance hit him with javelins.[36]

Knowledge of the composition of the enemy force could also prove extremely useful in taking advantage of an opponent's weaknesses, as Frontinus reports in the second book of his *Strategemata*. There, he provides the example of Scipio who, having acquired intelligence about the composition of the flank divisions of the Carthaginians, chose to attack the 'weaker' left flank consisting of 'Africans', while ordering his own left flank to retreat in the face of the attack of the elite 'Spaniard mercenaries'.[37] The two different battle formations at Tarsus (965) and Dorystolo (971) further prove the Byzantine commanders' adaptability, not just on different battlegrounds, such as the fields and meadows of Tarsus or the narrow strip of land flanked by woods and a marsh at Dorystolo, but against different enemies – the mixed infantry–cavalry army of the Tarsiots and the heavy infantry of Svyatoslav. However, whether or not Phocas and Tzimiskes depended on real-time intelligence to make up their battle plans or rather on experience from previous battles against both enemies, this still does not diminish the fact that knowledge of the enemy's composition could be crucial for the outcome of a battle.

Careful scouting and, perhaps, well-paid informants were the two main factors that gave warning of the Byzantines' concealed intentions and secret devices, and tipped the balance of battle in favour of Bohemond of Taranto outside Ioannina in 1082. Alexius Comnenus' decision to send skirmishing detachments to harass the Norman camp and gather intelligence regarding their numbers, and the commanding skills and fighting capabilities of their leader Bohemond, indicates Alexius' adaptability after his defeat at Dyrrachium the previous October. The emperor, 'fearing the first charge of the Latins', had a number of small and light chariots with spears fixed on top of them placed behind the first lines of his division at the centre, with infantrymen hiding underneath ready to emerge and be manoeuvred when the Norman cavalry charge was at striking distance from the Byzantine lines. In spite of this, however, 'as though he had foreknowledge of the Roman plan he [Bohemond] had adapted himself to the changed circumstances'. Bohemond's answer was to divide his forces into two major units and attack the flanks of the imperial army, thus engaging in a melee that quickly led the terrified Byzantines to flee the battlefield.

Bohemond's textbook tactical reconnaissance was applied a few months later by Alexius when he devised a plan similar to the one at Ioannina, his primary aim being to once again disrupt the Norman heavy cavalry charge. This time, the emperor had his men lay iron caltrops (τρίβολος, *trivolos*) in front of the centre of his formation, where he expected the Norman cavalry attack to take place. The course of the battle, however, was a repetition of what had taken place at Ioannina, with Bohemond finding out about the Byzantine plans, either by treason or simply by sending experienced scouts close to the enemy lines, and the result was another cavalry attack on the imperial army's flanks, which quickly melted away once again.[38]

Local knowledge of the terrain of operations was paramount for a general, especially for one like Alexius Comnenus who, although resilient and adaptable to every operational circumstance, had already been defeated three times in pitched battle by the Normans. In 1083, with Bohemond having reached the vicinity of Thessaloniki before marching south to besiege Larissa, the emperor decided to defeat the Norman cavalry by guile. Taking all the necessary precautions before a battle, as recommended by the *Praecepta Militaria* and the *De Rei Militari*,[39] Alexius interrogated a local man about the topography of Larissa and the surrounding areas, 'wish[ing] to lay an ambush there, for he had given up any idea of open hand-to-hand conflict; after many clashes of this kind – and defeats – he had acquired experience of the Frankish tactics in battle'.[40] The result was a triumph for the Byzantine emperor and his 'Vegetian' tactics, with his knowledge of the local terrain certainly playing a significant part in the outcome of the battle.

During the invasion of the island of Crete and the siege of Chandax in 961, Nicephorus Phocas was careful to send detachments of cavalrymen under Pastilas, the *strategos* of the Thrakesion, 'επί καταδρομήν και κατασκοπήν της νήσου' (to raid and spy on the island).[41] This could well have meant reconnaissance missions to collect intelligence about the supply and logistics of the Byzantine invading force, as Nicephorus would certainly have been aware of the time frame for the upcoming siege, along with raids to lay ambushes and seize prisoners for interrogation that might reveal crucial information on the military preparedness of the city's garrison. For a siege operation of this scale and importance, a general would have needed any information he could get his hands on, and Nicephorus Phocas was too experienced an officer to have overlooked the crucial role that intelligence would have played in this case.

The main question that arises at this point, however, is whether reliable – if there is such a thing as 'reliable' intelligence reports[42] – detailed and fast tactical intelligence relayed back to the commander of a field army, invading a foreign country or deployed to intercept an enemy, could be responsible for any long-term change in tactics and/or strategy. War is not an intellectual activity, but a savage and primitive one where humans and animals – and machines in the modern era – clash on the field of battle in a mayhem which could last hours, days and even months. During this mayhem, a general has always been required to come up with the best possible strategy and apply the best possible tactics in order to emerge victorious at a lesser cost than their adversary. Intelligence could provide the general with the best tactical background of information that would aid decision-making during battle. Intelligence is just one among dozens of different parameters that can influence decision-making, strategy and battle tactics, but unlike others, such as the weather or pure luck, intelligence was something a general could control.

In our sources, the term spy (κατάσκοπος) seems to apply invariably to watchmen, scouts, bandits and raiders into enemy territory sent to loot, take prisoners and gather information about the enemy. The Byzantine military treatises, however, occasionally use other terms to describe the role and tasks undertaken by these 'spies' when operating in enemy territory. There were the εκσπηλατόρες (Lat. *expilatores*, literally a robber or a plunderer, although in this context it probably means a scout) and the *trapezitai* (τραπεζίται),

> those who the Armenians call *tasinarioi* (τασινάριοι) . . . These men should be sent out constantly to charge down on the lands of the enemy, cause harm and ravage them . . . They should also capture some of the enemy and bring them back to the commanding general, so that he might obtain information from them about the movements and plans of the enemy.[43]

These men would have been primarily Armenians, settled in the region of Lesser Armenia, the Pontic frontier, and the regions of Cappadocia and Armeniakon 'from ancient times' – or more likely from the ninth century. Kekaumenus identifies these soldiers of the borderlands as χονσάριοι or χωσάριοι (*chonsarioi* or *chosarioi*) – which would develop into the term *hussar* in Western Europe – and assigns three distinct roles to these borderers: (1) watchmen,[44] (2) scouts/raiders, also called συνοδικοί (*synodikoi*),[45] and (3) scouts/guides[46] reconnoitring the invasion routes prior to the main army's advance. The general should constantly evaluate these soldiers based on their vigilance, efficiency and fidelity, and should always laden them with presents in order to keep their morale high and their loyalty to him. It is worth noting that the terms τραπεζίται and συνοδικοί are of Greek origin, although the latter is only attested in eleventh-century sources and not earlier, while τασινάριοι is of Armenian and χονσάριοι of Bulgarian origin (Χονσά meaning 'thieves'). Eventually, however, all of the above-mentioned terms came to mean both scouts and bandits.[47]

Further identified by the terms *akritai* (the borderer – from the Greek τὰ ἄκρα, 'the extremities') and *apelatai* (the one who drives away [the invaders] – from the Greek ἀπελαύνω, to 'drive away'), these border garrisons, scouts and watchmen were local men, who either served on the basis of the *strateia* or received *roga* by the central government. Some, having become impoverished, had resorted to plundering the regions on both sides of the borders.[48] In a world of regular border warfare, large-scale invasions, razzias and cross-border raids, these people represented the lightly armed irregular troops or militia turned brigands, whose bands Digenes Akritas aspired to join when he was still a teenager.[49] These were light infantry recruited from Armenians, Bulgarians and the native Byzantine population, which meant that they often were ethnically, linguistically[50] and religiously mixed, a fact epitomised by the legendary hero. The term *akrites* is derived from the Greek word ἄκρον (pl. ἄκρα), meaning border; similar border guards, the *limitanei*, were employed in the late Roman and early Byzantine armies to guard the frontiers.[51] In official Byzantine use, the term was used in a descriptive manner, being generally applied to the defenders as well as the inhabitants of the eastern frontier zone.[52] Their officers, however, were largely drawn from the local aristocracy, and it is interesting to note that the term *akrites* was used to describe the officer in charge of the *apelatai*, the 'Lord of the Marches', who would have held the aristocratic title of *patrician* and would have lived in a *kastron* (castle) dominating the region under his jurisdiction.[53] In the epic poem *Digenes Akrites*, the mother of the hero is descended from the noble families of the Kinnamades and the Doukades of Cappadocia, with twelve members of her family serving as generals, while the Muslim emir who kidnapped and eventually married her – Digenes' father – was also a nobleman from the *thughūr* called Musur, son of Chrysoverges, who commanded '3,000 chosen lancers' and subdued Syria, Kufa (in Iraq), Heraclea and Amorion, reaching as far inland as Ikonion.⁵⁴

Kekaumenus gives detailed advice to the Byzantine *akritai* on how to deal with the *toparchai* (local governors)⁵⁵ on the other side of the frontier; elsewhere, he advises the local *toparchai* on how to deal with the Byzantine commanders, as well as with the central government.⁵⁶ Kekaumenus wrote in the late 1070s but his information, based on family tradition, goes back at least two generations. In fact, his grandfather was a Byzantinised Armenian *toparches* in 'Greater Armenia';⁵⁷ therefore, he would have come into daily contact with the people who lived just beyond the frontier, either through local raids or through the marketplaces that would have been set up inside or in the vicinity of towns. Although the reality mirrored in Kekaumenus' *Strategikon* represents a militarised society that had been shaped as a result of the wars of the 'Reconquest' in the East, whose defence radically restructured the heavy presence of professional troops that overshadowed the older thematic militias, with the smaller *themata* grouped in five large regional commands headed by a *doux*,⁵⁸ both Kekaumenus' work and the earlier treatise *On Skirmishing*, each addressed to their own counterparts in the themes and their immediate subordinates, the *turmarchs*, give special importance to the pre-eminence of local diplomacy in the operational theatre of the East and the emphasis placed upon the initiative, and strategic and operational autonomy of the local commanders of the ακριτικά θέματα (the border themes).

The author of *On Skirmishing* recommends, in a typical border diplomacy fashion:

> He [general] ought also to have the businessmen go out. He should pretend to make friends with the emirs who control the castles in the border regions. He should also write to them and send men with gifts. As a result, with all his coming and going, the general might be able to get a clear picture of the plans and intentions of the enemy.⁵⁹

What this picture was and what kind of intelligence the general was advised to seek can be surmised from what immediately follows: 'He should find out how many men make up their army, how many horse and how many foot; he should find out about their commanders and the area in which they plan to attack.' Once again, what we see here is the importance placed upon strategic intelligence-gathering that could greatly assist a local commander in his decision-making – estimated numbers, composition of the army and projected invasion routes could prove invaluable for the outcome of a military campaign when 'real-time intelligence' was simply a fantasy. In a similar fashion, Kekaumenus advises his counterpart to be on peaceful terms with his neighbouring *toparches*, in case he provokes an alliance of local *toparchai* against him, which once again highlights the localised and fragmented nature of frontier diplomacy. What stands out most strongly from the reading of Kekaumenus' *Strategikon*, however, is the overcautious and cunning spirit underlying relations between the two sides:

> If your neighbouring *toparch* attempts to hurt you, do not become bold and overconfident but in a devious manner pretend you wish for peace and simplicity. Guard your country and make friends, if possible, from his country so that you will be able to learn his plans, and send to him private gifts.[60]

Trust between neighbours in the Byzantine–Muslim borders was a value that could not be taken for granted, but had to be earned by the smartest and most cunning of generals, as Kekaumenus' detailed examples – one including his own grandfather – vividly illustrate.

The soldiers manning the units of the *akritai* constituted an invaluable asset for the local authorities of the eastern themes; since they came from the border areas they were knowledgeable in the main routes, tracks and smaller paths, the *kleisourai* and the strongholds, the plains and the river valleys on both sides of the borders. They were also of mixed origin – Greek, Armenian and Arab – which is interesting in terms of their cultural acclimatisation; these *akritai* could speak the language of the people across the borders, most likely they prayed to the same God, they were in daily contact with them through trade and local markets and could even be related through marriage – once again, the example of Digenes Akritas is characteristic. Thus, these soldiers seem to have belonged to both categories of intelligence-gatherers; they were able to conduct espionage and commando operations in enemy territory and collect prisoners and invaluable information later relayed to the thematic authorities for further analysis. Their movement across the fluid 'no-man's land' between the empire and its neighbouring principalities was flexible and informal,[61] but also motivated by the real danger of being interrogated and possibly executed if caught acting as agents of the emperor. Nevertheless, their knowledge of the terrain, their local connections and network of informants and spies proved invaluable for an expeditionary force operating in the area, providing the commander with the necessary reconnaissance and tactical intelligence he needed to proceed with his campaign plans.

Notes

1. *The U.S. Army & Marine Corps Counterinsurgency Field Manual, U.S. Army Field Manual No. 3–24, Marine Corps Warfighting Publication No. 3–33.5* (2007), Chicago, IL: University of Chicago Press, p. 385.
2. A. Rolington (2013), *Strategic Intelligence for the 21st Century: The Mosaic Method*, Oxford: Oxford University Press; J. Keegan (2003), *Intelligence in War: Knowledge of the Enemy from Napoleon to Al-Qaeda*, New York: Pimlico.
3. Clausewitz, *On War*, p. 117.
4. Sun-Tzu [5th c. BC] (1963), *The Art of War*, trans. S. B. Griffith, Oxford: Oxford University Press, pp. 144–5.
5. Sun-Tzu, *Art of War*, p. 175.
6. Aeneas Tacticus (1928), *On Defence against Siege*, in The Illinois Classical Club (ed.), *Aeneas Tacticus, Asclepiodotus, Onasander*, The Loeb Classical Library, London: Heinemann, p. 87.
7. Constantine Porphyrogenitus, *Three Treatises on Imperial Military Expeditions* [hereafter *Three Treatises*] (1990), ed. J. Haldon, Vienna: Verlag der Österreichischen Akademie der Wissenschaften, especially treatise B, pp. 82–93. Cf. Vegetius, *Epitome of Military Science*, III.6, pp. 73–7.
8. Nizam al-Mulk (2012), *The Book of Government or Rules for Kings: The Siyar al-Muluk of Siyasat-nama of Nizam al-Mulk*, trans. H. Darke, London: Routledge, pp. 95–6. Cf. al-Harawi, XI, pp. 225–6.
9. *Strategikon*, VIII.2.49, p. 87. Cf. Vegetius, *Epitome of Military Science*, III.9, p. 85.
10. According to Plutarch, Alexander the Great asked the same questions to the Persian ambassadors at Pella about what sort of warrior the great king was: *Life of Alexander*, 5.2.
11. H. A. Murray (1943), 'Analysis of the Personality of Adolph Hitler with Predictions of his Future Behavior and Suggestions for Dealing with him Now and After Germany's Surrender', report delivered to the OSS, October 1943; W. C. Langer (1972), *The Mind of Adolf Hitler: The Secret Wartime Report*, New York: Basic Books. See also S. B. Dyson (2014), 'Origins of the Psychological Profiling of Political Leaders: The US Office of Strategic Services and Adolf Hitler', *Intelligence and National Security*, 29, pp. 654–74.
12. 'Generally, in the case of armies you wish to strike, cities you wish to attack, and people you wish to assassinate, you must know the names of the garrison commander, the staff officers, the ushers, gate keepers, and the bodyguards. You must instruct your agents to inquire into these matters in minute detail.' *Tu Mu*: 'If you wish to conduct offensive war you must know the men employed by the enemy. Are they wise or stupid, clever, or clumsy? Having assessed their qualities, you prepare appropriate measures.' Sun-Tzu, *Art of War*, p. 148.

Methods of Transmission: Reconnaissance, Intelligence

13. The only treatise that also refers to naval espionage is Leo's *Taktika*, which has a specific constitution 'On Naval Warfare': 'Therefore, it is necessary for you to observe with total accuracy the disposition of the enemy, and so prepare the outfitting of your dromons and the armament of your soldiers, their number and size and the rest of the equipment in proper fashion against the enemy. Also have small and fast dromons, not equipped for battle, but for scouting, [conveying] orders, and other needs that may occur.' Leo VI, *Taktika*, XIX.81, p. 534.
14. J. A. Richmond (1988), 'Spies in Ancient Greece', *Greece & Rome*, 45, p. 2.
15. 'Spies will enable a general to learn more surely than by any other agency what is going on in the midst of the enemy's camps; for reconnaissance, however well made, can give no information of any thing beyond the line of the advanced guard' (Jomini, *Art of War*, p. 269).
16. Keegan, *Intelligence in War*, p. 384.
17. N. Koutrakou (2000), '"Spies of Towns": Some Remarks on Espionage in the Context of Arab–Byzantine Relations (VIIth–Xth Centuries)', *Graeco-Arabica*, 7–8, p. 246.
18. Procopius, *De Bello Persico*, I.xxi, p. 196; F. Millar (1988), 'Government and Diplomacy in the Roman Empire during the First Three Centuries', *International History Review*, 10, pp. 345–77. Keegan also mentions the right of direct access by scouts to the commanding general, introduced by Caesar: *Intelligence in War*, p. 11.
19. *On Military Expeditions*, B, p. 86.
20. *On Skirmishing*, 2, p. 152; Leo VI, *Taktika*, XII.97, p. 270.
21. *On Skirmishing*, 1, p. 150. There is a useful comparison with the Mongol intelligence and warning system of the late twelfth century: J. T. Grubbs (2010), 'The Mongol Intelligence Apparatus: The Triumphs of Genghis Khan's Spy Network', *International Association for Intelligence Education*, pp. 5–7.
22. *On Strategy*, 8, p. 26; *De Ceremoniis*, p. 492; P. Pattenden (1983), 'The Byzantine Early Warning System', *Byzantion*, 53, pp. 258–99; H. S. Khalilieh (1999), 'The *Ribât* System and its Role in Coastal Navigation', *Journal of the Economic and Social History of the Orient*, 42, pp. 212–25; A. M. Fahmy (1956), *Muslim Naval Organisation in the Eastern Mediterranean from the Seventh to the Tenth Century AD*, Cairo: National Publication and Print House, pp. 55–76. On the use of fire signals at sea by the Byzantine navy in the mid-ninth century work of Syrianus Magistros, see I. Dimitroukas (ed.) (2005), Ναυμαχικά, Λέοντος ς΄, Μαυρικίου, Συριανού Μαγίστρου, Βασιλείου Πατρικίου, Νικηφόρου Ουρανού [Naumachika of Leo VI, Maurice, Syrianus Magister, Basil the Patrician, Nicephorus Ouranus], Athens: Kanaki, ch. 7, p. 118.
23. Richmond, 'Spies in Ancient Greece', p. 14.
24. The ancient Greeks had developed a sophisticated system of transmitting messages through fire and smoke signals, as indicated by Thucydides,

Aeneas Tacticus and Polybius; see Richmond, 'Spies in Ancient Greece', pp. 14–15. The author of *On Strategy* acknowledges that 'it is possible, as ancient authorities suggest, to report not only the approach of the enemy but their numerical strength in thousands, lighting the beacons once for each thousand men estimated'. Dagron, however, believes that this method was 'ni assez discrète, ni assez précise' (*Le traité*, pp. 253–4).

25. *On Strategy*, 7, p. 24; *On Skirmishing*, 2, p. 152.
26. Leo VI, *Taktika*, IV.26, p. 54; XII.42, p. 240; 97, p. 270; XVII.78, p. 426. Al-Ansari calls the scouting party *al-taliʾah* and defines it as 'a cavalry group which precedes the army for the collection and discovery of information [about the enemy]. As for its men: it is said that it is necessary to choose for the scouting party men of counsel and courage and wise in the experience (affairs) of wars' (al-Ansari, *Muslim Manual of War*, p. 80).
27. Leo VI, *Taktika*, IV.24, p. 54; *Praecepta Militaria*, V.12–16, p. 52.
28. *On Tactics*, 18, p. 290.
29. *Praecepta Militaria*, II.16–28, pp. 22–4.
30. Ibid., IV.35–6, p. 40; al-Ansari, *Muslim Manual of War*, p. 80.
31. Keegan, *Intelligence in War*, p. 20.
32. Probably using horses at prearranged watch posts situated in the interior, as the ones described by the author of *On Skirmishing* to transmit messages from the borders to the central authorities, which were situated at some three to four (Byzantine) miles apart: *On Skirmishing*, 1, p. 150. There is plausible modern evidence suggesting that a camel rider, unencumbered by baggage, could travel well over fifty miles per day: F. R. Trombley (2002), 'Military Cadres and Battle during the Reign of Heraclius', in G. J. Reinink and B. H. Stolte (eds), *The Reign of Heraclius (610–641): Crisis and Confrontation*, Leuven: Peeters, p. 245, n. 19.
33. *On Skirmishing*, 2, p. 152.
34. *Strategikon*, IX.5, p. 102. The author was also keen to stress the fact that the army should march in close formation to prevent the enemy from forming an estimate of their numbers: VIII.2.37–8, p. 86.
35. E. Peters (ed.) (1998), *The First Crusade: 'The Chronicle of Fulcher of Chartres' and Other Source Materials*, Philadelphia: University of Pennsylvania Press, II.5, p. 94.
36. *Strategikon*, preface of Book VII, pp. 64–5.
37. Frontinus, *Stratagems*, II.iii.4, p. 108. See also the examples given by Vegetius, *Epitome of Military Science*, III.26, p. 117; *Strategikon*, VIII.2/88, p. 90; Leo VI, *Taktika*, XX.184.
38. G. Theotokis (2014), *The Norman Campaigns in the Balkans, 1081–1108 AD*, Woodbridge: Boydell & Brewer., pp. 170–1.
39. *Praecepta Militaria*, IV.192–208, p. 50; Vegetius, *Epitome of Military Science*, III.9, pp. 84–5.
40. Theotokis, *Norman Campaigns in the Balkans*, p. 174.

41. *History of Leo the Deacon*, I.3, p. 62.
42. 'Many intelligence reports in war are contradictory; even more are false, and most are uncertain' (Clausewitz, *On War*, p. 117; see also pp. 223–4).
43. *On Skirmishing*, 2, p. 152; *On Tactics*, 18, p. 292.
44. Kekaumenus, p. 9.
45. Ibid.
46. Ibid., p. 17.
47. Dagron, *Le traité*, pp. 265–6.
48. Haldon, *Warfare, State and Society*, pp. 266–7.
49. *Digenes Akrites*, IV, pp. 82–6; V. Christides (1979), 'Arabic Influence on the Akritic Cycle', *Byzantion*, 49, pp. 94–109.
50. We see the emir speaking Greek with the three brothers (Digenes' uncles), although the latter would need an interpreter (δραγουμάνος) to ask for directions from a 'Saracen' peasant: *Digenes Akrites*, I, pp. 8, 14.
51. The *limitanei* were troops of the army corps stationed on the frontiers, who also served as garrisons made up of older legionary units and attached auxiliaries, and reinforced by other auxiliary and legionary cavalry units; see Haldon, *Warfare, State and Society*, pp. 67–9. These units of troops were not peasant militia in terms of their organisation and fighting abilities, but a functioning army in every respect: B. Isaac (1990), *The Limits of Empire: The Roman Army in the East*, Oxford: Oxford University Press, pp. 208–13, 287–8; B. Isaac (1988), 'The Meaning of the Terms *Limes* and *Limitanei* in Ancient Sources', *Journal of Roman Studies*, 78, pp. 125–47.
52. *Oxford Dictionary of Byzantium*, I, pp. 47–8, 127–8.
53. Kekaumenus, pp. 24, 26; *Digenes Akrites*, IV, pp. 138–40; *On Tactics*, 3, p. 264; 18, p. 292. Only the treatise *On Skirmishing* (3, p. 154) uses the term *akritai* to denote border guards in general.
54. *Digenes Akrites*, I, pp. 18–20.
55. Until the sixth century, the term used to denote the local magistrate in a broad sense and it was, sometimes, equated with a king. It reappeared in the period between the tenth and thirteenth centuries designating independent rulers and Byzantine local governors, who enjoyed substantial independence from the central government in Constantinople (*Oxford Dictionary of Byzantium*, III, p. 2095).
56. Kekaumenus, pp. 24–6, 76–7.
57. Ibid., p. 26.
58. J.-C. Cheynet (2003), 'Basil II and Asia Minor', in P. Magdalino (ed.), *Byzantium in the Year 1000*, Leiden: Brill, pp. 82–96; Haldon, *Warfare, State and Society*, pp. 84–94. 'Le stratège n'est pas encore devenu un dynaste personnellement soumis a l'empereur, il est resté son représentant direct; l'emir arabe, de son côté, n'est pas encore un "toparque" plus ou moins libre de son appurtenance' (Dagron, *Le traité*, p. 260).
59. *On Skirmishing*, 7, pp. 162–3.

60. Kekaumenus, pp. 24–6. See also C. Roueché (2000), 'Defining the Foreign in Kekaumenos', in D. C. Smythe (ed.), *Strangers to Themselves: The Byzantine Outsider*, Aldershot: Ashgate, pp. 203–14.
61. E. Jeffreys (2000), 'Akritis and Outsiders', in D. C. Smythe (ed.), *Strangers to Themselves: The Byzantine Outsider*, Aldershot: Ashgate, pp. 189–202. See the example of the Byzantine ambassador John the Sygkellos, who met with Greek *akritai* enquiring of him as to the emperor's health while – officially – on Muslim territory: *Theophanes Continuates*, pp. 96–7. In addition, Ibn Hawqal reports his personal experience travelling through Mesopotamia, and while still in Diyar-Rabiʿa, where he received useful directions from 'volunteers, those who formed the companies of brigands, both Muslims and Byzantines, due to their knowledge of the country and their experience with [mountain and valley] passages' (*Configuration de la terre*, I, p. 192).

6

Methods of Transmission of (Military) Knowledge (II): Espionage

> It has been customary from ancient times both among the Romans and the Persians to maintain spies at public expense; these men are accustomed to go secretly among the enemy, in order that they may investigate accurately what is going on, and may then return and report to the rulers. Many of these men, as is natural, exert themselves to act in a spirit of loyalty to their nation, while some also betray their secrets to the enemy.[1]

This chapter will explain the rather unconventional methods of procuring intelligence through espionage, an activity that was usually, but not necessarily, state-sponsored and which took place in times that preceded hostilities between states – the declaration of war being the key moment when we can draw a distinction between espionage and reconnaissance. It is the intention of this chapter to review the role of espionage in Byzantine foreign policy and to define and analyse the official and unofficial channels through which the Byzantines procured information that shaped their foreign policies and prompted them to become more adaptable to their enemies' strategies and tactics. I ask how the central authorities reacted to spies and espionage activity. Were their liberties and rights respected? What kind of information did they report back to their (pay)masters and in what way was this information processed?

Once I have identified the channels that transmitted intelligence used to provide guidance and direction to commanders in support of their strategic and tactical decisions, I will put the conclusions of this chapter into perspective. My aim is to paint a broader picture of what that says about the influence each culture had on its neighbours. Do we see nations that pursued a more defensive strategy and adapted more easily to the changing tactics of their enemies? Finally, can we say that certain cultures were more susceptible to tactical changes than others, and if so what were the deeper reasons behind this phenomenon?

Markets, Ports, Fairs, Taverns and Inns

In the field of international news, the dependence on the mercantile community has been marked since ancient times; merchants would be questioned by the local authorities on their city or port of origin about what kind of intelligence they were able to gather during their travels in or through enemy territory. It was not just regular movement and travelling that turned merchants into a natural source of information; it was also the places where they did business, where they socialised 'after hours', and the people they met there. Since an examination of the trade links between the Arabs and Byzantium would be beyond the scope of this study, it suffices here to say that commercial relations between the two worlds, although not always amicable, functioned relatively well since pre-Islamic times.[2]

Ports, markets and religious festivals were ideal places for intelligence-gathering, since it was in such places that the most diverse individuals would converge during the day – this was a place of work but also a place of socialising with others, not just locals but international tradesmen of various religions and nationalities.[3] The language barrier would not have been an issue in such hubs, as most of these men would have been proficient in Greek, Armenian and/or Arabic, and their frequent travels would have accustomed them to the given local traditions and way of life.[4] At ports, markets and festivals there would have been more or less unrestricted gossip about the political situation of the day, and rumours were passed on with surprising speed. As in the modern corporate world, where businessmen consider the establishment of networks abroad as essential, the general principle would have been more or less the same for a medieval tradesman as well.

Since antiquity, we can detect in military treatises a serious concern not over the members of the merchant class themselves, for whom distrust was mitigated since they were identified as non-combatants (ἄμαχοι, *amachoi*),[5] but rather over spies infiltrating their ranks and posing as ἔμποροι (merchants). In AD 365, fearing for his life, Procopius[6] avoided detection during his travels from Chalcedon to Constantinople, where he hoped to gather intelligence and hear the rumours circulating in the capital, due to his neglected personal hygiene and old clothes.[7] The sixth-century treatise *On Strategy* similarly advises the following to all spies assigned on a mission:

> Before leaving each spy should speak in secrecy about his mission to one of his closest associates. Both should agree upon arrangements for communicating safely with one another, setting a definite place and manner of meeting.

Methods of Transmission: Espionage

The place could be the public market in which many of our people, as well as foreigners, gather. The manner could be on the pretext of trading. In this way, they should be able to escape the notice of the enemy.[8]

Both the Byzantines and the Muslims sought to reduce and impose strict controls over all commercial activity in the eastern Mediterranean after the initial Muslim conquests, for fear of espionage. This is amply illustrated in the terms of the truce of Qinnasrin and the Baʿlabakk agreement of 637.[9] An anecdotal narrative by al-Baghdadi (1002–71) in his *Taʾrikh Baghdad* (History of Baghdad) relates how the Muslim attitude about city-building was influenced by Constantinople, and the degree to which the city was perceived as a model for imitation by the Abbasids. We read in an alleged conversation between Caliph al-Mansur (754–75) and the ambassador sent by Emperor Constantine V (741–75), regarding the building of Baghdad:

The Caliph asked Patrikios, 'What do you think of this city?' He answered, 'I found it perfect but for one shortcoming.' 'What is that?' asked the Caliph. He answered, 'Unknown to you, your enemies can penetrate the city anytime they wish. Furthermore, you are unable to conceal vital information about yourself from being spread to various regions.' 'How?' asked the Caliph. 'The markets are in the city,' said Patrikios. 'As no one can be denied access to them, the enemy can enter under the guise of someone who wishes to carry on trade. And the merchants, in turn, can travel everywhere passing on information about you.'[10]

It is important to note, however, that even though merchant activity within the empire was rigidly regulated since the era of Theodosius I, and merchants were forbidden to 'hold markets in places that lie beyond those that were agreed on in the treaty with that nation [Persia], so that they may not improperly spy into the secrets of another kingdom', there was a peculiar exemption:

Excepted here from, however, are those who accompany Persian ambassadors sent to Our Clemency at any time, and who have brought merchandise with them for trade. We do not deny them, out of kindness and out of respect of the embassy, the opportunity of trading beyond the places mentioned, unless they remain too long in any province under pretence of embassy, and do not accompany the ambassador on his return to his home. The punishment provided by this law justly falls on those who are bent on trade while they loiter, as well as on those with whom they trade.[11]

We know of a similar permission authorised by the emir of Egypt, Muhammad ibn Tugj al-Ihsid, in a series of correspondences that formalised the commercial relations between the nascent Ikhshidid dynasty in Egypt and the empire, sent some time between 935 and 944 during the reign of Emperor Romanus I.[12] It seems that, as Chrysos very aptly put it,[13] diplomatic missions were often turned into commercial caravans, thus providing ample opportunities for espionage, as I will discuss in more detail later on.

Naval espionage and the opportunity to obtain intelligence at major port-cities in the Mediterranean were also exploited by both empires.[14] In his *Kitab surat al-ard* (written c. 988), Ibn Hawqal complains that Byzantine merchants gathered intelligence while conducting their business at Muslim ports: 'They [Byzantines] sent their boats on the territory of Islam to engage in trade, while their agents roamed the country by taking the information secretly and by gathering information, after which they left.'[15] The jurist Abu Yusuf (d. 798), who served as chief judge (*qadi al-qudat*) during the reign of Harun al-Rashid, acknowledged the danger posed by merchants in transmitting information to the enemy.[16] Arabic-speaking infiltrators were sent by the Byzantines to the Egyptian port of Damietta in the Nile Delta before the Byzantine raid of 853,[17] while imperial agents (ἀκριβεῖς κατάσκοποι) – probably camouflaged as sailors or merchants – were also dispatched by the *protospatharius* Leo[18] to Tarsus, Tripoli and Laodicea to investigate whether the Muslims were aware of the Byzantine preparations for a naval expedition against Crete in 911.[19]

The late eleventh-century Italo-Norman chronicler Geoffrey Malaterra reports that the Normans sent Philip, son of Gregory the *patrikios*, to Muslim Syracuse to gather information about the enemy's army and fleet. He and his comrades were disguised as merchants and could roam around the port without attracting any unnecessary attention, 'for both he and all the sailors who were with him were most fluent in their language [Arabic] as well as Greek'.[20] Finally, Kekaumenus provides a vivid description of the cunning methods used to gain access to the Thessalian port-city of Demetriada, stressing the fact that ships coming to trade should not be trusted at any time, as they might pretend to come peacefully: 'We did not come here to fight [a war], rather to pay tolls and sell prisoners and other things we have from corsair activity.' But, in reality,

> the Hagarene ... after climbing over [διαβάντες, from the verb διαβαίνω: to stride, step across or pass over] the side of the walls, from where the locals had no suspicion, they climbed on top of the castle's battlements ... and they occupied the fortified city that was full of every goods immediately and without a battle.[21]

Methods of Transmission: Espionage

Spies used obscurity as their camouflage to avoid drawing attention from the local authorities. The real concern of many of the sources regarding information leaks to the enemy is reflected in the safety measures described by military writers to tackle this problem as efficiently as possible. Much of the evidence about spies in ancient Greece comes from the precautions recommended by Aeneas Tacticus to be taken following the outbreak of war or during the siege of a city. According to Aeneas, in order to prevent any information from being passed on to foreigners or enemy agents posing as merchants, no festivals are to be held outside of the city and no private gatherings are to be allowed during day or night.[22] Furthermore, 'no citizen or resident alien shall take passage on a ship without a passport [σύμβολο], and orders shall be given that ships shall anchor near designated gates'.[23] In order to enable local authorities to distinguish among friendly troops, agents or citizens from foreign lands and enemy infiltrators, several cities in ancient Greece had devised a series of verbal and written signs or signals called *synthemata* (συνθήματα), a common password that could easily be remembered (e.g. 'Athena' or 'Hermes Dolios') and tokens (σύμβολα) or *sphragides* (σφραγίδες, Lat. *bulla*).[24] The latter two are attested in mercantile activity and diplomatic missions throughout the centuries; a token was used as credentials to check the identity of a person, while *sphragides* (or *bullae*) were marks of authenticity and constituted proof of origin for an important document which, in addition, had to be kept confidential and away from prying eyes.

In the Muslim world, commercial and diplomatic contacts required what Islamic law calls the *aman*, or safe conduct.[25] Although no documentary evidence for the theory of the *aman* exists from the Arab–Byzantine world, we know that in later periods an *aman* was negotiated between a host (a sultan) and a group of visitors, such as an embassy or a group of merchants for a specific period – usually a year. If that group of people exceeded this specified period, then they would have to accept the status of *dhimmi* (non-Muslim citizen of an Islamic state). The tenth-century *Επαρχικόν Βιβλίον* (*Eparchikon Vivlion*, the 'Book of the Eparch') by Leo VI, written probably around the year of his death in 912, also places strict restrictions and regulations upon the guild life and mercantile activity in the empire's main cities and ports; for example, merchants coming from the Muslim world could not stay in the empire for more than three months.[26] Security considerations also prompted the imposition of a ban on the export of weapons and any other material related to warfare – this ban was extended by Tzimiskes in 971 to include several kinds of timber.[27] Qudama's short naval guide advises the city and port authorities to be vigilant for the possible infiltration by spies – the fear of spies in Egyptian ports was greatly intensified after the raid in Damietta[28] – and conduct thorough searches of every merchant leaving a Muslim port or city for any war supplies. According to Leo VI's sixty-third *Novella*, the person who ignored the ban on the export of weapons would have been punished by death.[29]

An ideal place to gather all kinds of intelligence about the enemy were the πανδοχεία or *funduqs*. These served as hostelry for travellers, but the institution took on new economic and social roles as, aside from catering to merchants' lodging needs and providing storage for their trade goods, they functioned as places of sales and governmental taxation.[30] As predecessors of modern hostels and inns, they were mainly situated alongside important roads, crossings and passes, and were places where anyone could meet and socialise with all sorts of people, including merchants and travellers who ate, drank and spent the night there. Here, one could recruit mercenaries, question witnesses, discuss contracts, conduct political negotiations and trade news and rumours – in a sense, these were the focal points of a town or a city where important and everyday people alike could meet after sunset and into the late hours. The important thing to bear in mind about these places is the diversity of people, trades, social and ethnic groups, and religions one could come across.[31] Naturally, as these people would, usually, have consumed copious amounts of wine and/or ale, 'their tongues would have gotten loose'.

Aeneas Tacticus makes special mention of the innkeepers. During a siege or emergency situation, 'even they' should not be allowed to receive any strangers without permission from the city authorities.[32] I have not come across any evidence in Byzantine primary sources regarding an incident involving people at a tavern, intoxicated or not, giving out secrets to enemy spies or agents, but we know, for example, from letters sent by Strasbourg spies from Breisach to their home town in 1417 that the city council had attempted not only to establish contact with the tavern-keepers, but even to send spies directly to them to catch up on whatever intelligence they could.[33] Perhaps the most famous incident of revealing top-secret military information comes from 1944. On the eve of the Normandy landings, a drunken American, Major-General Henry Jervis Friese, publicly took bets at a London hotel that the D-Day invasion would occur before 15 June. This was in spite of the real threat of Nazi agents operating in London pubs, bars and hotels where Allied troops lived and socialised.

Ecclesiastics

Ecclesiastics were so numerous and mobile, if we are to believe the saints' *Lives*, that their engagement in espionage was almost inevitable. These people would have had a legitimate reason to travel, either on pilgrimage or to visit other monastic centres, as well as to attend one of the numerous religious festivals that took place at the time. They would also most certainly have been respected by local officials and people in towns and villages, who would have offered them hospitality. The biography of St Paul the Younger reports on his regular correspondence with Emperor Constantine VII, and the monk's regular advice on matters of foreign policy. His deep knowledge of affairs in foreign countries, such as Bulgaria and the Arab emirates, reveals the widespread and sophisticated network of informants and contacts (monks, pilgrims, etc.) that delivered intelligence to the monastery on the Aegean island of Samos, where he had withdrawn.[34] Several incidents, however, point to the suspicion with which such people would have been confronted and the distrust that a lone traveller or a group of pilgrims would have aroused at border and road checkpoints. St Gregory Decapolites (d. 816), a saint of the late eighth and early ninth century from what is Jordan today, whose travels took him to Corinth, Rome and Constantinople, was accused by the local people and officials in Otranto of being a spy for the Byzantine government; the result was that he was publicly ridiculed and nearly lynched.[35] Further, we read of the suspicion and open persecution with which St Willibald, Bishop of Eichstatt in Bavaria (d. 787), and his group of followers were met when they landed in Syria on their way to Jerusalem via Cyprus for a pilgrimage:

> At that time [when they landed in Syria] there were seven companions with Willibald and he made the eighth. Almost at once they were arrested by the pagan Saracens, and because they were strangers and came without credentials they were taken prisoner and held as captives. They knew not to which nation they belonged, and, thinking they were spies, they took them bound to a certain rich old man to find out where they came from. The old man put questions to them asking where they were from and on what errand they were employed ... Then they left him and went to the court, to ask permission to pass over to Jerusalem. But when they arrived there, the governor said at once that they were spies and ordered them to be thrust into prison until such time as he should hear from the king what was to be done with them.[36]

Public awareness and fear of travellers and religious people acting as spies for foreign governments is more than evident in the reports we have of several similar incidents involving saints and pilgrims.[37] Clerics could, indeed, act as intelligence-gatherers or, as the example of a monk named Agapios of Mt Kyminas shows, take the role of active agents. Agapios in this case was passing along messages between the central government in Constantinople and Asotios of Ardanoutzin in Iberia, negotiating the annexation of that strategic town for the empire.[38] Other examples, however, point to the ever-present danger of spies disguising themselves as clerics to infiltrate into a city or port to collect information or deliver a message. Theophanes Continuates reports the story of John the Grammarian who, during his stay in Baghdad on a diplomatic mission, attached himself under disguise to a group of poor Iberian pilgrims in order to contact the renegade general Manuel in his residence in the city on behalf of Emperor Theophilus.[39] Perhaps the most famous story of monks acting as imperial agents dates from 522; then, according to Procopius, Nestorian monks were sent under the direct orders of Emperor Justinian to India to smuggle silkworm eggs hidden in rods of bamboo.[40] While under the monks' care, the eggs hatched, though they did not cocoon before arrival. The Byzantine Church was thus able to make fabrics for the emperor, with the intention of developing a silk industry in the empire, using techniques learned from the Sassanians.

Co-religionists that lived in the enemy country also formed a potential pool of spies dispatching information to the authorities across the borders. For Byzantium, this would mean the Christians living in Muslim territory, although we must bear in mind that a significant percentage of them would not have followed the Chalcedonian rite. The most notable example is that of Patriarch Theodore of Antioch, who was accused in 756 of 'frequently informing King Constantine [V] of the affairs of the Arabs through letters' and was sent into exile.[41] Since we can only speculate as to the contents of the patriarch's correspondence with the emperor, this may well have contained his personal accounts of the events in the caliphate in a crucial period for the Abbasid dynasty, which, coming from a native high up in the Church's hierarchy, could nevertheless have proved invaluable as a source of intelligence about the power struggle in Baghdad at the time.[42]

Travellers and Geographers

One can add to this category of intelligence-gathering the works of geographers and travellers. Their works constitute an invaluable pool of information about the geography, topography and the road network and bridges of the regions they describe; they provide both strategic and tactical intelligence about the wider political, social and economic spectrum of a neighbouring state or nation. Hence, we know that Alexander the Great did not omit to interrogate ambassadors from Persia who had come to Pella or merchants who would have travelled to the interior of the Persian Empire through Asia Minor. Furthermore, there would also have been people who travelled to Persia simply to see the country and/or to attend religious festivals, especially in the Greek-speaking part of Asia Minor along the Aegean coast.[43] There is an interesting parallel here with Julius Caesar's campaigns against the Gauls and the Germans. Caesar's accounts of these people seem to owe much to the work of Eratosthenes and Posidonius, who were in contact with and wrote descriptions of these people around the year 100 BC, while next to nothing was known about the people and the island of Britain. Lacking firsthand intelligence from merchants and travellers about the character of the British people, the topography of the island, its harbours and their approaches, and its size and population, Caesar took the bold decision to conduct a reconnaissance expedition. It is worth noting here that the Britons seem to have acquired better intelligence from merchants, who had alerted them beforehand to Caesar's invasion plans.[44]

A most important body of sources about the image of Byzantium in the Muslim world is the literature compiled by Muslim geographers and travellers, especially those that belong to the so-called 'Iraqi' school.[45] We know that several of the works of the late ninth and tenth centuries that included chapters containing valuable information on the lands of the Rum (Byzantines, lit. 'Romans'), including the organisation of the state and the army, were widely circulated and used by contemporary and later scholars. Examples include Ahmad al-Yaʿqubi (d. 897/8), who wrote the _Kitab al-buldan_ (_The Book of Countries_), which contains a description of the Maghreb;[46] Ibn Rustah (writing 903–13), who provided the most detailed description of Constantinople in Arabic up to that date;[47] and al-Masʿudi (896–956), an Arab historian and geographer, who was one of the first to combine history and scientific geography in a large-scale work.

Al-Masʿudi was very well informed about Byzantine affairs. He recorded the effect of westward migration upon the Byzantines, especially the invading Bulgars, and he was, of course, assiduously interested in Byzantine–Islamic relations. The geographic intelligence contained in his works is very important as he provides his reader with a description of the 'fourteen provinces called _band_ in that empire [Byzantium]', and notes down every significant fortified town in each province of the empire – including the Balkan themes – and the distances between them.[48]

For the ninth century in particular, we possess the works of Muslim geographers, such as Ibn Khurradadhbeh (d. 911), the chief spy of al-Mu'tamid (869–85) and author of the earliest surviving Arabic book of administrative geography, which contains some interesting information about the command structure of the Byzantine army:

> The patrikios commands 10,000 men; he has two turmarchs under his command, commanding 5,000 men each; each turmarch has under his orders 5 drungars in charge of 1,000 men each; under the command of each drungar are 5 comites in charge of 200 men each.[49]

Qudama ibn Ja'far, an Arab scholar and administrator for the Abbasid Caliphate and a Syrian Christian convert to Islam (c. AD 905), held several administrative positions in Baghdad before becoming a senior official in the caliphal treasury in the 920s. His *Kitab al-kharaj* (*Book on Taxation*), for which Qudama is primarily known, is a manual for administrators containing information about the structure and organisation of the state and army, along with invaluable details on the caliphate's neighbouring states, including Byzantium.[50] His description of the organisation of the Byzantine army along with a detailed account of the theme system, which also includes the generals of the various themes at that time, provides us by far with the best account we have about that period of Byzantine history. This information is repeated in the accounts of Ibn Khurradadhbeh (Qudama's father knew Ibn Khurradadhbeh personally, thus perhaps establishing the course of information flow) and another tenth-century Persian historian and geographer, Ibn al-Faqih al-Hamadhani. We read in al-Faqih's work:

> The province of Al Natulikus [Anatoliko], the meaning of which is 'the east'; and it is the largest of the provinces of the Romans; and its first boundary is Opsikion and Al Brakisis [Thrakesion], and its second the province of the Buccellarii and the seat of the imtratighus [*strategos*] is Marg Al Shahm; and its army consists of fifteen thousand men; and with him are three turmukhs. And in this province is Ammuriya [Amorion], which is at the present day waste, and Balis [Barbalissos] and Manbig [Hierapolis] and Mar'ash [Germanikeia], and that is the fortress of Burghuth.[51]

As the accounts of Ibn al-Faqih, Ibn Khurradadhbeh and Qudama ibn Ja'far resemble one another very closely, it has been assumed that they drew their information from a common source, a certain frontier official of the caliphate named al-Jarmi, who was captured by the Byzantines perhaps during the raid of Theophilus in 837.[52] He was eventually sent home after an exchange of prisoners in 845, after which date he wrote his work on the Byzantine Empire examining its leaders and officers, its road system and the right period for raids, which only survived until the following century. Treadgold has suggested that al-Jarmi had managed, somehow, to gain access to a manual examining Theophilus' military reforms of 840.[53] Rare information contained in his work includes the earliest record of the theme of Chaldia and the κλεισούραι of Seleukeia and Charsianon, and the latest mention of Cappadocia as a κλεισούρα before being upgraded into a theme. His knowledge of the boundaries of themes in Anatolia is also quite remarkable. The list of fourteen themes given by these three geographers is the best account we have on the history of the Byzantine thematic system before the period of the Macedonian dynasty; in fact, Constantine Porphyrogenitus wrote the earliest systematic account in Greek, and incorporated the list of precedence of Philotheos (see further down) about a century after al-Jarmi.

Ibn Hawqal is another tenth-century Muslim writer, geographer and chronicler, whose *Surat al-ard* (completed in 988) is perhaps the best full-length description of the lands of the Rum. Ibn Hawqal was not just the editor of a geographical survey like many of his predecessors; he was a prolific traveller who spent more than two decades of his life (943–69) exploring remote parts of Asia, Europe and Africa. His work includes a detailed description of Muslim Spain, Italy, Sicily and the 'land of the Romans', and as he notes in the preface to his work, he compiled his geography not simply from personal experience but from reading other authors, such as Ibn Khurradadhbeh,[54] and through a network of informants, whom he questioned, made them repeat their statements, and then cross-checked them with other informants to 'measure [their] veracity'.[55]

Where Ibn Hawqal examines the Mediterranean region, the reader can find a detailed description of the hierarchy of the Constantinopolitan court, the imperial palace and the prisons of the capital. The most significant piece of information regarding strategic and tactical intelligence contained in Ibn Hawqal's work concerns the geography of Asia Minor, the communications network, the towns and cities, and the economy of the region of Asia Minor, as seen through the eyes of a contemporary Muslim traveller. He mentions the invasion routes and distances between the Muslim *thughūr* and Constantinople, the main fortified places along the way, such as Mayyafariqin, Hisn-Ziyad and Tell-Arsanas, and the main towns such as Malatya, Charsianon, Tzamandos, Ankara, Nicomedia and

Chalcedon. All these towns and forts are accompanied by the author's calculation of the distances and the rate of march between them.[56] Here is what the author writes about the theme of Cibbyraeots and its capital:

> Attaleia is both a powerful fortress and an important rural township ... Attaleia is eight days' march to Constantinople, taking the road of the post service, and 15 days by sea, with a favourable wind. The territory that separates the two cities is fertile, well populated, and the traffic is uninterrupted all along the route, and the rural district of Attaleia is very flourishing and very productive up to the canal of Constantinople.[57]

Equally significant are Ibn Hawqal's views on the urban and rural economy, and state infrastructure of Byzantine Asia Minor in the middle of the tenth century, a period in which the cities and the general economy of the region had started to show some signs of revival after three centuries of continuous decline:

> Rich cities are few in their empire and their country, despite the extent of their territory and the continuity of life and its condition; indeed, the most notable part is formed by mountains, by the citadels of the fortresses, by troglodyte villages and hamlets to the houses cut into the rock buried underground ... Indeed, it is in a precarious situation; its strength is insignificant, its revenues are mediocre, its populations of humble condition, richness is rare, its finances are bad and its resources are scarce.[58]

Reflecting on the main questions that should be asked by a commander before an expedition and the kind of strategic intelligence an ambassador was expected to collect, we can see that the extracts taken from the three geographers, who used al-Jarmi's accounts and Ibn Hawqal's *Surat al-ard*, would have proved invaluable to a strategist planning an invasion of Anatolia. Indeed, we can see that in Ibn Hawqal's description of the region of Pamphylia, one of the invasion routes taken by Muslim forces into the Anatolian plateau, the author underlines the fact that the land is fertile and well populated. In addition, significant knowledge imparted includes the state of the roads, mountain passes and rivers, establishing whether an army can pass unhindered or not, whether fodder is available, who are the officers in every place and which of these are fortified. In passing, Ibn Hawqal acknowledges that, since good knowledge of the local terrain of operations is paramount for a general, any general should acquire the services of local border guards, either Byzantine or Muslim, who could provide him with local information and instructions about routes and pathways through the mountains and into Anatolia.

Methods of Transmission: Espionage

Ambassadors and Envoys

Invaluable strategic intelligence also came from ambassadors and envoys to the Muslim caliphate and smaller emirates neighbouring the Byzantine Empire. Byzantine diplomacy, especially when it came to its contacts with the Muslim world, lacked the modern concept of a resident ambassador as we understand it today. In the Middle Byzantine period, ambassadors were called upon to serve a variety of purposes, sometimes seeking to establish stability or trying to subvert or convert neighbouring rulers and princelings, thus bringing them closer to the imperial court which was the seat of power in the empire.[59] In general terms, their role was essentially 'reactive' and 'pre-emptive'. The central mechanism of Byzantine diplomacy responded to changing events rather than attempting to initiate them, while its design was clearly defensive in nature, aiming to repel any external threat rather than create favourable conditions for expansion.[60] When it came to the diplomatic relations with the Muslim caliphates, a state of war was considered to be the norm between the two powers and peace was very much an exception, although occasionally a truce was agreed between the two governments. It was only after the 780s that direct links between the two courts were established, the main concerns on both sides being the exchange of prisoners and the declaration of – or the threat of declaring – war rather than any major invasion.[61]

The instruments of this diplomacy were, of course, the ambassadors and envoys that followed a clearly established international set of patterns and rules.[62] As was the case in Byzantine lands, in the Muslim world foreign diplomatic missions were granted safe passage, or *aman*, in a manner similar to that applicable to the merchants and the rest of the travellers entering the *Dar al-Islam*, and they were also liable to the same restrictions on carrying weapons out of the country.[63] In theory, they were not to be harmed or maltreated in any way[64] and they were to be offered sumptuous hospitality in private residences for the duration of their visit, which could have extended for up to a year – although the example of Leo Choerosphactes' mission to Baghdad in 905 that lasted for two years shows that this rule was not strictly followed. Pomp and ceremony were no less well established in the Byzantine court, as testified by the tenth-century *De Ceremoniis*, in which the reception of Muslim ambassadors from Cordoba in the presence of the ambassadors of the emir of Tarsus is vividly described.[65] I will come back to the point of the participation and ranking of ambassadors in court ceremonials; before that, however, there is another key question that comes to mind: what were the liberties or restrictions imposed on ambassadors when on an official diplomatic mission in a foreign court, and were these respected by the authorities?

In principle, the freedoms and mobility of the ambassadors in Byzantium and the caliphate were not restricted but, rather, severely hindered. When an ambassador and his retinue from Baghdad arrived at the borders of Byzantium they were received by a βασιλικός (*vasilikos*), an imperial agent, who was to keep them under constant surveillance but who would not restrict their access to any places they wished to visit beyond the capital and the imperial palace. We have the example of the ambassadors of Caliph al-Wathiq (r. 842–7), who visited the cavern of the Seven Sleepers in Ephesus, or of Sayf ad-Dawla's envoy Ibn Shahram's visit to a temple some three days walk from Constantinople.[66] During their stay in the capital, they were hosted in a luxurious residence known as *mitaton*[67] and they participated in social and courtly festivities.[68]

The correlation between diplomatic duties and spying was well known in Byzantium. Kekaumenus repeatedly draws the attention of his readers to the wickedness of foreign envoys: 'Know that the incoming envoys are terrible (δεινοί) and cunning (πονηροί), and they make pretence even for the simplest of things.'[69] As early as the fourth century BC, several precautionary measures were recommended by Aeneas Tacticus in order to limit this kind of espionage, advising the city authorities to keep a close watch on all foreign ambassadors: 'Not everyone who wishes may converse with public embassies representing cities, princes, or armies, but there must always be present certain of the most trusted citizens who shall stay with the ambassadors so long as they remain.'[70] However, the main concern of authorities from ancient Greece to Byzantium seems not to have been so much the ambassadors themselves, who 'should be received honourably and generously, for everyone holds envoys in esteem', but rather 'their attendants, [who] should be kept under surveillance to keep them from obtaining any information by asking questions about our people'.[71] Frontinus, writing at the end of the first century AD, gives a number of examples of army officers acting as spies and infiltrating into a city disguised as slaves and attendants of a diplomatic mission; in one of these examples, a team of them purposely let loose a horse to run in the city and then chased it around in order to have the chance to inspect the city's fortifications.[72]

Ambassadors on a diplomatic mission could assume different roles depending on the circumstances, their instructions, and their qualifications and abilities as spies.[73] Regarding the last point, the author of *On Strategy* advises that envoys should undertake their mission eagerly, be naturally intelligent, be able to improvise and take advantage of the opportunities

Methods of Transmission: Espionage

presented to them, and be willing to risk their own life to achieve their mission. We have already mentioned what Nizam al-Mulk described as one of the major operational roles of an ambassador in his mission to a foreign court: the collection of information, either through direct observation or through second-hand sources, on the geography and topography of the enemy land, the state of the enemy army, economy and infrastructure, and the character of the enemy sovereign.

Constantine Porphyrogenitus reports in the *De Administrando Imperio* that the annual diplomatic embassies sent to the Patzinaks 'to keep the peace . . . and conclude conventions and treaties of friendship with them' would have brought with them invaluable strategic intelligence on the Patzinak state organisation and infrastructure, the economy, the army and the court.[74] In fact, the aforementioned work is not just a manual of foreign policy, diplomacy and the internal history and politics of the Byzantine Empire, but also, according to Jenkins, a 'comprehensive historical and geographical survey of most of the nations surrounding it'. This included not simply a collection of historical and topographical reports compiled by provincial governors and imperial envoys, but intelligence reports acquired by foreign agents and ambassadors. The information they provided upon questioning was highly scrutinised before being written down and sent to the imperial authorities.[75] It would have been these intelligence channels that provided the information for the chapter describing in detail the naval routes of the Rus' to the Black Sea and Constantinople and their famous *monoxyla*, or the numbers of horse, foot and different kinds of ships the nations of the Croats could muster at different periods in history.[76]

An envoy would have proved a valuable intelligence agent during the siege of a city, as he would have been the only person able to have limited access to the city and assess the enemy's situation. We are informed by Amatus of Montecassino that Robert Guiscard dispatched to the emir of Palermo a certain Peter the Deacon as an official ambassador who spoke fluent Arabic, but with secret instructions to investigate the city's defences.[77] During the siege of Dyrrachium by Bohemond of Taranto in 1107, and just before urgent negotiations took place that eventually led to the Treaty of Devol in September 1108, Byzantine envoys visited the Norman camp to broker a meeting between the count and the emperor. For Bohemond, it was crucial to conceal the desperate situation of his army after so many months of deprivations: 'When he [Bohemond] heard of their [envoys] approach he was afraid they might notice the collapse of his army and speak about it to the Emperor, so he rode out and met them at some distance from the camp.'[78]

A clever and cunning general could, potentially, mislead the envoys coming to negotiate a truce or an exchange of prisoners, or any spies that may have infiltrated his camp, into believing that his army was stronger and more numerous than in reality:

> If you believe that your forces are very weak compared to those of the enemy, then kill the spies or hold them in a secure fortress. If, however, you have a strong and impressive armament, fine equipment . . . then display your army in its orderly and impressive condition.[79]

A display of power and abundance of men, food and materiel could give the false impression to the enemy that winning the war would be much more difficult and costly than previously contemplated and they might even abandon the idea altogether. Frontinus devotes a special section of his *Strategemata* to 'how to produce the impression of abundance of what is lacking', where he recommends taking envoys and prisoners of war to the city's store houses, which would have been deliberately packed with food, to give the false impression of abundance of provisions to the besiegers.[80]

Other examples of intelligence-gathering by ambassadors include Theophanes' report of Frankish envoys being present in the imperial palace during the time of Nicephorus I's *coup d'état* that deposed Empress Irene in 802, surely a political event that was worth reporting back to Charlemagne's officials along with intelligence and gossip about the imperial court.[81] Ambassadors were also sent to assess the enemy's preparations for hostilities under the pretext of peace negotiations; hence, Emperor Artemius (r. 713–15) sent the eparch Daniel of Sinope to Damascus to 'διερευνῆσαι τα των Ἀράβων δε οπλιζομένων δια της κατά της Ρωμανίας κινήσεως και δυνάμεως αυτών' (investigate the arming and the numbers of the Arabs for their invasion of Romania), a mission to assess whether Caliph Walid was preparing for a large-scale expedition against the empire.[82] Almost two and a half centuries later, Constantine VII sent ambassadors to the Umayyad caliph of Cordoba for the second time in AH 338/July 949–June 950 to secure the neutrality of the Spanish Muslims in the face of a planned imperial expedition against Crete, or even an alliance against the Shiᶜa Fatimids of North Africa, an embassy that was viewed by the Spanish as having been sent to appraise their attitude towards the Cretan Arabs, to whom they were related.[83]

Finally, ambassadors could carry out secret operations in addition to the official mission for which they were dispatched overseas. Theophanes Continuates informs us about the secret negotiations between the Great

Khan of the Bulgars, Omurtag (814–31), and Michael II (820–9) regarding the dispatch of Bulgarian reinforcements to the Byzantine emperor against the rebellion of Thomas the Slav, after the latter had declared himself as rival emperor, transferred his forces from Anatolia to Thrace and besieged the capital in December 821.[84] Another typical example of a so-called παραπρεσβεία (false embassy) was the diplomatic mission of Leo Choerosphactes, a *magistros* (hence also known as Leo the Magister) and patrician in the imperial court and a high-ranking diplomat of Leo VI (he was related to his fourth wife Zoe). Leo was sent in 905 on an official mission to the emirs of Tarsus and Melitene, as well as to the Abbasid caliph in Baghdad, hoping to achieve a peace treaty. His orders, however, included a secret mission to establish a contact channel with the renegade general Andronicus Doukas, who had sought asylum at the Abbasid court after a failed rebellion against Emperor Leo the previous year.[85]

Imperial Court Ceremonial

As the Byzantine diplomacy of the period was called on to play a variety of roles, sometimes seeking stability and other times aiming to abase or convert the rulers of neighbouring regions into forming closer ties with Constantinople, Byzantine ceremonial played a prominent part in this 'game' of impressing local and foreign potentates. As different Byzantine ceremonial practices existed for every possible occasion and were loaded with multiple layers of meaning, their essence as a tool of contemporary diplomacy served as a 'teaser' and a 'deterrent', whose goal was to impress and intimidate both foreign and local guests, along with the citizens of the capital.[86]

Although the Arab community initially distanced itself from the pomp and ceremony displayed in the capital, Arab leaders quickly came to realise what a powerful tool of propaganda this was and came to adopt many of the elements of the grandeur and elegance they witnessed in the Byzantine capital, mixing it with elements that derived from the older, but still splendid, Sassanian ceremonial.[87] Hence, it was expected of both high-ranking foreign visitors and guests, including ambassadors and prisoners of war in the imperial courts of both capitals, to be invited to participate in religious celebrations on various holidays like the First of the Year in the Abbasid and Fatimid courts, and the Christmas and Easter Sunday ceremonies and banquets in Constantinople.

According to the *Kletorologion of Philotheos*,[88] a document recording the Byzantine lists of offices and court precedence from around 899, which was incorporated into the *De Ceremoniis* some four decades later, the ecclesiastical dignitaries were followed in the Christmas banquet by the 'Agarene friends', with the Eastern Muslims taking precedence over the Western ones coming from Sicily, Italy and Spain.[89] The fact that Muslims ranked second only to the ecclesiastical delegations sent from Rome, Antioch and Jerusalem highlights the importance with which the Byzantine government viewed its relations with the Abbasids. There are two facts I wish to underline at this point: the prominent place of Muslim diplomatic delegations in imperial banquets, sitting at the sixth table opposite the emperor and next to the Bulgarian 'friends', and the respect with which Muslim prisoners of war were treated in the ceremonies.[90]

We also know that both Byzantine and Muslim ambassadors were left to follow the customs of their own religion and were allowed to converse and socialise with members of the palace courts and other diplomats, although there would have been, of course, the necessary vigilance on the part of the agent keeping them under surveillance.[91] Embassies to foreign nations were composed of the elite of the high-ranking aristocracy, as proper education and good command of the language were essential skills for any diplomat. John the Grammarian was chosen to be sent to Syria in the ninth century because of his debating skills.[92] Al-Mas῾udi writes that while he was staying in Damascus in 946, he had the chance to converse personally with the imperial envoy John Anthypatos, who impressed everyone at court with his deep knowledge of history and philosophy.[93] Oratory skills and shrewdness were important for a foreign diplomat, not just to awe his hosts but also to save himself from embarrassing moments, such as in the case of the imperial ambassador to the Fatimid caliph al-Mu῾izz (953–75) who, during a conversation for which the ambassador had not been prepared, unwittingly revealed intelligence concerning the diplomatic relations between the empire and the Hamdanid emirate.[94]

Although kept under strict surveillance, it would have been possible for envoys to obtain intelligence from several different sources, from their everyday acquaintance with people in the palace to the simple observance of the language, customs, appearance and behaviour of officials and lay people. Attention to even the minutest detail could provide invaluable information to the authorities back home, as Nizam al-Mulk himself points out from his own experience after an audience with the ambassador from Samarqand.[95] They would also have had the opportunity to converse with co-religionists; for the Muslims, it would have been the mosque of Constantinople, the prison for Muslims in the imperial palace, or the baths and local markets – Liutprand, for instance, complained bitterly about his inability to see his 'friends' in the market of the capital. This was exactly the kind of strategic intelligence-gathering that an envoy was expected to

Methods of Transmission: Espionage

perform according to Nizam al-Mulk and al-Harawi, and what was certainly followed by Byzantine diplomatic missions as well.

However, ambassadors had to be careful not to be overly impressed by the reception staged by their hosts; after all, the purpose of the magnificent show put on by the courts in Constantinople and Baghdad was to awe their official guests, which explains why the following account of a diplomatic mission has survived in full. This rare account that we have on the reception of Byzantine envoys John Radenos and Michael Toxaras in Baghdad in 917 proves this point: even though by the late 920s the Abbasid Caliphate's authority was crumbling, peripheral powers were establishing themselves in Syria, Egypt and beyond, and the armies of the caliphs were unable to keep the peace and guarantee the borders of the empire, nevertheless they could still put on a magnificent show and look fearsome lining up in the corridors of the palace halls.[96] The envoys would have had the chance to view first-hand the different nations serving in the caliph's army, the different types of soldiers – the elephants allegedly 'caused much terror to the Greeks' – and, of course, their equipment with 'some ten thousand pieces of arms, to wit, bucklers, helmets, casques, cuirasses, coats of mail, with ornamented quivers and bows'.[97] The significant point here is that this is the image of the caliphate and its army that the Abbasid officials wished the envoys to witness – and that was certainly far from the reality.

Prisoners of War

A direct source of strategic intelligence about the enemy were the prisoners of war who, according to the author of *On Tactics*, 'sometimes captured together with their wives and children may prove more helpful than the spies'.[98] A raid to capture prisoners, either soldiers or civilians, is recommended by every Byzantine military treatise as one of the best methods to obtain intelligence about the strength and plans of the enemy.[99] The unit responsible for undertaking the task of capturing enemy prisoners for interrogation was, as we have seen, the corps of the *trapezitai* or *tasinarioi*, the rough border guards who knew their localities and the mountain passes and could conduct regular small-scale raiding operations for intelligence-gathering. Interrogating the prisoners was the most important and delicate part of the operation and had to be conducted by the general himself, because 'frequently very important and completely unsuspected information has been revealed by such questioning'.[100] The author of the treatise *On Tactics* takes the use of prisoners of war for intelligence even further, as he recommends to the general that he should 'give assurances of freedom for them [captives], their wives and children and then send them out to spy. After they have investigated how everything is going among their own people, they can return and report the facts.'[101] For the author, prisoners of war could act as spies in a much more efficient way considering their knowledge of the language, customs and their connections with the locals.

As the ninth century saw the stabilisation of the balance of power between Byzantium and the Abbasids, a new sort of diplomatic activity emerged between the two superpowers of the time, that was concerned mainly with the exchange of the large numbers of Christian and Muslim prisoners captured on both sides of the frontier.[102] Although irregular in nature, meetings concerning the exchange of prisoners frequently took place between the months of September and October at the mouth of the River Halys in Cilicia, where negotiations were conducted mainly by the local governors and several other intermediaries such as *qadis* (judges). Although it was not until the middle of the twentieth century that an international framework for the treatment of prisoners was agreed upon, the handling of prisoners on both sides of the Byzantine–Arab border was relatively humane.[103] This was despite the occasional maltreatment and/ or execution of Byzantine prisoners as a kind of psychological warfare, confirmed by numerous Arab sources, the equivalent of which can be seen in Byzantium as well during the reigns of Michael III and Basil I.[104] A radical change, however, took place after the enthronement of Leo VI. Was a 'paramount feeling of respect', as mentioned in the writings of Leo on Muslim warriors, the reason behind this change of attitude or, rather, the military exploits of the Muslims in the intervening period between the writing of the *Kletorologion* (899) and the *Taktika* (c. 900)?[105] Whatever the reason for this change and, especially, for the invitation of several high-ranking prisoners/hostages to partake in the Christmas and Easter Day celebrations, Leo's reign certainly marks a turning point during which we can see the first attempts to regulate the treatment of enemy prisoners based on the principles of the old Roman law and the Christian concepts of *humanitas et caritas*.[106]

A distinction needs to be made here between the prisoners captured during a campaign, who could provide mainly operational and tactical intelligence, as in the example of Nicetas Chalkoutzes that follows later on, and the more important high-ranking prisoners held captive in Constantinople, Baghdad or Aleppo, and the intelligence they could impart. Hence, al-Muʿtasim found the information provided by a captured Greek horseman of the garrison of *Qurra* (Koron, the seat of the *kleisourarch*[107] of Cappadocia) regarding the whereabouts of the Empreror Theophilus

Methods of Transmission: Espionage

on the east of the River Halys invaluable, as the Caliph was struggling to coordinate his invasion army with the northern one commanded by his general Afshin (838). According to Tabari, the captured cavalryman even provided the exact distance to the place where the emperor had pitched his camp 'behind the River Lamos (Halys), at a distance of four *pharsangs*'.[108] Important topographical information was provided by a prisoner of war to Umar, the emir of Melitene, before the battle at Bishop's Meadow (863), even though by that time the thematic forces sent by Michael III had already surrounded his much smaller army at Poson (Πόσων).[109] During a Byzantine expedition into the land of the Avars in the second half of the sixth century, Theophylactos Simocattes writes that due to a miscalculation the guides led the army into enemy land, where they quickly ran out of water and the soldiers resorted to drinking wine instead. 'On the third day the trouble intensified, and the whole army would have perished if a certain barbarian prisoner had not pointed out to them the River Helibacia, which was four *pharsangs* distant.'[110]

The Byzantines regarded any kind of information coming from prisoners – and deserters – as suspicious and treated it with caution, cross-checking it through other channels of information before assessing its value.[111] It is worth noting Leo VI's arming of Muslim prisoners of war to defend the capital from Symeon of Bulgaria in 896, after the disastrous defeat of the imperial army at Boulgarofygon. Desperate times called for desperate measures, and although the threat from the Bulgars was repelled, according to Tabari, Leo eventually 'recalled his order, removed them their weapons and scattered them in the [different] countries [of his empire], for fear that they might cause annoyance',[112] evidently showing the mistrust with which the Byzantines viewed the Muslim hostages. Finally, the example of Nicetas Chalkoutzes proves that the tactic of mistreating and keeping an envoy as a prisoner of war could backfire, as Chalkoutzes was forced to follow the Hamdanid invasion force into Cappadocia and Charsianon in 950. According to Skylitzes, Nicetas managed somehow (most likely though bribing his guards) to let Leo Phocas know about the emir's planned route home, thus allowing him to set up an ambush at the appropriate defile of the Taurus.[113]

There is another category of prisoner of war which includes a number of high-ranking Muslim captives that were fortunate enough, compared with their co-religionists, to spend their time in captivity in and around the imperial palace in Constantinople, able to practise their religion, take part in imperial banquets and even converse with the emperor himself. They belong to the same category of imperial 'guests' as the relatives or rivals of a potentate of a neighbouring country or a satellite state on the borders of the empire, who were held in the city or the palace as a potent form of diplomatic pressure, being awarded court titles and even married to members of the imperial family.[114] The deeper reason behind this tactic is obvious and has to do with propaganda, as the central authorities would have wished their prisoners to be at the heart of palace life and pomp and witness first-hand the magnificence of the capital, compared with other, less important, figures who, although they were also invited to the capital and the imperial banquets, would have been imprisoned further away from the palace or even in provincial prisons.[115]

The treatment of important figures as hostages did not abide to any international law, but there would have been some kind of universally accepted code of conduct between 'civilised nations' on how to treat noblemen and important figures in captivity. In fact, the example of Abu Firas confirms the case that the treatment of prisoners depended largely on how the enemy treated their high-ranking hostages; hence, it was only when the Byzantines learned that the Arabs restricted Byzantine prisoners with handcuffs that they forced Abu Firas, during his initial captivity at Charsianon, to wear handcuffs too as a reprisal.[116] Abu Firas (932–68) was a high-ranking member of the Hamdanid family and a cousin of Sayf ad-Dawla, who made him governor of Menbij (Hierapolis) – a focal point for the Byzantine–Hamdanid wars of the 950s–60s. A collection of his poems titled *al-Rumiyat* is one of our most important sources for the life of Sayf ad-Dawla and the Byzantine–Arab wars, a kind of personal 'war diary' with one of the pieces narrating his captivity in the capital in 962–6, where he met and conversed with Nicephorus Phocas.[117] His account confirms that noble Muslim prisoners of war were very well looked after; he writes about his glorious reception from Emperor Nicephorus himself, his stay at a luxurious lodging close to the palace, being provided with a personal servant, and his meetings with fellow Arab soldiers and co-religionists. Most interesting, however, is the part where he refers to his meetings with the emperor and the alleged dialogues he had with him, full of sarcasm and insults launched from both sides regarding not only matters of religion but also the fighting 'virtues' of both nations, where Abu Firas tries desperately to assert the fighting spirit and military superiority of his nation.[118] This vivid dialogue in Abu Firas' 'war diary' is certainly one of the most significant pieces of evidence that confirm the transmission of knowledge and ideas, despite the differences in religion and customs, between high-ranking military men of neighbouring nations like the Byzantines and the Arabs.

Another important prisoner in Constantinople was Harun ibn Yahya, who has provided us with perhaps the best description of the Byzantine capital from an Arab perspective.[119] Captured in Palestine after a Byzantine naval raid against the port-city of Ascalon sometime in the late ninth or early tenth century, he was initially transferred to Attaleia and then to Constantinople.[120] Despite being a prisoner of war, Harun was allowed to wander around the capital – presumably with an escort, although this is not clear in his work – and write his magnificent account of the buildings, monuments and churches of the capital, as well as of the customs, traditions and ceremonial of the Byzantine court. His description of Constantinople is relatively short but densely written and contains significant information about the topography and the fortifications of the city, its towers and its gates.[121] In his descriptions of the buildings of the capital, Harun includes several important landmarks of the period such as the Hippodrome, the Golden Gate, Justinian's Column and the Aqueduct. A specific building complex, however, drew the attention of our 'wandering prisoner' more than anything else, the imperial palace. This was a massive complex of buildings, courtyards, gardens and galleries with broad expanses that was surrounded 'on all its sides by a wall that is one parasang in perimeter'.[122] Harun writes about the three gates of the palace, the Hippodrome, the Mangana and the Sea Gate, which led to three large vestibules that came before the palace's main complex and private rooms. What is worth noting here is the different nations that made up the guards of the aforementioned gates, from African Christians, holding golden shields and lances, to guards of Khazar and Turkish origin, all of them sitting in courtyards paved with magnificent marble and golden decorations. Very important is Harun's description of the palace prisons, situated behind the Gate of Mangana. He mentions four different building blocks: one for the Muslims, one for the general population, one for the chief of the police and a separate one for the people of Tarsus; indeed, he is the only one who makes that kind of distinction between the different types of prisons, specifically referring to the one kept for the Tarsians.[123]

Harun's account also contains a unique and detailed description of the elaborate ceremonial of the traditional Christmas banquet and the annual religious procession from the palace to Hagia Sophia. In fact, some of Harun's details, such as his description of the musical organs playing during the banquet and the offering of dinars and dirhams to the Arab prisoners by the emperor, can be paralleled with Constantine VII's narrative details of the events in his *De Ceremoniis*.[124] The same has to be said about Harun's report of the procession from the Palace Gate to the Church Gate, which despite a few inaccuracies bears great similarities to the description found in the *De Ceremoniis*, except for one important, and unique in both Greek and Arab sources, detail regarding the participation of Arab prisoners in the procession and their ceremonial triple exclamation of 'may God prolong the life of the emperor for many years'.[125] Since Arab prisoners of war would have acted as 'representatives' of the caliph in ceremonial processions in the absence of an official delegation from Baghdad, Harun's detailed description of the Christmas banquet and imperial procession, emphasising the ceremonial magnificence, luxury and immense power of the emperor, has prompted speculations that he might have attended the ceremony himself as a prisoner and guest of Leo VI.

Other famous prisoners of war include al-Jarmi and Ioannes Kaminiates. For the former we have already discussed the extraordinary information provided by his work on the Byzantine army structure and organisation, and the history of the thematic institution in the first half of the ninth century, while it has been suggested that he probably had direct access to an official document of the reign of Theophilus. Kaminiates' Χρονικόν (*Chronicon*) not only illustrates in detail the military and siege tactics of the Muslim raiders (chapters 31–4), but also forms one of the most vivid and accurate accounts of the Muslim naval raids and slave trade of the period. Both these prisoners may have had the misfortune of spending a significant amount of time in captivity, admittedly under much better conditions than their fellow co-religionists, but their 'misfortune' turned them into eyewitness sources of events that were taking place at the seat of the enemy's power, transforming them into invaluable agents for the transmission of knowledge and strategic intelligence at a period when the Byzantine Empire had reached the peak of its glory.[126]

Several ordinary Muslim hostages and refugees, whom the Byzantines actively encouraged to apostatise and be baptised in return for a comfortable settlement in the empire, converted to Christianity. Having said that, it was more difficult for the authorities to convert Arab aristocrats held hostage in the *praetorium*, not only for fear of provoking retaliatory actions from the Abbasids but also because they would lose the ransom money paid for them. Constantine Porphyrogenitus gives us an account of the privileges enjoyed by these converts, who entered a military or civilian household, which included a significant amount of money and animals, and a three-year tax exemption from the *synone* (a monetary tax on cultivated land) and the *kapnikon* (a household/hearth tax), although it is unclear when this law was introduced.[127] It is noteworthy that the same source mentions the presence of 700 prisoners, probably Arab, in the Cretan expedition of 949, while al-Masᶜudi also notes a corps of 1,200 'Christianised Arab horsemen' in the Byzantine army in 943–4. Individual prisoners who decided to convert and settle in foreign lands do not constitute the only source for the transmission of knowledge and ideas over the Byzantine–Arab borders. Throughout the period that followed the Muslim expansion in Syria and the Middle East, several Arab and Arabic-speaking groups migrated into Byzantine Anatolia rather than accept Muslim domination, and their profound impact and influence on frontier society can be seen in the epic of Digenes Akritas and in the *Strategikon* of Kekaumenus.

Nomadic Tribes

The role of Arabs in Byzantine politics and the army dates back to pre-Islamic times, when the Syrian Desert and the Arabian Peninsula became an area of conflict for the two neighbouring superpowers of late antiquity, which clashed to the north (Mesopotamia) as well as to the south (Syria and Arabia) of their borders. The Lakhmid tribe had established itself as a buffer state against the Byzantine Empire (c. AD 300), centred on the city of al-Hira (south-west of Kufa) on the west bank of the Tigris, with al-Mundhir III (c. 505–54) launching raids as far north as Antioch. The Arab state that shielded imperial lands was ruled by the Ghassanids, a Christianised and Romanised Arab tribe from Yemen, who had migrated to Syria and Jordan in the third century and had set up their capital in Jabiya, in the Golan Heights.[128] As kings of their own people, the Ghassanid kings were acknowledged as *phylarchs*, native rulers of client frontier states, and given the high-ranking court title of 'patricians'. Nicolle, in his extensive study on the arms of the Umayyad era, lists numerous examples of both Byzantine and Sassanian influence on the equipment of the Arab warriors of the Syrian and Arabian deserts, and provides a powerful argument that the region was a melting pot of military cultures many centuries before the arrival of Islam, with strong influences from across Europe and Central Asia.[129]

With the Muslim conquests of the seventh century, the influx of Arab or Arabic-speaking tribes into Byzantine Anatolia proliferated. Either of Armenian or Kurdish origin, the Monophysite Mardaites, who inhabited the highlands of southern Anatolia, Isauria, Syria and the Lebanese Mountains, were used by the Umayyads as border guards in Syria and Cilicia. They are also attested, however, to have conducted guerrilla wars against their Muslim overlords over many decades, allegedly one of them reaching the outskirts of Jerusalem, and may have contributed to the lifting of the first great siege of Constantinople by Muᶜawiya.[130] In the 690s, Justinian II agreed with Abd al-Malik to relocate around 20,000 of them to the southern coast of Anatolia, as well as parts of mainland Greece, such as

Epirus and the Peloponnese, to serve in the thematic fleets of the imperial navy as part of his policy to restore population and manpower to areas depleted by earlier conflicts.[131] Conversely, it was Justinian II's policy of relocating people en masse after his successful campaign against 'Sklavinia and Bulgaria' in 688 that saw the Slavs, who were transferred from the Balkans to Asia Minor and the theme of Opsikion and on the frontier around Antakya, Apamea and Quros, taking Arab names.[132] According to Theophanes, a huge number of these Slavs, some 20,000 in all – although these numbers should be viewed with caution – deserted to the Arabs in 692. According to the same source, Mu°awiya used many of them during his invasions into Anatolia as guides and, most likely, military advisors.[133]

Other refugee groups include the Khurramites, an Iranian political and religious sect mixing Shi°a Islam and Zoroastrianism, who inhabited the Zagros Mountains before their expulsion by al-Mu°tasim around 834.[134] They were received with open arms by Theophilus, who enrolled them into a new *tagma* after they had converted to Christianity.[135] Their loyalty to their new lord, however, proved short-lived as they abandoned their position and fled at the Battle of Anzitene (838). After Theophilus recalled the Khurramite leader Theophobos to be punished, his troops proclaimed him emperor at Sinope. It was only in the following year that relations between the two were restored and the *tagma* was disbanded, with the Khurramites being dispersed throughout the themes of the East and West.[136] The integration of foreign nations and tribes into the Byzantine army and society took place without racial discrimination and without great difficulties or objections; at least this is the evidence from the sources. Indeed, throughout their history the Byzantines seem to have been much more reluctant to integrate Christian heretics, like the Manichaeans, into the Byzantine army than any other group of people. The result was the 'Romanisation' of foreign nations, their installation onto imperial lands – preferably previously uninhabited areas close to the frontiers to boost the local population and economy – their enrolment into the military codices as full-time units and the handing out of court titles to their leaders.[137]

At this point, I wish to examine in a little more detail the migration into Byzantine territory of the Arab nomadic tribe of the Banu Kilab, because this issue forms part of the wider topic of the nomadic migrations of the early tenth century and the threat these posed to the sedentary communities of the regions of Syria and northern Mesopotamia. Between the seventh and the beginning of the tenth century, most of the Bedouin tribes living in Syria and Mesopotamia had developed close economic ties with the urban populations of these regions under Byzantine or Abbasid administration, and they had become partly or fully sedentary. However, this process came to an abrupt end with a fresh influx of Bedouin tribes from the southern Arabian Peninsula, owing to the incursions of Ismaili Qarmatian tribes from Yemen to Syria, Palestine and Mesopotamia after 902.[138] Consequently, several tribes found the opportunity to settle in areas like Edessa and Harran (Banu Numayr), Mosul (Banu Uqayl) and Palestine (Banu Tayy) and replenish the numbers of preexisting tribes in the wider region, eventually leading to the 're-nomadisation' of the Jazira, and the mounting pressure upon the remaining sedentary societies by the middle of the century. Increasing Bedouin raids to agricultural districts and the conversion of newly reclaimed agricultural lands into pasture for their animals severely compromised the fiscal integrity of Byzantine and Abbasid – and later Hamdanid – rule. This was a problem that put increasing pressure on Constantinople and, after 935, on the Aleppan government, but for the Byzantines it was the Hamdanid emirate that would play the role of buffer state and eventually neutralise these tribes in the 950s.

It is in this socio-political background that in 935/6 the Arab tribe of the Banu Habib from northern Mesopotamia fled to the Byzantine Empire and converted to Christianity. Ibn Hawqal identifies as the main cause of their migration the high taxation and depredations of the newly established Hamdanid regime of the Diyar-Rabi°a, after Nasir ad-Dawla's victory over a caliphal army that same year (935), reinforced by Bedouins from the Banu Habib, that was sent to eject him from Mosul:[139] 'It was then that the Hamdanids fell on the country and had the people undergo all kinds of harassment and refinements of tyranny and arbitrary injustices. They imposed new rights and surveyed extraordinary taxes unknown until then.'[140] This desertion should also be seen in the light of the recent advances of the imperial armies on the Taurus frontier, more specifically the campaigns of Tzimiskes to capture Melitene in 927–34. Ibn Hawqal writes that the Banu Habib were 12,000 soldiers with their families, all cavalry, and describes them as well armed and very experienced. They were converted to Christianity, enrolled in the military codices and settled along the eastern border themes with plots of land. They were treated just like any other thematic cavalrymen and their forces formed the garrisons of the five new themes of Melitene, Charpezicium, Chozanon, Armosata and Derzene. In typical Byzantine fashion, Romanus I distributed among their leaders several high-ranking military titles.[141]

The Byzantines proved more than willing to welcome several thousand disaffected Bedouins into their newly conquered territories as garrison troops, as they had done with the Armenians in Lesser Armenia, the Pontic frontier and the regions of Cappadocia and Armeniakon not so long ago. By the middle of the tenth century, the Byzantine government was becoming increasingly aware of the military importance of the Bedouins. These nomadic troops were not mentioned by either Leo VI in his *Taktika* (c. 900) or the anonymous author of the *Sylloge Taktikorum* (c. 930), whereas the treatises of the middle and later tenth century identify them with the term *Arabitai* (Αραβίται) to distinguish them from the rest of the Muslim units (Αγαρηνοί, *Agarenoi*). This distinction certainly highlights the emerging role of these lightly armed troops and their skirmishing tactics in the Byzantine–Arab wars of the period, as reflected in the works commissioned by Nicephorus Phocas (c. 969) and Nicephorus Uranus (c. 991).[142] The migration of such a vast number of troops from across the frontiers and their integration into the socio-economic and military establishments of the empire would certainly have played a major role in the cross-border transmission of ideas. Further, their experience would have been greatly appreciated in the changing strategic environment of the middle of the tenth century.

Renegades, Apostates and Deserters

Several high-ranking apostates or renegades deserted to the enemy for various reasons, ranging from personal convictions and ambitions to persecution after falling out with the central authorities. Perhaps the most characteristic example of a Muslim prisoner who made a career in the Byzantine army as a convert was Anemas, the son of Kouroupas, emir of Crete, who was taken hostage by Nicephorus Phocas in 961; we later see him distinguishing himself at the Battle of Dorystolon as the emperor's bodyguard commanding an elite unit of *kataphraktoi*.[143] On the opposite side, there is the famous Leo Tripolites, whose Arabic name was Rashiq al-Wardami, but is more commonly known in Arab sources by his sobriquet, *Ghulam Zurafa* (Servant of Zurafa), evidently the name of his first Muslim master. Probably a Mardaite from Attaleia who was captured by the Arabs, he converted to Islam and became a commander of the fleet that raided and captured Thessaloniki in 904 and, along with another renegade captain called Damian, defeated the Byzantine fleet of Admiral Himerios in 912.[144]

Two cases of officers mounting a rebellion against the emperor and seeking the help of the Arabs were Euphemius and Andronicus Doukas. A victim of the machinations of an Arab minister of Leo VI called Samonas – to whom I will turn below – Andronicus Doukas had sought asylum in the Abbasid court after a failed rebellion against Emperor Leo the previous year and was contacted in secret, as we have already seen, by Leo

Methods of Transmission: Espionage

Choerosphactes in 905. Euphemius was appointed admiral of the theme of Sicily in 826, when he mounted a rebellion against the authority of Michael II and proclaimed himself emperor, probably wishing to exploit the fluid political situation in the capital after the loss of Crete to the Andalusian Muslims in the same year. Fully aware that he could not withstand the counterattack of the imperial forces, he appealed to the Muslims of Africa, who dispatched a fleet in the summer of 827, thus initiating the Arab invasion of the islands of Sicily and Malta.[145]

An important example of the value of information provided by enemy defectors can be found in 1107–8 during Bohemond of Taranto's siege of Dyrrachium. After the arrival of Alexius Comnenus near the port-city of Dyrrachium, the emperor summoned for advice three 'Westerners' who had defected to the Byzantine army in previous years and were familiar with the 'Frankish' battle tactics. According to Anna Comnena, one of them was a veteran of the 1081 Norman invasion of Illyria and Greece, named Peter of Aulps, who served as a senior commander under Robert Guiscard and had defected to the imperial army during the siege of Antioch by the Crusader armies in June 1098.[146]

Aware of the danger posed by deserters, the author of the *Strategikon* recommends two stratagems to avoid any intelligence from leaking to the enemy. First, either a deserter should be sent intentionally to the enemy carrying false information or, if the officers suspect that certain soldiers are likely to desert, they should be given false intelligence to deliver to the enemy general. Second, the general could take advantage of the enemy's suspicion of incoming deserters:

> Letters ought to be sent to deserters from our side that have joined the enemy in such a way that the letters will fall into the enemy's hands. These letters should remind the deserters of the pre-arranged time for their treachery, so that the enemy will become suspicious of them, and they will have to flee.[147]

If we compare this recommendation with what Anna Comnena reports about the incident involving Peter of Aulps and a conspiracy to send false letters to the Norman camp at Dyrrachium, then the similarities become obvious: 'He [Peter] composed letters which were apparently answers to some of Bohemund's most intimate friends and were conceived on the assumption that the others had already written to him, wooing his friendship and revealing the tyrant's secret intentions.'[148]

Finally, the author of the *Strategikon* directs the attention of his readers to the 'peoples [of our army] akin to the enemy' and advises the general to send away any units of troops in his army that belong to the same race as the enemy 'to avoid their going over to the enemy at a critical moment'.[149] Written at the beginning of the seventh century, this recommendation reads all too familiar if we bear in mind the desertion of the Cuman troops that came into contact with their kin in Alp-Arslan's vanguard the day before the fateful battle at Manzikert on 26 August 1071.

Rogue or disaffected officials in the imperial court or the provincial aristocracy could also act as inside agents, providing valuable intelligence to the enemy in secret. We have already seen the sensitive diplomacy exercised by Byzantium in the Armenian borderlands with the caliphate, and specifically the regions of Taron and around Lake Van, and the manipulation of local politics and family connections by imperial agents. In the same vein, Constantine Porphyrogenitus mentions the example of Krikorikios of Taron (died c. 930), who pretended to be on good terms with the Byzantine emperor, but

> acted at the pleasure of the chief prince of the Saracens . . . and everything that the Romans were planning in secret against their Saracen adversaries he would divulge to Syria, and would always keep the commander of the faithful informed secretly through his letters of what was going on among us.[150]

It was not only rogue or disaffected officials who could act as inside agents; personal friendships or family relations of officials could procure information to an 'enemy' agent for personal reasons. For example, we know that Agathias' Persian excursus in his *Histories* contains information not found anywhere else, as the chronicler was drawing on information procured for him by his friend and interpreter Sergius working on the Persian royal annals.[151] Likewise, during the last quarter of the tenth century one observes a remarkable increase in Arab individuals switching sides and becoming integrated into the Byzantine elite by gaining court titles in exchange for recognition of the emperor's supremacy.[152]

The case of Samonas, however, deserves more attention because of a strange episode that occurred in 904. Samonas' career in the imperial court is a typical case of a eunuch coming from humble origins, who was nonetheless promoted to become one of the most influential officials of the palace.[153] He was an Arab from Melitene (born c. 875) and entered the household of Stylianos Zaoutses (the father-in-law of Emperor Leo VI) as a captive. After Zaoutses' death in 899, he was promoted rapidly into Leo's personal service, receiving the highest court title of patrician in

906. Between 900 and the year of his downfall in 908, he seems to have been particularly involved in security and intelligence issues, becoming essentially Leo's trusted right-hand man as his chief of intelligence, also playing an active role as an agent to reveal a plot against Leo VI by Andronicus Doukas and Eustathius Argyrus.

The bizarre episode narrated by the continuators of George Monachus and Theophanes involves Samonas' alleged desertion to the Abbasids, with him being prevented from crossing the River Halys and deserting to the caliphate only at the last moment by Constantine Doukas, the son of the head of the noble family of the Doukades. Desertion by disaffected or disgraced officials of the imperial court was not something new, but it was the strange circumstances under which this incident occurred that has prompted speculations about the real events of that year. In fact, Jenkins has argued that the *Scriptores post Theophanem* were uncritically biased against the Macedonian dynasty and in favour of the provincial aristocratic families, such as the Doukades and the Argyroi.[154] Instead, according to the *Vita Euthymii*, Samonas should be viewed as a loyal chief spy and conspirator against attempts to overthrow Leo VI, whose machinations in compromising Doukas' stay in Baghdad portray a man of great intelligence and cunning.[155]

Professional Spies

So far, I have examined several sources of information and intelligence-gathering, ranging from merchants and travelling priests to ambassadors and deserters. Nevertheless, 'actual spies are the most useful. They go into the enemy's country and can find out exactly what is going on there and report it all back to those who sent them.'[156] We saw that the term spy (κατάσκοπος) seems to apply invariably to all watchmen, scouts, bandits and raiders into enemy territory sent to loot, take prisoners and gather information; however, κατάσκοποι or αληθείς των κατασκόπων[157] can also refer to secret agents (spies) dispatched to spy and gather intelligence regarding the strategic and operational plans, and military preparedness of the enemy. These were professional agents, as opposed to merchants or travelling monks, who were to:

> carry out a necessary function in providing us with such information about the enemy as may be useful for us to know either to gain some advantage or avoid injury ... [For example] any preparations for war against us or expeditions against any other neighbouring peoples or, on the other side, expeditions by some of them against the enemy.[158]

The same author advises that public markets should be used as meeting places for spies and other agents carrying intelligence, where they can pretend to trade goods, thus avoiding drawing attention to their discussions. Then, valuable information about 'the enemy's plans against us and of the situation in their country' could be passed on in secret.

Intelligence, literacy and familiarity with the customs, language and geography of the country to which they were sent to conduct espionage were deemed paramount for any spy, including some basic knowledge of the political context and diplomatic relations between their country and that of the enemy. What is more important for the context of this study, however, is familiarity with and appreciation of the military equipment and weaponry of the enemy, which would allow for the delivery of the best possible reports to their paymasters.[159] In order to avoid issues of conflicting loyalties, spies should not be of the same race as the enemy, and their families should reside in the country of their employers, 'so that love for family will keep them from remaining permanently with the enemy or from getting involved in any plots against their own people'.[160] It is clear that the issue of desertion and double agency – passing on information to the enemy in secret – was a major worry for the authors of military treatises.

These authors were also gravely concerned about the threat of enemy spies infiltrating a camp because of all the non-combatants escorting an army on campaign. To counter this possibility, they counted on the comradeship and familiarity of individual soldiers in an army unit to detect them.[161] In fact, there is a stratagem on how to uncover enemy agents appearing in the works of Vegetius, the *Strategikon* and Leo's *Taktika*: whenever a general suspected that a spy had infiltrated the camp, he would signal through the trumpets for all soldiers to go into their tents, and since there was only a handful of soldiers quartered in each tent, they and their officers presumably would have been able to detect any intruder they did not recognise.[162] The Byzantine spy system seems to have been rather defensive and reactionary to the Arab espionage and intelligence-gathering tactics. As enemy agents could loom anywhere in the camp, a general had to 'spread rumours among the enemy that [they] are planning one thing; then go and do something else . . . plans about major operations should not be made known to many, but to just a few and those very close'.[163]

Taking precautions against the leaking of information to enemy spies was not enough; deception was the key to turning the tables in one's favour, and the point raised by many tacticians was not just to detect and expose enemy agents but to transform them into double agents, in what we would identify today as counter-espionage.[164] Spies should be given assurances for their freedom and safety, and they should be dispatched back to their country to spy on behalf of the Byzantine authorities. As these people would have prior knowledge of the language, customs and geography of the enemy country, including a local network of friends and acquaintances, they would be able to observe the enemy's movements, learn their plans and report back to Constantinople.

Perhaps the most untrustworthy category of people, however, were the deserters. The authorities could not easily determine where their true loyalties lay and 'often enough they ha[d] been sent by the enemy deceitfully to plot against their hosts'.[165] What was recommended in this case was that either 'suspected defectors should be told the opposite of what we intend to do, so that we may use them to deceive',[166] ergo unwittingly turning them into double agents, or a general could send his own defectors to the enemy camp or country to deliver fabricated intelligence in order to confuse the enemy – these are called 'expendable agents' by Sun Tzu.[167] The result will be that 'either they [the enemy] will not believe it [false intelligence] and become careless or they will believe it and take the wrong action. Your intentions, therefore, will be accomplished, whereas theirs will end up achieving nothing.'[168]

Ports, markets and religious festivals have been ideal places for espionage for millennia, since it was in places like these that diverse people mingled and conducted their business. There are many examples in our sources where the opportunity to obtain intelligence at major port-cities in the Mediterranean was exploited by both empires. All military authors of the period are highly concerned with the activities of the ἔμποροι and with the spies that would infiltrate their ranks to collect intelligence. In order to curtail the flow of information across the borders, central authorities resorted to placing severe restrictions on the activities of merchants, with mixed results for the effectiveness of blocking intelligence from reaching enemy agents.

Strategic intelligence was also acquired from ambassadors, envoys and their staff, along with reports from travelling laymen and ecclesiastics. These were social groups that were respected and treated amicably by the authorities, although exceptions did apply as we have seen. As such, some had privileged but strictly controlled access to sites and people that even included the imperial court and the emperor. Envoys and ambassadors, along with prisoners of war who were intentionally set free to report back to their countrymen, assumed the task of intelligence-gathering during sieges, as they would have been the only people with limited access to the city, able to assess the enemy's situation. There should be a distinction, however, between the operational and tactical intelligence provided by prisoners captured during a campaign and the intelligence that could be obtained by high-ranking and other important prisoners held captive in Constantinople, Baghdad or Aleppo. Finally, we have several recorded cases of Muslim and Christian apostates, which included entire tribes, that could also have provided imperial officials with first-hand strategic intelligence about the political developments in the Muslim states neighbouring the empire in the East and the state of the enemy's army, economy and infrastructure.

Notes

1. Procopius, *Wars*, vol. I, I.xxi, p. 197.
2. Al-Mas'udi (d. 956 in Egypt) records naval trade between Constantinople, Syria and Egypt in the Umayyad period and later. See al-Mas'udi (1989), *The Meadows of Gold: The Abbasids*, trans. and ed. P. Lunde and C. Stone, London: Routledge, VIII, pp. 75–85. Ibn Khurradadhbeh reports the existence of active merchant traffic between China and Constantinople in the Abbasid period, see M. Canard, (1964), 'Les relations politiques et sociales entre Byzance et les Arabes', *Dumbarton Oaks Papers*, 18, p. 50. There is also the official correspondence between al-Ma'mun and Theophilus about the extension of commercial relations between the two states, with clarifications regarding trade and the movement of caravans between the empire and Aleppo in 969 after the reduction of the latter to the status of client emirate, see Vasiliev, *Byzance et les Arabes*, 1, pp. 119–21; Canard, *Hamdanides*, pp. 835–6. The classic work on this topic is P. Crone (1987), *Meccan Trade and the Rise of Islam*, Oxford: Oxford University Press. See also P. Crone (2007), 'Quraysh and the Roman Army: Making Sense of the Meccan Leather Trade', *Bulletin of the School of Oriental and African Studies*, 70, pp. 63–88. Asa Eger supports the idea of the open involvement of the people of the *thughūr* in frontier trade and exchange based on the patterns of major *thughūr* cities and villages in the lowland areas and the network of fortified road stations or *fanadiq* (sing. *funduq*) (Asa Eger, *Islamic–Byzantine Frontier*, pp. 287–9). Durak also argues for the existence of vibrant trade across the Byzantine–Islamic frontier, despite its militarisation; see K. Durak (2008), 'Commerce and Networks of Exchange between the Byzantine Empire and the Islamic Near East from the Early Ninth Century to the Arrival of the Crusaders', PhD dissertation, Harvard University.
3. Generally, on the methods of conducting espionage, see B. Walter (2011), 'Urban Espionage and Counterespionage during the Burgundian Wars

(1468–1477)', *Journal of Medieval Military History*, 9, pp. 132–43; I. Arthurson (1991), 'Espionage and Intelligence from the Wars of the Roses to the Reformation', *Nottingham Medieval Studies*, 35, pp. 134–54.
4. *On Strategy*, 42, p. 124.
5. Leo VI, *Taktika*, I.7, p. 12.
6. Not to be confused with the sixth-century historian. He was, rather, related through his mother to Emperor Julian. In 358, during the reign of Constantius II, he was sent as an envoy to the Sassanian court. Ammianus reports that a baseless rumour spread, according to which Julian had ordered Procopius to take the purple in case of his death. After Julian's death in 363, and fearing Jovian's wrath, Procopius went into hiding in the Chersonese in the Crimea: *Oxford Dictionary of Byzantium*, III, p. 1731.
7. Ammianus Marcellinus (1948), *History*, trans. J. C. Rolfe, London: Heinemann, vol. II, XXVI.vi.4–6, pp. 598–9.
8. *On Strategy*, 42, p. 122. Cf. Kekaumenus, p. 33. The author of the treatise *On Skirmishing* introduces the term πραγματευτής (business representative) as the agent tasked with improving diplomatic relations with the neighbouring *toparches* (local governor): *On Skirmishing*, 7, pp. 162–3.
9. W. E. Kaegi (1986), 'The Frontier: Barrier or Bridge?', in G. Vikan (ed.), *17th International Byzantine Congress. Major Papers*, New York: A. D. Caratzas, pp. 281–2, 284 n. 11, 293.
10. J. Lassner (1970), *The Topography of Baghdad in the Early Middle Ages: Text and Studies*, Detroit, MI: Wayne State University Press, pp. 58 and 61.
11. *Codex Iustinianus*, 4.63.4.
12. M. Canard (1936), 'Une lettre de Muhammad ibn Tugj al-Ihsid émir d'Égypte à l'empereur romain Lecapene', *Annales de l'Institut d'études orientales de la Faculté des lettres d'Alger*, 2, pp. 189–209.
13. E. Chrysos (1993), 'Η Βυζαντινή διπλωματία ως μέσο επικοινωνίας' [Byzantine Diplomacy as a Means of Communication] in N. Moschonas (ed.), *Η Επικοινωνία στο Βυζάντιο* [Communication in Byzantium], Athens: National Research Foundation, p. 401.
14. For naval intelligence-gathering in Byzantium and the Muslim world, see V. Christides (1997), 'Military Intelligence in Arabo-Byzantine Naval Warfare', in K. Tsiknakes (ed.), *Το Εμπόλεμο Βυζάντιο, (9ος–12ος αι.)* [Byzantium at War, (9th–12th c.)], Athens: National Research Foundation, pp. 269–81; Christides, 'Two Parallel Naval Guides, pp. 52–103, especially pp. 90–2.
15. Ibn Hawqal, *Configuration de la terre*, I, p. 193.
16. Abu Yusuf Yaʿqub (1979), *Kitab al-kharaj* [Islamic Revenue Code], trans. A. A. Ali, Lahore: Islamic Book Centre, pp. 386–7.
17. Christides, *Conquest of Crete*, p. 164; Christides, 'Two Parallel Naval Guides', p. 82.
18. He is identified by the author of the *De Ceremoniis* as άρχων της Κύπρου (*archon tes Kyprou*), a term which could denote the lord magistrate of the

island of Cyprus and was, for centuries, synonymous with *megistanas* and *dynatos* (the powerful): see *Oxford Dictionary of Byzantium*, I, p. 160.
19. *De Ceremoniis*, p. 657.
20. Malaterra, *Deeds*, 4.2.
21. Kekaumenus, p. 34.
22. Aeneas Tacticus, *On Defence against Siege*, p. 53.
23. Aeneas Tacticus, *On Defence against Siege*, p. 55.
24. F. S. Russel (1999), *Information Gathering in Classical Greece*, Ann Arbor: University of Michigan Press, pp. 182–6; Richmond, 'Spies in Ancient Greece', pp. 6–10.
25. Bonner, *Jihad in Islamic History*, pp. 84–94; Fahmy, *Muslim Naval Organisation*, p. 18.
26. Koutrakou, 'Spies of Towns', p. 262. See also E. Christophilopoulou (1935), *Το Επαρχικόν Βιβλίον του Λέοντος του Σοφού και αι Συντεχνίαι εν Βυζαντίω* [The Book of the Eparch by Leo the Wise and the Guilds in Byzantium], Athens: Pournara Publications.
27. Canard, 'Relations politiques', p. 52.
28. Fahmy, *Muslim Naval Organisation*, p. 18; Christides, *Conquest of Crete*, p. 164.
29. Leo VI, *Novellae Constitutiones*, p. 107, cols 561–4; M. Hamidullah (1961), *Muslim Conduct of State*, Lahore: Islamic Book Centre, pp. 313, 316–19; Christides, 'Two Parallel Naval Guides', p. 91; Christides, 'Military Intelligence', pp. 269–71.
30. O. R. Constable (2003), *Housing the Stranger in the Mediterranean World: Lodging, Trade, and Travel in Late Antiquity and the Middle Ages*, Cambridge: Cambridge University Press, pp. 40–67.
31. B. Kumin (1999), 'Useful to Have, but Impossible to Govern: Inns and Taverns in Early Modern Bern and Vaud', *Journal of Early Modern History*, 3, pp. 153–203. See also A. Tlusty (2002), 'The Public House and Military Culture in Germany, 1500–1648', in A. Tlusty and B. Kumin (eds), *The World of the Tavern: Public Houses in Early Modern Europe*, Aldershot: Ashgate, pp. 136–59.
32. Aeneas Tacticus, *On Defence against Siege*, p. 55.
33. Walter, 'Urban Espionage', pp. 136–7.
34. *Vita St-Pauli Junioris* (1892), ed. H. Delehaye, *Annalecta Bollandiana*, 11, pp. 19–74 and 136–81, especially p. 72; N. Koutrakou (1995), 'Diplomacy and Espionage: Their Role in Byzantine Foreign Relations, 8th–10th Centuries', *Graeco–Arabica*, 6, p. 130.
35. C. Mango (1985), 'On the Re-reading of the Life of St. Gregory the Decapolite', *Βυζαντινά*, 13, p. 637. I did not have the chance to go through F. Dvornik (1926), *La vie de Saint Grégoire le décapolite et les slaves macédoniens au IXe siècle*, Paris: Champion, p. 58.
36. C. H. Talbot (1954), *The Anglo–Saxon Missionaries in Germany, Being the Lives of SS. Willibrord, Boniface, Leoba and Lebuin Together with the*

Hodoepericon of St. Willibald and a Selection from the Correspondence of St. Boniface, London: Sheed and Ward, p. 162.

37. Koutrakou, 'Spies of Towns', pp. 263–4; Koutrakou, 'Diplomacy and Espionage', pp. 129–31.
38. *DAI*, 46, pp. 216–18.
39. *Theophanes Continuates*, p. 119; J. Rosser (1976), 'John the Grammarian's Embassy to Baghdad and the Recall of Manuel', *Byzantinoslavica*, 37, pp. 168–71.
40. Procopius, *De Bello Gothico*, vol. 2, IV.17, pp. 546–7.
41. Theophanes Confessor, *Chronicle*, p. 594.
42. Koutrakou, 'Diplomacy and Espionage', pp. 137–8.
43. Richmond, 'Spies in Ancient Greece', pp. 3–4.
44. Caesar (1917), *The Gallic Wars*, trans. by H. J. Edwards, London: Heinemann, IV.20–2, pp. 205–13.
45. N. M. El Cheikh (2004), *Byzantium Viewed by the Arabs*, Cambridge, MA: Harvard University Press, pp. 5–13; C. Galatariotou (1993), 'Travel and Perception in Byzantium', *Dumbarton Oaks Papers*, 47, pp. 221–41.
46. M. A. al-Malik Farra (1981), *A Critical Edition of 'Kitab Al-Buldan' by Al Ya'qubi, Ahmad Ibn Abi Ya'qub (Ishaq) B. Ja'far B. Wahb B. Wadih Al-Kitab Al-Abbasi*, London: British Library Research and Development Department.
47. A. Ibn Rustah (1892), *Kitāb al-aᶜlāk an-nafīsa*, ed. M. J. de Goeje, Bibliotheca Geographorum Arabicorum 7, Leiden: Brill.
48. Vasiliev, *Byzance et les Arabes*, 2.II, pp. 395–409.
49. Ibn Khurradadhbeh (1889), *Kitāb al-masālik wa'l-mamālik*, ed. M. J. de Goeje, Bibliotheca Geographorum Arabicorum 6, Leiden: Brill. I was able to find the translation of this particular extract from Ibn Khurradadhbeh's work in Haldon, *Warfare, State and Society*, p. 110.
50. S. E. Bonebakker (1986), 'Kudāma', in *Encyclopedia of Islam*, vol. 5, Leiden: Brill, pp. 318–22; Qudama ibn Jaᶜfar (1889), *Kitab al-kharaj*, ed. M. J. de Goeje, Bibliotheca Geographorum Arabicorum 6, Leiden: Brill, pp. 196–9, 256.
51. E. W. Brooks (1901), 'Arabic Lists of the Byzantine Themes', *Journal of Hellenic Studies*, 21, p. 74.
52. For al-Jarmi, see al-Masᶜudi's *Kitab al-tanbih wa'l-ashraf*, his second work finished shortly before his death (in Brooks, 'Arabic Lists', p. 70).
53. W. Treadgold (1983), 'Remarks on the Work of al-Jarmī on Byzantium', *Byzantinoslavica*, 44, pp. 211–12.
54. Ibn Hawqal, *Configuration de la terre*, p. xiv.
55. Ibn Hawqal, *Configuration de la terre*, p. ix.
56. Ibid., pp. 190–2. He uses the term *pharsang*, which is a historical Persian unit of itinerant distance comparable to the European league (1 *pharsang* equals about 5 kilometres).
57. Ibid., p. 196.

58. Ibid., pp. 194–5.
59. J. Shepard (1992), 'Byzantine Diplomacy, A.D. 800–1204: Means and Ends', in J. Shepard and S. Franklin (eds), *Byzantine Diplomacy*, Aldershot: Ashgate, pp. 41–71; A. Kazhdan (1992), 'The Notion of Byzantine Diplomacy', in J. Shepard and S. Franklin (eds), *Byzantine Diplomacy*, Aldershot: Ashgate, pp. 3–21.
60. H. Kennedy (2004), 'Byzantine–Arab Diplomacy in the Near East from the Islamic Conquests to the Mid Eleventh Century', in M. Bonner (ed.), *Arab–Byzantine Relations in Early Islamic Times*, Aldershot: Ashgate, pp. 81, 90.
61. Kennedy, 'Byzantine–Arab Diplomacy', p. 82.
62. Chrysos has identified seven different types of embassies; see Chrysos, 'Η Βυζαντινή διπλωματία ως μέσο επικοινωνίας', pp. 399–407.
63. Bonner, *Jihad in Islamic History*, pp. 84–93; M. Canard (1949–51), 'Deux épisodes des relations diplomatiques Arabo–Byzantines au Xe siècle', *Bulletin d'études orientales*, 13, pp. 51–69, especially 52–4; Canard, 'Relations politiques', pp. 36–9.
64. The embassy of Liutprand of Cremona is an obvious exception to this set of rules.
65. *De Ceremoniis*, pp. 570–93.
66. Canard, 'Relations politiques', p. 40; Vasiliev, *Byzance et les Arabes*, 2.II, pp. 295–6.
67. Lat. *metatum*, denoted a temporary lodge in the countryside used by soldiers, couriers and/or civil servants in transit; see *Oxford Dictionary of Byzantium*, II, p. 1385. Today, this term refers to a shepherd's lodge in areas of the southern Peloponnese, the Cyclades and Crete.
68. S. Patoura-Spanou (2005), 'Όψεις της Βυζαντινής διπλωματίας' [Facets of Byzantine Diplomacy], in S. Patoura-Spanou (ed.), *Διπλωματία και Πολιτική, Ιστορικές Προσεγγίσεις* [Diplomacy and Politics, Historical Approaches], Athens: National Research Institute, p. 151.
69. Kekaumenus, p. 13.
70. Aeneas Tacticus, *On Defence against Siege*, pp. 54–6.
71. *On Strategy*, 43, p. 124.
72. Σέξτος Ιούλιος Φροντίνος [Sextus Julius Frontinus] (2015), *Στρατηγήματα* [*Strategemata*], ed. G. Theotokis, trans. V. Pappas, Athens: Hellenic Army Press, I.i.3, p. 8. See also I.ii.1, pp. 17–19.
73. On the qualities and skills of Byzantine diplomats, see D. Nerlich (1999), *Diplomatische Gesandtschaften zwischen Ost- und Westkaisern 756–1002*, Bern: P. Lang, pp. 107–21; N. Drocourt (2008), 'La diplomatie médio-byzantine et l'antiquité', *Anabases*, 7, pp. 57–87.
74. *DAI*, 1, p. 48.
75. *DAI*, pp. 10–12. See also T. Lounghis (1990), *Κωνσταντίνου Ζ' Πορφυρογέννητου De Administrando Imperio (Προς τον ίδιον υιόν Ρωμανόν). Μια μέθοδος ανάγνωσης* [Constantine VII Porphyrogenitus' *De Administrando*

Imperio (To My Own Son Romanus): A Reading Method], Thessaloniki: Vanias; Shepard, 'Imperial Information and Ignorance', pp. 107–16.
76. See chapters 9 and 31 of the *DAI* respectively.
77. Amatus, *History of the Normans*, V.24, p. 142.
78. Anna Comnena, *The Alexiad*, XIII.ix, p. 244.
79. Leo VI, *Taktika*, XVII.91, p. 432. Cf. *Strategikon*, IX.2, p. 95.
80. Frontinus, Στρατηγήματα [*Strategemata*], III.xv.1–2, 4, pp. 250–2. Cf. Onasander, *Strategikos*, X.3, p. 417; Leo VI, *Taktika*, XX.87, p. 566.
81. Theophanes Confessor, *Chronicle*, pp. 654–5.
82. Ibid., p. 534.
83. Ibn Idhari in Vasiliev, *Byzance et les Arabes*, 2.II, pp. 218–19. See also Vasiliev, *Byzance et les Arabes*, 2.I, p. 331.
84. *Theophanes Continuates*, pp. 49–57. Lemerle and Kyriakes believe that the story was later tampered with for reasons of imperial propaganda, and they suggest that it was Michael who had contacted Omurtag first, as the history of *Georgius Monachus* makes clear: see *Georgius Monachus* in *Theophanes Continuates*, p. 788; P. Lemerle (1965), 'Thomas le slave', *Travaux et mémoires*, 1, pp. 255–97, especially pp. 279–81; Κυριάκης, Βυζάντιο και Βούλγαροι, pp. 125–6; Koutrakou, 'Diplomacy and Espionage', p. 140 n. 56.
85. Vasiliev, *Byzance et les Arabes*, 2.I, pp. 190–1; G. Kolias (1939), *Léon Choerosphactès, magistre, proconsul et patrice*, Athens: Verlag der 'Byzantinisch-neugriechischen jahrbücher', pp. 54–5; P. Magdalino (2004), 'In Search of the Byzantine Courtier: Leo Choirosphaktes and Constantine Manasses', in H. Maguire (ed.), *Byzantine Court Culture from 829 to 1204*, Washington, DC: Dumbarton Oaks, p. 147 n. 36.
86. C. Angelidi (2013), 'Designing Receptions in the Palace (*De Cerimoniis* 2.15)', in A. Beihammer, S. Constantinou and M. Parani (eds), *Court Ceremonies and Rituals of Power in Byzantium and the Medieval Mediterranean, Comparative Perspectives*, Leiden: Brill, pp. 465–86; L. V. Simeonova (1998), 'In the Depths of Tenth-Century Byzantine Ceremonial: The Treatment of Arab Prisoners of War at Imperial Banquets', *Byzantine and Modern Greek Studies*, 22, pp. 75–104. The classic work on the topic is N. Oikonomides (1972), *Les listes de préséance byzantines des neuvième et dixième siècles*, Paris: Ed. du Centre national de la recherche scientifique.
87. El Cheikh, *Byzantium Viewed by the Arabs*, pp. 152–62; Canard, 'Le cérémonial fatimite', pp. 356–7.
88. J. B. Bury (1958), *The Imperial Administrative System in the Ninth Century: With a Revised Text of the 'Kletorologion' of Philotheos*, New York: Burt Franklin. For dealing with the reception of foreign embassies, see *De Ceremoniis*, pp. 393–410.
89. Bury, *Imperial Administrative System*, pp. 155–6.
90. Ibid., pp. 157–8.
91. Al-Harawi, X, pp. 224–5; Canard, 'Relations politiques', p. 40.

92. *Theophanes Continuates*, pp. 95–6.
93. Mas'udi (1967), *al-Tanbih wa'l-ishraf*, 2nd edn, Baghdad, p. 261.
94. 'That you should treat him [the envoy] in a nice way, make him welcome, have extended interviews with him, multiply the questions on all things, ask about the situation of his master and revenues from its territory. Excite him somewhat with harsh words, as this reveals the bottom of the heart and shows his hidden thoughts.' See al-Harawi on the 'secret interrogation' of an ambassador by the prince: XI, p. 225. On the embassy to al-Mu'izz, see S. M. Stern (1950), 'An Embassy of the Byzantine Emperor to the Fatimid Caliph Al-Mu'izz', *Byzantion*, 20, pp. 239–53. Cf. Liudprand of Cremona's banquet conversation with Nicephorus Phocas: Liudprand of Cremona (1993), *The Embassy to Constantinople and Other Writings*, ed. J. J. Norwich, London: J. M. Dent, especially ch. 11, p. 182. See also N. Koutrakou (1995), '*Logos* and *Pathos* between Peace and War: Rhetoric as a Tool of Diplomacy in the Middle Byzantine Period', *Thesaurismata*, 25, pp. 7–20, especially pp. 9–10.
95. The ambassador reached the conclusion that the vizier was a Rafidi, based on the ring on his finger. Rafidi is used in a derogatory way to identify Shi'a Muslims and those who, in general, reject legitimate Islamic authority and leadership; see Nizam al-Mulk, *Book of Government*, p. 97.
96. G. le Strange (1897), 'A Greek Embassy to Baghdad in 917 A.D.', *Journal of the Royal Asiatic Society of Great Britain and Ireland*, pp. 35–45; Kennedy, 'Byzantine–Arab Diplomacy', p. 89. Cf. the vivid description of the imperial palace and throne by Liudprand, who came to Constantine on an embassy from Berengar II in 949, in his *Antapodosis*: Liudprand of Cremona, *Embassy to Constantinople*, VI.5, p. 153. See also the reception of Olga, the Queen of the Rus', in 957 as described in the *Russian Primary Chronicle*, pp. 82–3.
97. Le Strange, 'Greek Embassy to Baghdad', pp. 41–2.
98. *On Tactics*, 18, p. 292.
99. For a detailed presentation of the treatment of prisoners of war through the writings of the tacticians of the Early and Middle Byzantine periods, see D. Letsios (1992), 'Die Kriegsgefangenschaft nach Auffassung der Byzantiner', *Byzantinoslavica*, 53, pp. 213–27.
100. *Strategikon*, VII.5, p. 97.
101. *On Tactics*, 18, p. 292.
102. Kennedy, 'Byzantine–Arab Diplomacy', pp. 87–8.
103. Patoura, *Αιχμάλωτοι*, pp. 22–4; M. Campagnolo-Pothitou (1995), 'Les échanges de prisonniers entre Byzance et l'Islam aux IXe et Xe siècles', *Journal of Oriental and African Studies*, 7, pp. 1–56. For a detailed examination of the political and religious laws regarding the treatment of prisoners in the Islamic world, see M. Khadduri (1955), *War and Peace in the Law of Islam*, Baltimore, MD: Johns Hopkins University Press; R. A. Khouri al Odetallah (1983), 'Άραβες και Βυζαντινοί. Τό Πρόβλημα τῶν Αἰχμαλώτων

Πολέμου' [Arabs and Byzantines: The Problem of Prisoners of War], PhD thesis, Aristotle University of Thessaloniki, pp. 23–7; M. Munir (2010), 'Debates on the Rights of Prisoners of War in Islamic Law', *Islamic Studies*, 49, pp. 463–92, especially pp. 485–7 and the detailed bibliography provided in n. 1. I have not had the chance to go through M. Lykaki (2016), 'Οι Αιχμάλωτοι Πολέμου στη Βυζαντινή Αυτοκρατορία (6ος–11ος αι.): Εκκλησία, Κράτος, Διπλωματία και Κοινωνική Διάσταση' [Prisoners of War in the Byzantine Empire (6th–11th c.): Church, State, Diplomacy and Social Dimensions], unpublished PhD thesis, University of Athens.

104. Vasiliev, *Byzance et les Arabes*, 2.I, pp. 222–40.
105. Toynbee, *Constantine Porphyrogenitus*, pp. 382–3; Simeonova, 'Byzantine Ceremonial', pp. 79–80.
106. 'Do nothing unholy, shameful, or inhumane to the prisoners. If you have not been treated unjustly, do not act unjustly' (Leo VI, *Taktika*, XIX.39, p. 518). See also the letter written by Nicolaos Mysticus to Caliph al-Muqtadir in July 922: *Letters*, 102, p. 376. Cf. what the Prophet is reported to have said about the treatment of prisoners of war: Munir, 'Debates', pp. 485–7.
107. A *kleisourarch* was the commander of a *kleisoura*.
108. Vasiliev, *Byzance et les Arabes*, 1, pp. 296–7; Bury, 'Mutasim's March', p. 123.
109. *Theophanes Continuates*, p. 181; Huxley, 'Bishop's Meadow', p. 445.
110. *History of Theophylact Simocatta*, VII.6, p. 185.
111. *Strategikon*, IX.3, p. 97; Onasander, *Strategikos*, p. 421; *On Strategy*, 41, p. 120; Leo VI, *Taktika*, XVII.92, p. 432; Kekaumenus, pp. 10–11. To my knowledge, the oldest example of prisoners of war deliberately sent to be caught and, thus, mislead the enemy army with false information comes from the Battle of Kadesh (1274 BC), fought between the Egyptian forces of Ramses II and the Hittite king Muwattalli II. As Ramses and the Egyptian advance guard were about 11 kilometres from Kadesh, they captured two nomads: 'When they [the nomads] had been brought before the Pharaoh, His Majesty asked, "Who are you?" They replied "We belong to the king of Hatti. He has sent us to spy on you." Then His Majesty said to them, "Where is he, the enemy from Hatti? I had heard that he was in the land of Khaleb, north of." They of Tunip replied to His Majesty, "Lo, the king of Hatti has already arrived, together with the many countries that are supporting him . . . They are armed with their infantry and their chariots. They have their weapons of war at the ready. They are more numerous than the grains of sand on the beach. Behold, they stand equipped and ready for battle behind the old city of Kadesh."' J. Tyldesley (2000), *Ramesses II: Egypt's Greatest Pharaoh*, London: Penguin, pp. 70–1.
112. Tabari in Vasiliev, *Byzance et les Arabes*, 2.II, pp. 11–12. See also Haldon, 'Strategies of Defence', pp. 149–62.
113. John Skylitzes, *Synopsis of Byzantine History*, pp. 233–4.

114. Armenian princelings – as mentioned in the 'Armenian chapters' of the *DAI* and Symeon of Bulgaria – are some characteristic examples of this case; see Shepard, 'Byzantine Diplomacy', pp. 59–60.
115. Canard, 'Relations politiques', p. 46; al Odetallah, 'Άραβες και Βυζαντινοί', pp. 81–2; Simeonova, 'Byzantine Ceremonial', p. 90. For the participation of 'Agarene' prisoners from the *praetoriun* at the Christmas banquet, see *De Ceremoniis*, p. 743.
116. Al Odetallah, 'Άραβες και Βυζαντινοί', pp. 103–4.
117. Abu Firas' extracts for the wars between 950–8 can be found in Vasiliev, *Byzance et les Arabes*, 2.II, pp. 349–71. For his captivity in the capital, see al Odetallah, 'Άραβες και Βυζαντινοί', pp. 103–6; Patoura, *Αιχμάλωτοι*, pp. 95–7. Cf. the dinner dialogue of Phocas with Liudprand of Cremona: *Embassy to Constantinople*, ch. 11, p. 182.
118. Canard, 'Relations politiques', p. 46; M. Canard and N. Adontz (1936), 'Quelques noms des personnages byzantins dans une pièce du poète Abû-Firâs', *Byzantion*, 11, pp. 451–60.
119. Vasiliev, *Byzance et les Arabes*, 2.II, pp. 380–9; A. A. Vasiliev (1932), 'Härün-ibn Yahya and his Description of Constantinople', *Seminarium Kondakovianum*, 5, pp. 149–63.
120. Patoura, *Αιχμάλωτοι*, p. 104, n. 74.
121. 'It is a large city twelve parasangs by twelve, the parasang is, what they say, of a mile and a half. It is surrounded by the sea from the side of the East; from the West Coast, it extends the land through which passes the road of Rome. It has a fortified enclosure'; see Vasiliev, *Byzance et les Arabes*, 2.II, p. 383. Cf. Odo of Deuil's description of Constantinople, its buildings and fortifications, some two and a half centuries later (October 1147). Ironically, Odo considers that the Land Wall of Constantinople 'is not especially strong, and the towers are not very high, but the city trusts, I think, in its large population and in its ancient peace'. Odo of Deuil (1949), *La croisade de Louis VII, roi de France*, 4 vols, ed. H. Waquet, Paris: Académie des inscriptions et belles-lettres, vol. 3, pp. 44–6.
122. Vasiliev, *Byzance et les Arabes*, 2.II, p. 384.
123. Vasiliev, *Byzance et les Arabes*, 2.II, pp. 384–5; *De Ceremoniis*, p. 592; Muqaddasi in Vasiliev, *Byzance et les Arabes*, 2.II, pp. 422–3; El Cheikh, *Byzantium Viewed by the Arabs*, pp. 144–5; Patoura, *Αιχμάλωτοι*, p. 106.
124. Vasiliev, *Byzance et les Arabes*, 2.II, pp. 384–5; *De Ceremoniis*, II.15, p. 592.
125. Vasiliev, *Byzance et les Arabes*, 2.II, p. 391 and n. 5.
126. Simeonova, 'Byzantine Ceremonial', p. 82; Patoura, *Αιχμάλωτοι*, pp. 107–8; El Cheikh, *Byzantium Viewed by the Arabs*, pp. 154–9.
127. *De Ceremoniis*, pp. 694–5. For a useful comparison with the integration of outsiders in Islamic society, see E. Landau-Tasseron (2006), 'The Status of Allies in Pre-Islamic and Early Islamic Arabian Society', *Islamic Law and Society*, 13, pp. 6–32.

128. I. Shahid (1991), 'Ghassān', *Encyclopedia of Islam*, II: C–G. Leiden: Brill, pp. 462–3.
129. D. Nicolle (1997), 'Arms of the Umayyad Era: Military Technology in a Time of Change', in Y. Lev (ed.), *War and Society in the Eastern Mediterranean, 7th–15th Centuries*, Leiden: Brill, pp. 9–100; Trombley, 'Military Cadres', pp. 253–8. Crone argues that the trade in leather and other pastoralist products, a tradition which she ascribes to the Meccans, could make sense on the assumption that the goods were destined for the Roman army, which is known to have required colossal quantities of leather and hides for its equipment; Crone, 'Quraysh and the Roman Army', pp. 63–88.
130. *Oxford Dictionary of Byzantium*, p. 1297; M. Moosa (1969), 'The Relation of the Maronites of Lebanon to the Mardaites and Al-Jarājima', *Speculum*, 44, pp. 597–608.
131. Treadgold, *Byzantine State and Society*, pp. 327–32; Haldon, *Byzantium in the Seventh Century*, pp. 70–3; Charanis, 'Transfer of Population', pp. 140–54.
132. Asa Eger, *Islamic–Byzantine Frontier*, p. 291. See also F. Husayn (2012), 'The Participation of Non-Arab Elements in the Umayyad Army and Administration', in F. Donner (ed.), *The Articulation of Early Islamic State Structures*, London: Routledge, pp. 279–80.
133. Theophanes Confessor, *Chronicle*, pp. 511 and 560–1; Koutrakou, 'Spies of Towns', p. 255; Charanis, 'Transfer of Population', p. 143.
134. E. Venetis (2005), 'Ḵorramis in Byzantium', *Encyclopaedia Iranica*, available at: www.iranicaonline.org/articles/korramis-in-byzantium (accessed 26 June 2018).
135. The *tagma* was a professional military unit of the Byzantine army, established during Constantine V's reign (741–75). *Theophanes Continuates*, p. 112; J.-C. Cheynet (1998), 'Theophile Thèophobe et les Perses', in S. Lampakis (ed.), *Byzantine Asia Minor*, Athens: National Research Institute, pp. 39–50.
136. Treadgold, *Byzantine State and Society*, pp. 441–3; *Oxford Dictionary of Byzantium*, pp. 2067–8; I. S. Kiapidou (2003), 'Battle of Dazimon, 838', *Online Encyclopaedia of the Hellenic World, Asia Minor*, Athens: Foundation of the Hellenic World.
137. Canard, 'Relations politiques', p. 43; Haldon, *Warfare, State and Society*, pp. 117–28; Nicolle, *Crusader Warfare*, I, pp. 165–7.
138. Cappel, 'Byzantine Response to the Arab', pp. 113–32; J. L. Krawczyk (1985), 'The Relationship between Pastoral Nomadism and Agriculture: Northern Syria and the Jazira in the Eleventh Century', *JUSUR*, 1, pp. 1–22; J. F. Haldon and H. Kennedy (1980), 'The Arab–Byzantine Frontier in the Eighth and Ninth Centuries: Military Organization and Society in the Borderlands', *Zbornik Radova Vizantoloski Institut*, 19, pp. 115–16; F. von Sievers (1979), 'Military, Merchants and Nomads: The Social Evolution of the Syrian Cities and Countryside during the Classical Period, 780–969/164–358', *Der Islam*, 56, pp. 230–3.

139. Ibn Zafir in Vasiliev, *Byzance et les Arabes*, 2.II, pp. 120–1; Kennedy, *The Prophet*, p. 268.
140. Vasiliev, *Byzance et les Arabes*, 2.II, p. 420.
141. According to Treadgold, they may have been divided into five units of 2,400 each to be distributed to the aforementioned new themes; W. Treadgold (1992), 'The Army in the Works of Constantine Porphyrogenitus', *Rivista di Studi Bizantini e Neoellenici*, 29, pp. 128–30; Treadgold, *Byzantium and its Army*, pp. 111–12.
142. *On Skirmishing*, 7, p. 162; *Praecepta Militaria*, II.101–11, p. 28; *On Tactics*, 10, p. 280.
143. *Oxford Dictionary of Byzantium*, p. 96.
144. *Theophanes Continuates*, pp. 366–8; al-Masᶜudi in Vasiliev, *Byzance et les Arabes*, 2.II, p. 38; Vasiliev, *Byzance et les Arabes*, 2.I, pp. 163–81; *Oxford Dictionary of Byzantium*, II, p. 1216.
145. A. Metcalfe (2009), *The Muslims of Medieval Italy*, Edinburgh: Edinburgh University Press, pp. 10–12.
146. Anna Comnena, *The Alexiad*, XIII.iv, pp. 235–7. Cf. Vegetius, *Epitome of Military Science*, III.9, p. 84–5.
147. *Strategikon*, VIII.1/11, 28, pp. 80–81. Cf. Frontinus, Στρατηγήματα [*Strategemata*], I.ii.3, p. 18.
148. Anna Comnena, *The Alexiad*, XIII.iv, pp. 235–6.
149. *Strategikon*, VII.1/15, p. 69.
150. *DAI*, 43, p. 188.
151. A. Cameron (1970), *Agathias*, Oxford: Oxford University Press, p. 39; A. Cameron (1969), 'Agathias on the Sasanians', *Dumbarton Oaks Papers*, 23, pp. 67–183.
152. See the detailed footnote which includes several examples from the 970s to 990s in A. Beihammer (2012), 'Strategies of Diplomacy and Ambassadors in Byzantine–Muslim Relations in the Tenth and Eleventh Centuries', in A. Becker and N. Drocourt (eds), *Ambassadeurs et ambassades au coeur des relations diplomatiques: Rome, Occident médiéval, Byzance, VIIIe s. avant J.-C. – XIIe s. après J.-C*, Metz: Centre de recherche universitaire Lorrain d'histoire, Université de Lorraine, pp. 383–4, n. 42.
153. R. Janin (1936), 'Un ministre arabe à Byzance: Samonas', *Echos d'Orient*, 36, pp. 307–18; L. Rydén (1984), 'The Portrait of the Arab Samonas in Byzantine Literature', *Graeco–Arabica*, 3, pp. 101–8. For the alleged episode of Samonas' escape from Constantinople to his home town of Melitene in 904, see R. J. Jenkins (1948), 'The "Flight" of Samonas', *Speculum*, 23, pp. 217–35.
154. Jenkins, 'The "Flight" of Samonas', pp. 218–22.
155. *Vita Euthymii* (1888), ed. C. de Boor, Berlin, p. 26.
156. *On Tactics*, 18, p. 292.
157. A term, not much different from κατασκοποι, used to define the different agents gathering intelligence.

158. *On Strategy*, 42, p. 122. Cf. Kekaumenus, pp. 5, 9.
159. Walter, 'Urban Espionage', pp. 138–9.
160. *On Strategy*, 42, p. 122.
161. Leo VI, *Taktika*, I.7, p. 12.
162. Vegetius, *Epitome of Military Science*, III.26, p. 118; *Strategikon*, IX.5, p. 105; Leo VI, *Taktika*, XVII.89, p. 430.
163. *Strategikon*, VIII.I/8, p. 80; 2/72, p. 88; Leo VI, *Taktika*, XX.8, p. 540. Leo VI also urges the general not to lead his troops through inhabited areas, especially if his army is rather small, 'to avoid being observed by enemy spies' (IX.21, p. 160).
164. For purposes of comparison, see the detailed and highly accurate study by E. S. Gürkan (2012), 'The Efficacy of Ottoman Counter-intelligence in the 16th Century', *Acta Orientalia Academiae Scientiarum Hungaricae*, 65, pp. 1–38.
165. *Strategikon*, VIII.I/41, p. 82.
166. *Strategikon*, VIII.I/11, p. 80.
167. Sun-Tzu, *Art of War*, p. 146.
168. Leo VI, *Taktika*, XX.15, p. 542.

7

Tactical Changes in the Byzantine Armies of the Tenth Century: Theory and Practice on the Battlefields of the East

This chapter will focus on the tactical changes that took place in the units of the Byzantine army in the tenth century. The most useful primary sources for identifying these changes are the following military treatises: the Στρατηγική Έκθεσις Και Σύνταξις Νικηφόρου Δεσπότου (*Praecepta Militaria* of the Emperor Nicephorus Phocas) (c. 969),[1] the anonymous *Sylloge Taktikorum* (c. 930),[2] a short tenth-century work entitled Σύνταξις οπλιτών τετράγωνος έχουσα εντός καβαλλαρίους (better known by its Latin title: *Syntaxis Armatorum Quadrata*)[3] and the c. 969 Περί παραδρομής του κυρού Νικηφόρου του βασιλέως (better known by the title bestowed on it by its 1985 editor, Dennis: *On Skirmishing*).[4] These works provide crucial information on how armies should be organised and deployed on the battlefield up to the period when they were compiled.

I will discuss the recommendations of the authors of the treatises regarding the marching, battle formations, armament and battlefield tactics of the Byzantine army units and I will ask whether they reflect any kind of innovation or tactical adaptation to the strategic situation in the East. The main point that I wish to raise in this section of the study concentrates on how far we can say that 'theory translated into practice' on the battlefields of the period at Hadath (954), Tarsus (965), Dorystolon (971), Alexandretta (971), Orontes (994) and Apamea (998). How successful were the Byzantines at adapting to the changing military threats posed by their enemies in the East, according to the evidence we can discern from careful study of the writings of the tacticians of the period?

Tactical Changes: The Double-ribbed Hollow Infantry Square

Beginning his discussion on the infantry corps and the necessary equipment that should be borne by the στρατιώτας (soldiers), the author of the Στρατηγική Έκθεσις (infra *Praecepta Militaria*) proceeds with the brief but nevertheless detailed description of their battle formation:

Tactical Changes: Theory and Practice

The formation of the aforementioned foot soldiers is to be a double (διττή) square, called by the ancients a four-sided formation, which has on each side three units (παραταγάς), so that all four sides have twelve units in total. They must be set apart as much as to allow between twelve and fifteen cavalry men to go in and out [of the square].[5]

Hence, the infantry was supposed to march into battle in twelve taxiarchies of a thousand men each, forming a square that had three units on either side with a specific number of intervals – either two or three depending on the numbers of cavalry and infantry units[6] – to allow for the unencumbered access in and out of the square of the cavalry units fighting alongside them. The *Praecepta*, however, is not the only text that mentions this specific battle formation. Chapter 47 of the *Sylloge Taktikorum* mentions an infantry square punctuated by eight to twelve intervals in total, with the author probably being familiar with the text of the *Syntaxis Armatorum Quadrata* and copying from it the diagram of the square.[7] In fact, the author of the *Sylloge Taktikorum* goes into such detail as to calculate the exact measurements of the openings of the intervals and the total manpower for each one of the infantry units. The text provides us with a more detailed – more encyclopaedic than practical – description of the square adopted by Nicephorus, and although this certainly preceded the version incorporated in the *Praecepta*, we can say that the latter is a more refined and realistic approach to the realities of warfare in the tenth century. Is it, however, possible to trace any innovation in these two descriptions of the Byzantine infantry's formation of the tenth century? If so, how does this relate to the battlefield tactics of their enemies?

A quadrilateral formation was not something new, as square formations were adopted by both the ancient Greeks and the Romans when pitching camp, marching through hostile terrain, when fearing encirclement and in cases of emergency in general.[8] It is to be noted that this hollow square with regular intervals on all sides that worked as both a base and a place of refuge and regroupment for the cavalry during battle cannot be found in any of the military treatises prior to the tenth century – indeed, neither the *Strategikon* (c. 600) nor the *Taktika*[9] (c. 900) make any mention of this type of infantry square, before its detailed examination by the *Syntaxis Armatorum Quadrata*, the *Sylloge* and the *Praecepta*.

McGeer has suggested that this fairly recent formation of the Byzantine tacticians was inspired by the standard Byzantine ground plan of temporary military encampments that was most likely developed during the campaigns in Mesopotamia by the Domestic of the Scholae, John Curcuas, in the years 922–44.[10] Although the exact period for the development of this formation is hard to prove based on only the information found in the primary sources, the link between the plan for the Byzantine camp and Nicephorus' battle formation seems quite attractive. Indeed, according to the evidence from the *Taktika*, from as early as the sixth century, the preferred ground plan for Byzantine camps was a square or rectangle with two entrances on each side and four major roads dividing the camp into nine sectors. Other versions include three entrances on the east and west side and two on the north and south, with five roads dividing the camp into twelve sectors, depending on the number of taxiarchies the general had at his disposal, and the ratio between the infantry and cavalry units.[11]

A fundamental question to be asked, however, is how can we link the ground plan of a tenth-century camp to the battle formation of the Byzantine infantry. The answer lies in the writings of Nicephorus Phocas, in the chapter where he examines the encampment of the army:

> They [the soldiers] must keep their places in the camp exactly as they set to deploy in battle formation, so that, in the event of a sudden report of the enemy, they will be found ready as though in battle formation . . . Eight intervals must be left open in the army's encampment so that three chiliarchs (χιλίαρχοι)[12] have two intervals. These must be in the shape of a cross on the four sides of the encampment . . . two roads from the east to west and two from north to south.[13]

There are several advantages for any army in using a square formation: (1) the enemy cannot attack the camp from all sides without dividing their forces, thus leaving the soldiers inside feeling more secure that they will not be enveloped;[14] (2) a square can be seen as a place of refuge for any soldier or unit that has retreated from the battlefield and can withdraw inside the formation to regroup; (3) the square formation offers far fewer opportunities to a soldier who wishes to desert his comrades before or during battle. On account of these defensive advantages, it seems straightforward why the Byzantines preferred the square formation for their encampments and marching formations, especially in enemy territory, and they had adapted it to their battle formations by the middle of the tenth century.

Tactical Changes: The Position of the Cavalry in Mixed Formations

Other evidence pointing to a change in the tactical formations of mixed armies in the tenth century involves the position of the cavalry in the armies of the period and the role of the infantry in relation to that. Until this period, the classic formation of a mixed army set the infantry units at the centre – where the general also stood – while the cavalry took its place at the flanks of the entire formation.[15] The reason behind this was that the core troops of the army were deployed in the centre to receive the enemy attack and/or deliver the key counterattack that would decide the outcome of the battle, while the smaller but more flexible cavalry units on the sides would protect the flanks of the centre division – essentially the weakest points of a heavy infantry formation – and look for ways to envelop the enemy.

The *Taktika* also refers to what should happen in the event that the cavalry has to withdraw from the battlefield: 'They [the cavalry] should seek refuge to the rear or behind the battle line, but they should not go further than the wagons. If they still cannot hold out, they should dismount and defend themselves on foot.'[16] Thus, Leo advised his mounted troops to retire behind the infantry lines for cover and stay close to the wagons, and in a case of dire emergency that might involve the entire army being encircled, they should fight on foot, probably by forming a square or rectangle.

By the middle of the tenth century, however, the cavalry was positioned inside the square infantry formation – in essence a camp converted into a battle formation – manoeuvring through the intervals on the sides to open the battle. Therefore, the tactical initiative was retained by the cavalry, which remained the *force de frappe* in the hands of the Byzantine generals. 'If it should happen that the enemy hits our cavalry units hard and repels them, God forbid, they should retire inside the infantry formation for protection',[17] which clearly designates the square as a place of refuge for the cavalry units in case of needing to retreat from the battlefield, regroup or launch another attack.

The tactical manoeuvre of retiring the cavalry behind the infantry or close to an encampment for protection is not new.[18] However, it is testament to the ingenuity of the tacticians of the period in question that they combined centuries-old wisdom in military affairs to the circumstances of their time. This transformed the infantry formation into a sort of mobile base for the cavalry, while the latter still retained their follow-up role as troops that would deliver the knock-out blow, and secure the booty and prisoners.[19]

Tactical Changes: The Size and Formation of the Infantry Taxiarchy

Having examined the mixed formations, the author of the *Praecepta* proceeds with the study of the individual taxiarchies that formed the key elements of the infantry square, and the way the foot soldiers were deployed in the formation and in what numbers:

> The heavy infantry men should be deployed in a two-fold formation (αμφίστομος), putting two infantry men in the front [of the taxiarchy] and two in the back. And in between them there should be three light archers, so that the depth of the unit would be seven men.[20]

The immediate change that we notice in the text involves the depth of the taxiarchy, a significant development that ran against the advice of previous military manuals. Treatises like the *Strategikon* and the *Taktika* recommend twenty lines of foot soldiers that would include some sixteen spearmen supported by four rows of archers immediately behind them,[21] keeping a steady ratio between the heavy infantry and the archers of 4:1.[22]

The *Sylloge* follows up on the comments of the aforementioned authors in chapter 43, titled 'Infantry Formations According to Them [the Greeks and the Romans]': 'The depth of an infantry unit should not exceed sixteen rows (ορδίνοι), while it should not fall below seven.'[23] That is because in a unit that is deeper than sixteen men, the javeliners, the archers and the slingers would be impeded by the number of infantrymen directly in front of them. Similarly, if it has less than seven rows it does not have the necessary depth to withstand a simultaneous attack from the front and the flanks. The ideal depth of a phalanx in a mixed formation, however, according to the author of the *Sylloge*, is ten men, where the first and last four rows consist of infantrymen, thus forming a double-faced formation (αντίστομον), while between them there should be two rows of archers, slingers and javeliners mixed together.[24]

Following his description of the infantry taxiarchy, Phocas explains why he decided to deviate from the ancient dicta that wanted the infantry phalanx to be deeper; he seems familiar with the depth of the Macedonian phalanx – an exemplar formation for the ancient and Byzantine tacticians – and he admits that 'we do find the ancient Macedonians making their phalanx sixteen men deep, occasionally twelve or ten'.[25] Although precise rules regarding the depth of a Greek phalanx varied from period to period, armies usually formed their phalanx 'eight shields' deep, as it was common among the Spartans and the Athenians before the 370s BC, with the former increasing the depth to twelve men after the radical structural changes of the fourth century BC.[26] The Thebans habitually fought in very deep formations – depths of 25 to 50 shields can be seen in the sources. Otherwise, only forces which vastly outnumbered their opponents adopted average depths greater than eight, doubling the length of the file to sixteen shields.[27]

Even though Phocas' direct source remains a mystery, the organisation and deployment of the Macedonian phalanx on the battlefield is examined by a number of military treatises, from the authors of the *Taktika* and the *Sylloge Taktikorum* to Onasander, Aelian, Asclepiodotus and Arian, all of whom may have influenced the author of the *Praecepta Militaria* as they constituted the classical education of a Byzantine officer. It is possibly for this reason that Phocas feels the need to explain his decision to break with the well-established tradition of the Macedonian phalanx on the battlefield: 'In our own day, however, such formations are no longer employed and this type of phalanx is impractical. When compared with the wars of the ancients, even the offspring of Hagar have greatly reduced the depth of their formations.'[28] This last excerpt is a characteristic example of the author's realistic approach to warfare in the East in the tenth century.

Another important change that we notice both in the *Sylloge* and the *Praecepta* is the term 'double-faced' that refers not only to the mixed cavalry–infantry formation of the army, but also to the battle formation of each taxiarchy that formed the infantry square. A double-phalanx of heavy infantrymen facing the front and back of the formation was not an innovation of the period, as the author of the *Strategikon* affirms:

> Assuming that the files are sixteen men deep and hostile forces appear both in front and to their rear, if the enemy approaching the front is getting very close give the command: 'Divide in the middle. Form double phalanx.' The first eight men halt. The other eight face about and move back, thus forming a double phalanx.[29]

Our authors' insistence, however, in putting heavy infantrymen at the back of each formation as well as to the front, a 4–2–4 formation in the *Sylloge* and a 2–3–2 in the *Praecepta*, is a strong indication that this was a recent development based on the experience acquired in the wars of the mid-tenth century in the operational theatres in Cilicia and Mesopotamia. Furthermore, both manuals are careful to put experienced men in the first and last lines of the taxiarchy, as well as in the first and last file. The two pentecontarchs (commanders of forty-nine men) stood at the wings of each 100-man line; the first line consisted entirely of dekarchs (commanders of nine men) while the ουραγός (rearguard) was formed of pentarchs (commanders of four men) and tetrarchs (commanders of three men). These were experienced and reliable soldiers who could boost the morale and fighting abilities of the men in front of them and could deal effectively with any emergency coming from the rear of the formation.[30]

The *Sylloge Taktikorum* is the first of the military treatises of our period that refers to a new unit that was established around the beginning of the tenth century, the infantry corps of the *menavlatoi*. The term *menavlion*[31] has been identified as a heavy javelin or spear, designed for thrusting and not casting. It comes from the Latin *venabulum* (μέναυλον),[32] one of the three terms used to describe a heavy javelin – the other two were the *verutum* (βηρύττα)[33] and *martiobardulus* (μαρτζυβάρβουλον),[34] all of which are used by Vegetius and the author of the *Taktika*.[35] This particular type of spear was made of 'hard wood (oak, cornel) and just so thick as hands can wield them'.[36] Its shaft had a length of between eight to ten cubits, which according to Schilbach works out between 2.7 and 3.6 metres.[37] The operational role of this new unit is described in the *Sylloge Taktikorum*:

> There is the tagma of the so-called menavlatoi, numbering 300, all shield-bearers; first, they should be made to stand in the front ranks of the intervals [between the units]; when the enemies approach within a bow-shot range, they [*menavlatoi*] should proceed through the intervals and towards the enemy units and deploy either in a straight line or in a wedge formation as it has been said on chapter 46, at a distance of between 30 to 40 orgyai (οργυαί) [from the rest of the army]. Their purpose is to stab boldly the horses of the kataphrakts with their menavlia.[38]

This was a unit of men on foot – separate from the twelve main units of the heavy infantry and skirmishers – numbering some 300 men. It was deployed in the front ranks of the intervals between the infantry units that faced the front (κατά μέτωπον), but once the enemy attack approached within bow-shot distance they would manoeuvre out of the square and deploy at a distance of some 55 to 72 metres in front of the main army, either in a linear or wedge formation. Their operational role was to receive the enemy attack, which was to come from the heavy cavalry, by kneeling and anchoring the butts of the *menavlia* to the ground, aiming their weapons at an angle against the enemy horses.

As the *menavlatoi* were a newly established unit in the Byzantine army of the tenth century, its role and tactics on the battlefields of the East would have evolved through trial and error throughout the second and third quarter of the century. Hence, it should not come as a surprise to find the author of the *Praecepta* introducing significant changes in the numbers and operational role of this corps of infantrymen. First, the total number of the *menavlatoi* quadrupled from 300 to 1,200, which reveals how important they were in receiving the attacks of enemy heavy cavalry. We can identify a change in the structure of the unit as well, as the *Praecepta* also incorporates them into the taxiarchies of the main army; according to Phocas, each of the twelve chiliarchs/taxiarchs should have a complement of 100 *menavlatoi*.

Tactical Changes: Theory and Practice

The original position of each unit of *menavlatoi* is not specified in the *Praecepta*, although the author refers only to the extra slingers, archers and javeliners that were to deploy at the intervals of the infantry taxiarchies; thus, we understand that this was not supposed to be their starting point. However, we read in the *Taktika* of Nicephorus Uranus that

> the menavlatoi must be at the ready in the back [of the taxiarchy] and on whichever side they see the enemy kataphraktoi attacking, those menavlatoi must immediately move out through the aforementioned intervals and take their places in front of the infantry formation.[39]

Therefore, once the point of attack of the enemy units was located, the *menavlatoi* manoeuvred through the intervals – probably divided into two units of fifty men, led by two *pentecontarchs* on either side of the taxiarchy – and deployed in front of the rest of the infantry, thus making it eight deep. Phocas, however, was careful to highlight that the *menavlatoi* must be deployed in front of the infantry, 'by no means isolated from them, but instead closely ranked with them'.[40] This statement reveals that in the decade or so between the writing of the *Sylloge* and the *Praecepta* the deployment of a separate corps of *menavlatoi* projected at a distance of some 55–70 metres in front of the main army had obviously failed to produce any satisfactory results.

This unit of heavy spearmen was active in campaigns throughout the century.[41] In relation to the introduction of the corps of the *menavlatoi* and the increase in the depth of the first ranks of the heavy infantry, we notice a significant change that was unveiled by Uranus in his *Taktika* at the turn of the century. In chapter 56, Nicephorus records: 'It is necessary to combine two files of infantry men and make them into one; that is, one file must move into the next one and the seven men must become fourteen and thicken the formation.'[42] What the author suggests is a simple, fast and effective infantry manoeuvre that would double the depth of the phalanx, excluding the two extra lines of *menavlatoi* in front of the *oplitai*, with only a marginal loss of space – the total length of the phalanx would be reduced by just one file.

This extract from the *Taktika* may read as an original manoeuvre based on Uranus' experience in the East and the Balkans, but the basic idea behind it greatly resembles the one described by Aelian and the author of the *Strategikon*:

> The depth of the phalanx is, thus, doubled when the second file is thrown into the first, so that the leader of the second file takes his place behind the leader of the first; the second man in the second file becomes fourth in the first; the

third in the second becomes sixth in the first file and so on, till the whole second file be inserted in the first.[43]

The depth of the files may be increased or doubled. Assume that the troops are standing four deep and the commander wants to double that to correspond to the depth of the enemy's line and to make his own stronger in preparation for the charge. The command for this is 'Enter' [*Intra*]. And the files become eight deep.[44]

Similar manoeuvres feature in the changing of the formation of the Athenian cavalry of the fifth century BC, from a 10 × 10 square to a 20 × 5 rectangle. The manoeuvre required: (1) a standard parade formation in order for (2) the files to open sufficiently for the rear five-man unit of each ten-man file to (3) move left and forward and take its place next to the unit that was just in front of it. If we take the three stages of this manoeuvre in reverse order and we assimilate the five-man units of cavalry to a single infantryman in Uranus' description, then both manoeuvres appear almost identical.[45]

Tactical Changes: The Role of the Armenians and the Rus'

In the opening lines of the *Praecepta*, the author refers to the qualities necessary for a foot soldier serving in the imperial army. Youth, stature, strength and agility in the use of their weapons are the desirable qualities underlined by Phocas. He also feels the need to comment on the nationality of the foot soldiers: 'It is both best and necessary to choose foot soldiers [στρατιώτας] from the Byzantines and the Armenians, heavy infantry men [ὁπλίτας] large in stature and no more than forty years of age.'[46] This recommendation is coupled with what we read just a few paragraphs later: 'Inside the aforementioned intervals [of the infantry square], if there are javeliners, whether Rhos (Ρώς) or either foreigners . . .'[47]

The authors of the *taktika* works were always keen to stress the qualities of a soldier regarding his age, stature, physique and agility in the use of the sword and lance.[48] Compared with the *Sylloge Taktikorum* and Leo's *Taktika*, none of the authors makes any recommendations regarding the nationality of the foot soldiers, with the possible exception of the treatise *On Skirmishing* (c. 950), which refers to the sentries posted on the Byzantine borders as being of Armenian origin – although this can hardly be seen as a recommendation but rather as a simple remark.[49] Does the mention of the two nationalities (Rus' and Armenian) reveal any changes in the army's tactics and organisation? Is the employment of these two types of foot soldiers by the Byzantines in this period of expansion related to a possible reaction to the battle tactics employed by their enemies in the East?

Tactical Changes: Theory and Practice

Armenians had been serving in the Byzantine army as early as the sixth century, sometimes voluntarily as refugees or adventurers migrating from the East, at others forcibly removed from their homes in Cilicia and Syria and settled on imperial lands, mostly in Thrace, Macedonia and Cyprus. In the middle of the tenth century, the number of Armenians in the empire increased significantly due to the Byzantine government's need to repopulate the various regions that were captured by the Arabs, such as Melitene, Tarsus, Adana and Antioch, cities that suffered significant losses in population after the expulsion or execution of their inhabitants.[50] Hence, by the middle of the century, a whole new range of military districts under independent commanders had been established, initially by upgrading the former *kleisourai* to the status of themes, as well as the incorporation of new regions as *themata*.[51] These were called 'frontier' (ακριτικά) or Armenian (Αρμενι[α]κά) themes, not only because of their predominantly Armenian population, but also to differentiate them from the older and larger Byzantine (Ρωμαϊκά) themes of the interior.[52]

The Armenians played a prominent role in the operations against the Arabs in northern Mesopotamia, Cilicia and Syria. The role of the Armenian contingents of Mleh, a patrician and commander of the predominantly Armenian regions around Tzamandos and Lycandos (914–34), in the military campaigns of John Curcuas – himself an Armenian – can hardly be overestimated, especially in the capture of Melitene and its environs in 934.[53] The *De Ceremoniis* and Leo the Deacon maintain that Armenians constituted a large percentage of the army that was mustered to invade Crete in the reigns of both Leo VI and Romanus II, while Nicephorus Phocas settled a great number of them on the island after its conquest in 961.[54] Finally, the army of 50,000 mustered by Bardas Phocas against Sayf ad-Dawla in 954 consisted of Armenians, Turks, Rus', Bulgars, Slavs and Khazars.[55] It was not only the ready and abundant supply of Armenian soldiers, however, that made them attractive employees by the Byzantine state. Their warlike qualities were famous and, even though the Byzantine chroniclers' antipathy towards them is evident,[56] we should not underestimate their valour and fighting ability on the battlefields of Mesopotamia and Cilicia.[57]

The role of the infantry units of the middle of the tenth century, as reflected in the *Praecepta Militaria* written in a period of intense military activity in Asia Minor and the Balkans, demonstrates the empire's need for professional soldiers with discipline, high morale, uniformity in training and tactical specialism to be deployed alongside the elite *tagmatic* units – a sharp contrast to the view of foot soldiers as a mere rabble to be used for the defence of towns and castles in the preceding centuries.

Armenian soldiers represented the best foot soldiers the empire had at its disposal in ample supply in the tenth century. Their mention, however, in that particular excerpt of the *Praecepta* should not be taken as an indication of a possible reaction to the battle tactics employed by their enemies in the East, but rather as an up-to-date account of the Byzantine army.

The course of the employment of Rus' soldiers in the Byzantine army developed somewhat differently. Prior to the arrival of the famous Varangian Guard to the rescue of Basil II in 988, contacts between the empire and the Rus' existed even before the first siege of Constantinople in 860.[58] The key dates, however, for the future of Rus' soldiers in Byzantine service are the years 868, 911 and 941, which mark the treaties that were drawn between the Byzantine emperors and the Rus' great princes, following the sieges of the imperial capital. From this period onwards, we find scores of Rus' participating in every major operation, although I should point out that these troops were employed as individuals and never formed a separate and distinct unit before 988.

These elite Rus' troops were primarily heavily armed foot soldiers that fought in close formation, shield by shield. The fact that they are mentioned in the *Praecepta* should not come as a surprise because the Byzantines were quick to spot the advantages of employing these heavily armed infantrymen into their ranks. Indeed, the late tenth-century treatise *On Tactics* mentions these troops forming elite units of heavy infantry, probably spearmen or javeliners, escorting the emperor and performing special duties during the campaign.[59]

It would be very interesting if we were able to reconstruct the Rus' fighting tactics, particularly against the heavy infantry of the Iranian Daylami. The latter were regarded as the elite infantry of the Arab armies (the Tulunids, the Ikhshidids, the Hamdanids, the Mirdasids and the Fatimids – although not before the reign of al-Aziz)[60] and had very similar equipment to the Rus', meaning large battle-axes and swords, accompanied by two-pronged spears (*zupin* or *mizraq*).[61] Their ethnic background – coming from the Elburz Mountains of north-western Iran – made them excellent fighters in mountainous and broken terrain where the cavalry could not operate, and perhaps the employment of the Rus' would have been an answer to these sturdy foot soldiers. The relatively small numbers of the Rus' prior to the arrival of the Varangian Guard, however, would not have allowed their deployment in the main Byzantine army units – the taxiarchies – but rather in smaller and distinct units performing special duties, bearing in mind that the Byzantines used to keep both indigenous and foreign units ethnically and geographically coherent.

Tactical Changes: Theory and Practice

Tactical Changes: Greek Fire and Other Incendiaries

An interesting passage from the *Praecepta* is the one mentioning the use of Greek fire inland:

> The commander of the army should have with him small cheiromaggana [χειρομάγγανα], three elakatia [ηλακάτια], a swivel tube with liquid fire and a hand-pump [χειροσίφουνα], so that, if the enemy is using the same deployment in equal strength, our men can gain the upper hand over the foe and break them up by using both the cheiromaggana and the artificial liquid fire [σκευαστού και κολλητικού πυρός].[62]

There are three important terms in this passage: χειρομάγγανα, χειροσίφουνα and σκευαστού και κολλητικού πυρός. Beginning with the last of the three, the term 'artificial liquid fire'[63] is used to describe Greek fire,[64] as almost all of the authors of military treatises who mention this weapon are using adjectives that derive from the verb σκευάζω (to manufacture).[65] Greek fire was an unstable and highly flammable substance, which was produced, almost certainly, out of crude oil or a distillation of it that was obtained from wells very close to the surface in the regions of the north Black Sea coast, in the Caucasus region between the Black and Caspian Seas, and in areas of northern Mesopotamia.[66] The use of naphtha – Persian word for petroleum, originally denoting oil- or chemical-based incendiary substances in Arabic – and other incendiaries in warfare are attested since the fifth century BC; thus, the innovation in the introduction of Greek fire by Kallinikos in the third quarter of the seventh century lies in the method of projecting this liquid rather than in the substance itself.

Recent studies have shown that the essential mechanism for projecting Greek fire was an adaptation of a Graeco-Roman pump, a device that could be divided into three parts: (1) a small hearth or brazier that was used to heat the oil in its container before battle, with prolonged heating producing longer ranges, (2) a swivel tube, through which the oil was projected in any direction against the enemy, and (3) a bronze siphon (σιφών) or pump, by means of which some of the pressure to project the oil was obtained. Hence, the second of the terms used by Phocas – the χειροσίφουνα (*cheirosiphona*) – comes from the plural form of the last element of Kallinikos' device, the bronze pump, with the addition of the prefix χειρό- (hand-held). Finally, the term μάγγανα (*mangana*) describes other launching devices like catapults, *ballistae* and *gerania* (Γ-shaped cranes turning in a circle) that were used to throw containers of Greek fire and other incendiaries against the enemy.[67]

Greek fire was primarily used in naval and siege warfare, but the use of this weapon in pitched battle – by a land army at their enemies marching against them – is not reported by any Byzantine primary source before the *Praecepta*. We may trace the period of this 'tactical innovation' to the writings of Leo VI:

> You [admiral] should also employ the other method, with small *siphones* throwing [the fire] by hand which are held behind iron *skoutaria* [shields] by the soldiers. These are known as *hand-siphones* and were recently invented by our Majesty. They also throw processed fire into the faces of the enemy.[68]

This is the earliest written evidence for such hand-held *siphones* – small devices designed to project Greek fire against the crews of enemy ships. The author of the *Sylloge Taktikorum* also makes reference to *strepta*, 'which shoot clearly by machine the liquid fire that is also called brilliant by the many, and the so-called *cheirosiphones* which our majesty have now devised'.[69] Therefore, we can conclude that the move to smaller hand-held projectiles of Greek fire used directly against enemy crews had taken place some time in Leo VI's reign. However, was Nicephorus' recommendation of bringing these devices/projectiles into battle, 'in case the enemy is deployed in equal strength', a tactical innovation by the emperor? Or was he influenced by the employment of similar weapons by the Arabs?

Muslim armies had been using various devices that could project incendiary substances as early as the beginning of the ninth century.[70] One was by means of what was known as a *zarraqa*, a bronze piston pump – very similar to the Greek siphon – from whose nozzle a jet of burning liquid was projected, while a box full of naphtha was attached to it by pipes.[71] The most common use of incendiaries in Islamic history, however, took the form of containers of naphtha. These ranged from small grenades called *karaz*, made out of glass and clay, to the larger *qidr* (pots or clay containers) to be thrown from a siege machine.[72] Muslim armies made much use of the *manjaniq* or mangonel, a machine which involved the swinging of a beam or the movement of a counterpoise to strike and propel a missile with great force, and the *arrada*, a ballista which hurled missiles through a torsion of ropes.[73]

A number of Muslim sources also refer to the *naffatun*, a special unit of troops – predominantly Iranian Daylami –[74] that was attached to each corps of archers in the armies of the Abbasid period, wore fire-proof suits and threw incendiary materials – naphtha grenades and naphtha emitted from a siphon.[75] As early as the tenth century, they are reported

Tactical Changes: Theory and Practice

by al-Harthami to be used in the opening stages of pitched battle, where the main attack from the Muslims was preceded by volleys of arrows and naphtha. Similarly, in the mid-twelfth century, al-Tarsusi gives us a very detailed description of the different types of naphtha and their use in land and naval warfare.[76] In the mid-ninth century, al-Tabari reports encounters between 'fire hurlers' (*naffatun*) and Byzantines in Upper Mesopotamia in AH 249 (863–4), while the same chronicler mentions the use of oil and flammable substances by Musa ibn Bugha in Qarzim against Daylami infantry in a land battle.[77] Finally, Ibn al-Athir reports a confrontation between Yaqut, governor of Shiraz, and the Buwaihid Imadaddaula in AH 322 (944), where he clearly mentions the use of naphtha grenades against troops on the ground.[78]

We now know that Byzantine and Muslim naval preparedness and technology were on an equal footing, and the Muslims of the tenth century used Greek fire – or their version of incendiary mixture – as efficiently as the Byzantines did.[79] In fact, the great tradition of the Muslim world regarding incendiary weapons and the number of examples cited here of their use in pitched battles in Mesopotamia and the Taurus frontiers long before the writings of Nicephorus Phocas, point to the fact that the Byzantines were the ones influenced by their enemies in the use of hand-held *siphones* of Greek fire in pitched battle.

Whether these weapons, recommended by Phocas to be taken by a commander in battle, had any real effect upon the enemy is very difficult to know with certainty. It seems more likely that they were deployed mainly for psychological reasons rather than to inflict significant casualties on the enemy. Indeed, more than twelve centuries before the compilation of the *Praecepta*, we read in Sun-Tzu: 'In general, fire is used to throw enemies into confusion so that you can attack them. It is not simply to destroy enemies with fire.'[80] Frontinus also notes in his *Strategemata* (late first century AD) the use of carts filled with pitch and sulphur, which were set on fire and driven against the enemy to cause panic in its ranks.[81] We should note that some three centuries later, several Muslim military treatises report not only the use of light canons in battle against the Mongols, but that the 'Egyptians' also had a cavalry force specially equipped with a canon (*midfa*) and crackers (*sawarikh*), which were used to frighten the enemy – mainly the horses – and cause confusion in the enemy ranks.[82] Joinville, recounting the deeds of the French king Louis IX during the Seventh Crusade (1248–54), in his *Life of St Louis* (completed in 1309), has the following to say about the psychological effects of the use of Greek fire against the besiegers of Mansourah in December 1248:

This is what Greek Fire was like: it came straight at you as big as a vinegar barrel, with a tail of fire behind as long as a big lance. It made such a noise as it came that it seemed like the thunder of heaven; it looked like a dragon flying through the air. It gave so intense a light that in the camp you could see as clearly as by daylight in the great mass of flame which illuminated everything.[83]

Tactical Changes: Structure and Deployment of Cavalry Units

The regular cavalry units, which made up the bulk of the Byzantine army's cavalry force, were grouped in small tactical units of fifty men, each called a *bandum*, and deployed in battle formations of a hundred men across (two *banda*) and five men deep. According to the *Syntaxis Armatorum Quadrata* and the *Praecepta*, the first two rows comprised heavy cavalrymen, with the following two consisting of mounted archers. In order to make the formation double-faced, the last row deployed heavy cavalry as well.[84] The double-face formation was not an innovation for the cavalry units as it was for the infantry; rather, it was recommended by the author of the *Strategikon* for the four divisions of the second line of cavalry troops: 'Make these divisions double-fronted in order to meet attacks from the rear.'[85]

These comments about the depth of the cavalry *bandum* indicate a significant change in the unit's deployment on the battlefield. Indeed, the author of the *Praecepta* suggests decreasing the number of lines from the previous eight (or even ten),[86] recommended in the *Strategikon* and the *Taktika*, to just five as it would not impede its fighting efficiency on the field of battle. As a matter of fact, it is the author of the *Strategikon* who acknowledges that the ancient authorities considered the depth of four men to be sufficient for a cavalry *tagma*, 'as greater depth [was] viewed as useless and serving no purpose'.[87] In reality, the Spartan cavalry μόρα (*mora* – a division of the Spartan army, varying in strength) deployed a 24 × 4 formation – a variation on the more common 12 × 8 rectangle – during the Peloponnesian War,[88] the Athenians made use of formations of 10 × 10 and 20 × 5 based on the ten φυλαί (*phylai* – races, tribes; also a military contingent furnished by a tribe) created by Cleisthenes in the late sixth century BC,[89] while the Theban and Boeotian ἴλαι (*ilai* – military subdivisions of a cavalry *mora*) formations were usually 10 × 5.[90]

Furthermore, the proportion of heavy cavalry to mounted archers had changed by the middle of the tenth century. The *Strategikon* recommends that a third of all troops in each division should be archers stationed at the flanks of the formation, while the rest should be lancers placed at the centre.[91] Yet three centuries later, the author of the *Sylloge Taktikorum* describes a 2–1–2 formation with only the middle of the five lines composed of mounted archers, with Phocas doubling this number a generation later to a 2–2–1 formation, where the third and fourth rows had a full complement of two archer *banda* each, and the ουραγός line of tetrarchs – lancers – made the formation double-faced.[92] Phocas also insists that the five-man depth and the ratio of lancers to archers had to be maintained whatever the number of men available, contrary to the *Strategikon*'s recommendation that 'it is necessary to regulate the depth of formation according to the type of unit'.[93]

The military treatises of the tenth century are the first to attest to the 'reintroduction' of the *kataphraktoi* unit and their τρίγωνος παράταξις (*trigonos parataxis*) – a tactical formation described as a triangular one, but resembling rather a trapezium – which was simple in design and easy to create on the battlefield. The *kataphraktoi* were by far the most elite unit of the Byzantine army when it came to training, experience and, of course, equipment.[94] Described in every detail by Phocas, following closely but not copying the author of the *Sylloge*, it stood twelve rows deep, with the first one numbering twenty men and with every line increasing this number by four men – two on either side; hence, the twelfth row had sixty-four men, raising the number of men for the entire formation to 504.

Phocas also gives an alternative formation, in case the commander did not have the necessary number of men at his disposal, with a total of 384 men divided into twelve rows, with the first row having ten men instead of twenty and the last one fifty-four instead of sixty-four.[95] This, however, was a mixed formation that included a certain number of mounted archers and javeliners, who played a specific tactical role in the τρίγωνος παράταξις (wedge cavalry formation). Although the *kataphrakts* formed the first four rows, from the fifth row to the twelfth and in the middle of the formation were mounted archers enclosed within the surrounding *kataphraktoi*, while the two horsemen on each wing alternated between lancers and archers.

It is in the latter part of the work where we can point out the difference with the writings of the *Sylloge*'s author:

> The first row and the second, and the third and fourth, and also the twelfth and the eleventh, and the tenth and the ninth should all be composed of kataphrakts; the rest four rows [in the middle] should be composed of archers and lancers.[96]

Thus, what Phocas did was to move the archers to the centre of the triangular formation 'where they [archers] could be protected by them [*kataphrakts*]'.[97] Does this change in the structure of the triangular *kataphrakt* formation signal any kind of Byzantine adaptation to the reality of warfare in the East? Was this an attempt to secure the flanks of the formation from enemy attacks, and is it possible to link this change to the introduction of the two *banda* of *prokoursatores*, as I will describe below?

Overall, the wedge cavalry formation (τρίγωνος παράταξις) was not an innovation of the tacticians of the tenth century; rather, it has its roots in ancient Greek cavalry tactics and can be traced as far back as the seventh century BC. According to Asclepiodotus, paraphrased two centuries later by Aelian: 'It is said that the Scythians and the Thracians invented the wedge formation, and that later the Macedonians used it, since they considered it more practical than the square formation.'[98] Apparently, it was King Phillip II of Macedonia who adopted this formation from the Thracians in the first years of his reign, probably by seeing the Thracian cavalry in action, as these people were neighbours of the Macedonian tribes for many centuries.[99]

We can trace the origins of the wedge formation further back in time if we consider that a wedge (τρίγωνος) is, practically, half of another famous formation in ancient Greece. This was practised by the southern neighbours of the Macedonians – the Thessalians – since the seventh century BC and was known as the rhomboid (ῥομβοειδεῖ) formation: 'The half of the rhomb is called a wedge, taking a triangular form; so that the wedge is discoverable in the very formation of the rhomb.'[100] Finally, the famous predecessor of the Byzantine wedge was the Roman *cuneus*, described by Vegetius as primarily an infantry formation:

> A *cuneus* is the name for a mass of infantry who are attached to the line, which moves forward, narrower in front and broader behind, and breaks through the enemy lines, because a larger number of men are discharging missiles into one position.[101]

The rhomboid formation may have had a few advantages over the wedge, as it could change direction more easily as a single entity, with the πλαγιοφύλακες (*plagiofylakes*) – the officers on the left and right angles of the rhomb – taking command of the entire unit in the place of the ἴλαρχος (*ilarchos*) – the squadron commander who was at the fore-point.[102] Both formations, however, were suitable for shock tactics:

> For the front of the wedge formation is narrow, as in the rhomboid, and only one-half as wide, and this made it easiest for them to break through, as well as brought the leaders in front of the rest, while wheeling was thus easier than in the square formation, since all have their eyes fixed on the single squadron-commander, as in the case also in the flight of cranes.[103]

An issue that has been raised by modern historians regarding the effects of a cavalry attack on a tight infantry formation should be mentioned at this point. In the past, numerous ancient and medieval historians have maintained that 'it was an axiom that cavalry could not make a frontal attack on an unbroken line of heavy-armed spearmen' and that no horse would run into a mass of foot soldiers to face certain death.[104] The cavalry charge was mainly a psychological weapon that aimed at frightening the enemy soldiers and creating gaps in their formation that could then be penetrated. If a charge failed to break the line of infantry, the cavalry could often then either retreat and renew its charge, or advance the last few yards to engage in single combat. In contrast, there is the opinion that was raised initially by Worley and, more recently, by Sears and Willekes that horses in their final stage of cavalry charge would resemble more a group of stampeding cattle that could trample virtually anything in its path. They maintain that on open and favourable ground, cavalry can charge infantry, even if the infantry is in good order.[105]

Judging by the way the authors of the tenth-century treatises arranged the archers and lancers within the triangular formation, they presaged an attack on three stages:[106] (1) the approach, during which the mounted archers would attempt to open up gaps in the ranks of the opposing infantry; (2) the moment of impact and the ensuing melee, when the first four rows of lancers would press on with their assault to cut their way through the enemy ranks; (3) the pursuit, when the lancers and archers in the wings would be better suited to follow up with the retreating enemy units. Hence, without negating the stampeding effect of attacking horses over an enemy infantry formation, I believe that the biggest chance for horsemen to break through the enemy's formation and engage in a melee was to thin down the front ranks of the enemy. This inevitably required the presence of archers, such as those included by both authors of *taktika*, to shoot volleys of arrows against the enemy concentrated on a specific part of their formation to create a gap that would be taken advantage of by the cavalry.

Tactical Changes: The Third Line of Cavalry

The military manuals of the tenth century record another change in the tactical formation of the Byzantine cavalry of this period. The third line of cavalry was first reported in chapter 46 of the *Sylloge Taktikorum*, and was also introduced by Nicephorus Uranus and the author of the treatise *On Skirmishing* under its Arabic name (*saqah*), which clearly demonstrates the influence of the Arabs on the Byzantine commanders and tacticians.[107]

For centuries, the traditional model for the deployment of cavalry was the disposition of the entire cavalry force into two lines, along with the necessary units of flank guards, out-flankers and rear-guard units, as set down in detail in the *Strategikon* and paraphrased in the *Taktika* some three centuries later.[108] The author of the *Strategikon* emphasises the following:

> To form the whole army simply in one line facing the enemy for a general cavalry battle and to hold nothing in reserve for various eventualities in case of a reverse is the mark of an inexperienced and absolutely reckless man.[109]

What follows, clearly shows the degree of Byzantine adaptability to the battle tactics of their enemies:

> Just as the Avars and Turks line up today keeping themselves in that formation, and so they can be quickly called to support any unit that may give way in battle. For they do not draw themselves up in one battle-line only, as do the Romans and Persians, staking the fate of tens of thousands of horsemen on a single throw.[110]

Our author regards the drawing-up of the entire cavalry force in one battle line as unwieldy, unable to manoeuvre in rough terrain, difficult to manage and coordinate in battle, and lacking the necessary support in case it finds itself outflanked or counter-attacked while pursuing a fleeing enemy. Conversely, deployment in two lines bears the following advantages: the morale and confidence of the men in the first line is higher, knowing that their rear is protected from outflanking enemy manoeuvres, while it is also less likely that they will run away or desert their post. Furthermore, the support line can work as a place of refuge and a rallying point for the retreating soldiers, and can reinforce the first line when necessary. Finally, according to our author, an army arrayed in two lines can repel an enemy who is not only equal but also superior in numbers.

Breaking with the traditional model, which reflected military experience gained while fighting invading armies in the Balkans and the East in the sixth century, the compiler of the *Sylloge Taktikorum* suggests the addition of a third line of cavalry and outlines its role during battle:

> The third and last line [παράταξις] is the *saka*, which is deployed a bow-shot behind the second [βοηθός] line, being the same as the first [πρόμαχος] line in terms of numbers, units and all the rest, and if there are enough kataphrakts, the middle unit should be a triangular formation, if there are not enough, the middle unit should be like the rest.[111]

Tactical Changes: Theory and Practice

From what we have concluded so far, a number of major changes in the battlefield deployment of the Byzantine infantry and cavalry forces took place in the middle of the tenth century that could be considered a tactical response to the tactics employed by the Arabs as experienced by great officers in the East, such as John Curcuas, Nicephorus Phocas, John Tzimiskes and Nicephorus Uranus. The corps of infantry would march into battle in twelve taxiarchies of a thousand men each, forming a double-ribbed square that had three units and a specific number of intervals on each side to allow for the unencumbered access in and out of the square of the cavalry units fighting alongside them. Each infantry taxiarchy was also transformed to a large extent, not only regarding the depth of the phalanx formation, now reduced to 7 + 1 men (heavy and light infantrymen including the *menavlatoi*), but also in the deployment of the men in each file. This now involved a two-fold formation, where the heavy infantrymen were put in the first two rows as well as at the back, to defend from potential attacks from behind and any attempts to encircle them. Other evidence includes the position of the cavalry in a mixed formation which, by the middle of the century, was positioned inside the square infantry formation – in essence a camp converted into a battle formation – manoeuvring through the intervals on the sides to open the battle while also being able to fall back into the square in case they were repulsed by the enemy. Changes in the battle formation of the elite cavalry unit of the *kataphraktoi* further demonstrates a Byzantine tendency to adjust to the tactics of the enemy, as the increase of the proportion of archers in the *kataphrakts'* wedge shows. Finally, a great innovation of the period was the introduction of a third line of cavalry which had the crucial role of providing cover from any enveloping manoeuvres, identified in our Greek treatises under its Arabic name, *saqah*.

Notes

1. Nicephorus Phocas' *Military Praecepts* and an incomplete version of Nicephorus Uranus' *Taktika* can be found in E. McGeer (1995), *Sowing the Dragon's Teeth: Byzantine Warfare in the Tenth Century*, Washington, DC: Dumbarton Oaks.
2. *Sylloge Tacticorum, quae olim 'inedita Leonis Tactica' dicebatur* [hereafter *Sylloge Taktikorum*] (1938), ed. A. Dain, Paris: Société d'édition les belles lettres; G. Chatzelis and J. Harris (2017), *A Tenth-Century Byzantine Military Manual: The 'Sylloge Tacticorum'*, London: Routledge.
3. A. Dain (1967), 'Les stratégistes byzantins', *Travaux et mémoires*, 2, p. 367; McGeer, 'Syntaxis Armatorum Quadrata', pp. 219–29.

4. G. T. Dennis (2008), 'The Anonymous Byzantine Treatise *On Skirmishing* by the Emperor Lord Nicephoros', in G. T. Dennis (ed.), *Three Byzantine Military Treatises*, Washington, DC: Dumbarton Oaks, pp. 143–239; G. Dagron and H. Mihaescu (1986), *Le traité sur la guérilla ('De velitatione') de l'Empereur Nicéphore Phocas (963–969)*, Paris: Editions du Centre national de la recherche scientifique.
5. *Praecepta Militaria*, I.39–46, p. 14.
6. Ibid., I.46–52, p. 14.
7. McGeer, 'Syntaxis Armatorum Quadrata', p. 228.
8. Xenophon, *Anabasis*, III.4.19–23; Aelian, XVIII, pp. 80–1; *On Strategy*, 15, p. 46.
9. The author of the *Taktika* writes about a τετράγωνον πλινθίου σχῆμα (four-sided, rectangular formation), where the horses and the baggage train should be placed in case of an emergency: *Taktika*, XIV.20, pp. 302–3.
10. McGeer, *Sowing the Dragon's Teeth*, pp. 259–64.
11. Vegetius, *Epitome of Military Science*, III.8, pp. 79–83; *Strategikon*, B.22, pp. 158–62; *On Strategy*, 28.1–44, pp. 86–8; *On Tactics*, 1.1–190, pp. 246–54.
12. The term used to translate to the 'commander of a thousand men'. However, the number of troops under a chiliarch's (or *drungarie*, as he was mostly known after the seventh century) command varied over the centuries from 1,000 (before 840 and after 959) to as low as 400 (840–959); see Treadgold, *Byzantium and its Army*, pp. 93–105, especially p. 97 table 4.
13. *Praecepta Militaria*, V.23–6, p. 52.
14. *On Tactics*, 1.18–29, p. 246.
15. Vegetius, *Epitome of Military Science*, II.15, p. 46; *Strategikon*, XII.B.13, p. 144; Leo VI, *Taktika*, XIV.61, p. 326.
16. Leo VI, *Taktika*, XIV.62, p. 328.
17. *Praecepta Militaria*, II.94–8, p. 28. The *Sylloge* adds that the cavalry units should either enter the square – 'where they were before [the battle begun]' – or remain outside and on the sides of the infantry units and continue fighting alongside them: *Sylloge Taktikorum*, 47.19, p. 90.
18. Procopius (1916), *Histories*, trans. H. E. Dewing, 6 vols, London: Heinemann, vol. 2, IV.3.15–25, p. 231; IV.17.24–9, p. 369; IV.19.8–9, p. 379. See also Frontinus' description of the same tactic applied by Hannibal: Στρατηγήματα [*Strategemata*], II.3.9, p. 110.
19. *Praecepta Militaria*, II.76–7, p. 26.
20. Ibid., I.62–5, p. 16.
21. Although the *Strategikon* notes that the archers could be posted inside the infantry formation, alternating one heavy infantryman with one archer: *Strategikon*, XII.B.12, p. 144.
22. *Strategikon*, XII.B.12, p. 143; Leo VI, *Taktika*, XIV.59–60, pp. 324–6. Asclepiodotus notes a number between eight and sixteen for the depth of a phalanx; see Asclepiodotus (1928), *Tactics*, in The Illinois Classical Club

(ed.), *Aeneas Tacticus, Asclepiodotus, Onasander*, The Loeb Classical Library, London: Heinemann, II.1, pp. 250–2; VI.2, p. 274.
23. *Sylloge Taktikorum*, 43.1, p. 66.
24. Ibid., 45.16–17, pp. 73–4.
25. *Praecepta Militaria*, I.65–70, p. 17.
26. H. van Wees (2004), *Greek Warfare, Myths and Realities*, London: Bloomsbury Academic, p. 185; W. K. Pritchett (1971), *Ancient Greek Military Practices*, Berkeley: University of California Press, pp. 135–6.
27. For the doubling of the phalanx, see Asclepiodotus, *Tactics*, X.17–18, pp. 308–10.
28. *Praecepta Militaria*, I.75–80, p. 17.
29. *Strategikon*, XII.B.16, p. 147. See Frontinus' terminology regarding Alexander's battle formation at Gaugamela (331 BC): '*aciem in omnem partem spectantem ordinavit, ut circumventi undique pugnare possent*'; Frontinus, *Stratagems*, II.iii.19, p. 118. Cf. Onasander, *Strategikos*, X.2, p. 410.
30. *Praecepta Militaria*, I.8–10, p. 12; *Sylloge Taktikorum*, 45.10–13, pp. 72–3; Leo VI, *Taktika*, IV.6, p. 50. We should also note the difference in the terminology: the author of the *Sylloge* refers to the phalanx as αντίστομον (45.17), while Nicephorus identifies it as αμφίστομον (I.78). For an analysis of these two terms, see *Sylloge Taktikorum*, 42, p. 65; *On Strategy*, 31, pp. 94–6; Aelian, 37–42, pp. 134–44; Asclepiodotus, *Tactics*, XI.4, pp. 316–18.
31. For the origin, construction and use of the *menavlion*, see J. F. Haldon (1975), 'Some Aspects of Byzantine Military Technology from the Sixth to the Tenth Centuries', *Byzantine and Modern Greek Studies*, 1, pp. 32–3; T. G. Kolias (1988), *Byzantinische Waffen. Ein Beitrag zur byzantinischen Waffenkunde von den Anfängen biz zur lateinischen Eroberung*, Vienna: Verlag der Österreichische Akademie der Wissenschaften, pp. 194–5; E. McGeer (1986–7), 'Μεναύλιον – Μεναύλατοι', *ΔΙΠΤΥΧΑ*, 4, pp. 53–8; McGeer, *Sowing the Dragon's Teeth*, pp. 209–11, 267–72; T. Dawson (2007), '"Fit for the Task": Equipment Sizes and the Transmission of Military Lore, Sixth to Tenth Centuries', *Byzantine and Modern Greek Studies*, 31, pp. 4–6; M. P. Anastasiadis (1994), 'On Handling the *Menavlion*', *Byzantine and Modern Greek Studies*, 18, pp. 1–10.
32. Leo VI, *Taktika*, VI.27, p. 96; IX.71, p. 182; XI.22, p. 204; XIX.14, p. 508.
33. Vegetius, *Epitome of Military Science*, II.15, p. 47; Leo VI, *Taktika*, VI.22, p. 92.
34. Vegetius, *Epitome of Military Science*, I.17, pp. 16–17; Leo VI, *Taktika*, VII.3, p. 106.
35. Haldon, 'Byzantine Military Technology', p. 33.
36. *Praecepta Militaria*, I.119–21, p. 18.
37. There is an ongoing debate over the length of the *menavlion*: Anastasiadis, 'On Handling the *Menavlion*', pp. 3–4; McGeer, 'Μεναύλιον', pp. 54–5; T. Dawson (2002), 'Suntagma Hoplon: Equipment of Regular Byzantine Troops, c. 950–c.1204', in D. Nicolle (ed.), *A Companion to Medieval Arms*

and Armour, Woodbridge: Boydell & Brewer, p. 83, n. 20; Dawson, 'Fit for the Task', pp. 7–10.
38. *Sylloge Taktikorum*, 47.16, p. 89.
39. Nicephorus Uranus, *Taktika*, 56.106–10, p. 94.
40. Ibid., I.105–6, p. 18.
41. The unit is clearly mentioned in the early 990s treatise, acting against an enemy that occupied a mountain pass: *On Tactics*, 20.130, p. 300.
42. Nicephorus Uranus, *Taktika*, 56.111–15, p. 94.
43. Aelian, 28, pp. 114–15. Cf. Asclepiodotus, *Tactics*, X.17, p. 308.
44. *Strategikon*, XII.B.16, pp. 148–9. Cf. *On Strategy*, 31, pp. 94–6.
45. L. J. Worley (1994), *Hippeis, the Cavalry of Ancient Greece*, Oxford: Oxford University Press, pp. 75–7. Cf. the changing of formation of the late fifth-century Spartan cavalry *moira* from a 12 × 8 to a 24 × 4 rectangle, again with the stages taken in reverse order: Worley, *Hippeis*, pp. 90–1.
46. *Praecepta Militaria*, I.1–4, p. 12; Nicephorus Uranus, *Taktika*, 56.1–2, p. 88.
47. *Praecepta Militaria*, I.51–3, p. 14; Nicephorus Uranus, *Taktika*, 56.56–8, p. 90.
48. *Sylloge Taktikorum*, 36.1, p. 58; 47.15, p. 89; Leo VI, *Taktika*, IV.1, pp. 46–7.
49. *On Skirmishing*, 2.11–13, p. 152.
50. Canard, *Hamdanides*, pp. 736, 820–3; Bosworth, 'City of Tarsus', pp. 278–9; Dagron, G. (1976), 'Minorités ethniques et religieuses: l'immigration syrienne', *Travaux et mémoires*, 6, pp. 193–6.
51. Haldon, *Warfare, State and Society*, p. 84.
52. For the organisation of these themes, see Dagron, *Le traité sur la guérilla*, pp. 239–57; H. Ahrweiler (1969), 'Recherches sur l'administration de l'empire Byzantin aux IXe–XIe siècles', *Bulletin de correspondance hellénique*, 84, pp. 45–7; H. Ahrweiler (1965), 'La frontière et les frontières de Byzance en Orient', in H. Ahrweiler (ed.), *Byzance: les pays et les territoires*, London: Variorum, pp. 216–18; N. Oikonomides (1974), 'L'organisation de la frontière orientale de Byzance aux Xe–XIe siècles et le Taktikon de l'Escorial', in M. Berza and E. Stănescu (eds), *Actes du XIVe Congrès international des études byzantines*, Bucharest: Editura Academiei Republicii Socialiste România, I, pp. 285–302.
53. Canard, *Hamdanides*, pp. 731–6; Whittow, *Orthodox Byzantium*, pp. 314–20.
54. The theme of Sebasteia was requested to provide 1,000 Armenians for the Cretan expedition of 911, while the military settlements at Prine and Platanion contributed 500 each – all cavalry: *De Ceremoniis*, I, p. 652. For the 949 expedition, the eastern *tagmata* provided 1,000 Armenians, while the Thrakesion was asked for 600 – again, all cavalry: *De Ceremoniis*, I, pp. 666–7. For the settlement of Armenians in Crete, see *History of Leo the Deacon*, II.8, p. 80: 'Ἀρμενίων τε καὶ Ῥωμαίων καὶ συγκλύδων ἀνδρῶν φατρίας ἐνοικισάμενος' (Then he pacified the entire island, settling it with bands of Armenians, Romans, and other rabble). See also, Tsougarakes, *Byzantine Crete*, pp. 73, 238.

55. Mutanabbi (1950), *Poem on Hadath*, in A. A. Vasiliev (ed.) and M. Canard (trans.), *Byzance et les Arabes, 867–959*, Brussels: Institut de philologie et d'histoire orientales, 2.II, p. 331; Canard, *Hamdanides*, p. 779.
56. They were keen to highlight the Armenians' unruliness and unreliability: *On Skirmishing*, 2, p. 152; Leo the Deacon, IV.7, p. 113. See also the comments by Bar Hebreus in *The Chronography of Gregory Abû'l Faraj, the Son of Aaron, the Hebrew Physician, commonly known as Bar Hebraeus: Being the First Part of his Political History of the World* (2003), trans. E. A. Wallis Budge, 2 vols, London: Gorgias, p. 168.
57. Their tenacious defence of a pass against Sayf ad-Dawla's Iranian–Daylami troops in 953 has been highlighted by contemporary Muslim sources; see Vasiliev, *Byzance et les Arabes*, 2.I, p. 349 n. 6; *Histoire de Yahya*, p. 771. Asochik portrays the Armenians as valiant and faithful soldiers in two imperial campaigns in the Balkans in 971 and 986: Asochik (1883–1917), *Histoire universelle*, trans. E. Dulaurier and F. Macler, Paris: Les Presses Universelles, 2.3.44–5, pp. 126–7. See also E. McGeer (1995), 'The Legal Decree of Nikephoros II Phokas Concerning Armenian *Stratiotai*', in T. Miller and J. Nesbitt (eds), *Peace and War in Byzantium*, Washington, DC: Catholic University of America Press, pp. 123–37.
58. Theotokis, 'Rus, Varangian and Frankish Mercenaries', pp. 126–56.
59. 'Let him [the emperor] also have some Rus and malartioi'. *Malartioi*, according to Dennis, were later (eleventh century) referred to as *kontaratoi* (spearmen): *On Tactics*, 10.37–40, p. 280; 19.35, p. 294.
60. Bosworth, 'Buyids', pp. 149–51; Beshir, 'Fatimid Military Organization', p. 42; Lev, 'Fatimid Army', pp. 171–97; Bacharach, 'African Military Slaves', pp. 471–95; McGeer, *Sowing the Dragon's Teeth*, pp. 233–6.
61. Bosworth, 'Buyids', pp. 149–51.
62. *Praecepta Militaria*, I.150–5, p. 20.
63. The term κολλητικού can also be translated as sticky, probably reflecting the qualities of many incendiary liquids like naphtha.
64. For a selected bibliography on Greek fire, see A. Mayor (2006), *Υγρόν Πυρ, Δηλητηριώδη Βέλη και Σκορπιοί–Βόμβες, Βιολογικά και Χημικά Όπλα στον Αρχαίο Κόσμο* [Greek Fire, Poisonous Arrows and Scorpion Bombs, Biological and Chemical Weapons in the Ancient World], trans. Annita Gregoriadou, Athens: Enalios; J. R. Partington (1999), *A History of Greek Fire and Gunpowder*, London: Johns Hopkins University Press, pp. 1–41; J. F. Haldon and M. Byrne (1977), 'A Possible Solution to the Problem of Greek Fire', *Byzantinische Zeitschrift*, 70, pp. 91–9; H. R. Ellis-Davidson (1973), 'The Secret Weapon of Byzantium', *Byzantinische Zeitschrift*, 66, pp. 61–74; A. Roland (2008), 'Secrecy, Technology, and War: Greek Fire and the Defense of Byzantium, 678–1204', in J. France and K. De Vries (eds), *Warfare in the Dark Ages*, Aldershot: Ashgate, pp. 655–79; K. Korres (1995), *'Υγρόν Πυρ', Ένα Όπλο της Βυζαντινής Ναυτικής Τακτικής* ['Greek Fire': A Weapon in Byzantine Naval Tactics], Thessaloniki: Vanias;

M. Mercier (1952), *Le feu gregéois, les feux de guerre depuis l'antiquité, la poudre à canon*, Paris: Paul Geuthner; J. H. Pryor and E. M. Jeffreys (2006), *The Age of the ΔΡΟΜΩΝ: The Byzantine Navy c. 500–1204*, Leiden: Brill, appendix VI, pp. 607–31.

65. A. Dain (1940), *Appellations grecques du feu grégeois*, Paris: Mélanges Ernout, p. 123; Korres, 'Υγρόν Πύρ', p. 67; Aeneas Tacticus, *Our Defence against Siege*, 35, p. 182.
66. For example, *DAI*, 53, p. 284.
67. Leo VI, *Taktika*, XV.41, 44, 45, p. 370; I. P. Stephenson (2006), *Romano–Byzantine Infantry Equipment*, Stroud: Tempus, pp. 133–40.
68. Leo VI, *Taktika*, XIX.64, pp. 528–9.
69. *Sylloge Taktikorum*, 53.8, p. 102–3.
70. A. Y. al-Hassan and D. R. Hill (1992), *Islamic Technology: An Illustrated History*, Cambridge: Cambridge University Press, p. 108; Ellis-Davidson, 'Secret Weapon of Byzantium', p. 65; Partington, *History of Greek Fire*, pp. 25–7.
71. al-Hassan and Hill, *Islamic Technology*, p. 108. On the use of flamethrowers in Muslim warships, see V. Christides (1984), *The Conquest of Crete by the Andalusians (ca. 824–961)*, Athens: Academy of Athens, appendix III.
72. al-Hassan and Hill, *Islamic Technology*, p. 109; Nicolle, *Crusader Warfare*, II, p. 243.
73. al-Hassan and Hill, *Islamic Technology*, p. 108; Nicolle, *Crusader Warfare*, II, pp. 244–5; V. J. Parry (1970), 'Warfare', in P. M. Holt, A. K. S. Lambton and B. Lewis (eds), *The Cambridge History of Islam*, Cambridge: Cambridge University Press, II, p. 831.
74. Hamblin, 'Fatimid Army', pp. 164–70; Beshir, 'Fatimid Military Organization', p. 42; Nicolle, *Crusader Warfare*, II, p. 243.
75. Mercier, *Le feu gregéois*, pp. 42–68; al-Hassan and Hill, *Islamic Technology*, p. 106; Partington, *History of Greek Fire*, p. 22; Parry, 'Warfare', p. 831.
76. Al-Harthami (1964), *Mukhtasar siyasat al-hurub*, ed. ʿAbd al-Raʾuf ʿAwn, Cairo: Silsilat kutub al-turath, ch. 24, pp. 41–4; al-Tarsusi (1948), *Tabsira arbab al-lubab*, in C. Cahen (trans.), 'Un traité d'armurerie composé pour Saladin', *Bulletin d'études orientales* 12, pp. 145–8.
77. Al-Tabari, *History*, 35, pp. 11, 151–2.
78. Mercier, *Le feu gregéois*, p. 51.
79. Christides rejects Haldon and Byrne's view in 'A Possible Solution to the Problem of Greek Fire' that only the Byzantines succeeded in using Greek fire in naval and siege warfare; see Christides, *Conquest of Crete by the Andalusians*, appendix II; Christides, 'Two Parallel Naval Guides', p. 92.
80. Sun-Tzu, *Art of War*, 12, p. 164.
81. Frontinus, II.IV.17 (On Creating Panic in the Enemy's Ranks), pp. 130–1.
82. al-Hassan and Hill, *Islamic Technology*, p. 113. See also D. Ayalon (1978), *Gunpowder and Firearms in the Mamluk Kingdom: A Challenge to a Mediaeval Society*, London: Frank Cass.

83. J. de Joinville (1963), 'Life of St-Louis', trans. M. R. B. Shaw, *Joinville and Villehardouin: Chronicles of the Crusades*, London: Penguin, p. 216.
84. *Praecepta Militaria*, IV.80–5, p. 44; McGeer, 'Syntaxis Armatorum Quadrata', pp. 219–29.
85. *Strategikon*, II.4, p. 26.
86. Ibid., II.6, pp. 27–8; Leo VI, *Taktika*, XVIII.143, p. 494.
87. *Strategikon*, II.6, p. 27. Cf. Asclepiodotus, *Tactics*, VII.4, p. 280.
88. Very little is known about the organisation of the Spartan cavalry in the Peloponnesian War – the first such cavalry force after the disbandment of the Spartan cavalry following the second Messenian War (685–68 BC). Historians have presumed that the infantry model was followed; see Worley, *Hippeis*, pp. 89–91.
89. Judging, however, by the formation of the Syracusans – who were following Athenian practices – in the Sicilian campaign of the Peloponnesian War (415–13 BC), it was not unusual for an eighty-man squadron to form in 10 × 8, 8 × 10, 16 × 5 and 20 × 4 rectangles; see Worley, *Hippeis*, pp. 74–6, 100–1.
90. Worley, Hippeis, pp. 61–2.
91. *Strategikon*, II.3, p. 26.
92. *Praecepta Militaria*, IV.39–47, p. 40; *Sylloge Taktikorum*, 46.3, p. 77.
93. *Praecepta Militaria*, IV.56–60, p. 42; *Strategikon*, II.6, p. 28.
94. For the defensive and offensive equipment of the Byzantine *kataphraktoi*, see *Praecepta Militaria*, III, pp. 34–8; McGeer, *Sowing the Dragon's Teeth*, pp. 69–70, 214–17; Dawson, 'Suntagma Hoplon', pp. 81–90; O. Gamber (1968), '*Kataphrakten*, Clibanarier, Normanreiter', *Jahrbuch der Kunsthistorischen Sammlungen in Wien*, 64, pp. 7–44.
95. *Praecepta Militaria*, III.1–17, p. 34.
96. *Sylloge Taktikorum*, 46.7, pp. 78–9.
97. *Praecepta Militaria*, IV.135–40, pp. 45–6.
98. Asclepiodotus, *Tactics*, VII.3, p. 278; Aelian, XVIII, pp. 80–1.
99. Worley, *Hippeis*, p. 157; P. Rance (2004), 'Drungus, δρούγγος, and δρουγγιστί: A Gallicism and Continuity in Late Roman Cavalry Tactics', *Phoenix*, 58, pp. 96–130, especially pp. 120–30.
100. Aelian, XIX, p. 88.
101. Vegetius, *Epitome of Military Science*, III.19, p. 97.
102. Asclepiodotus, *Tactics*, VII.2, p. 276.
103. Asclepiodotus, *Tactics*, VII.3, p. 278; Aelian, XVIII, pp. 80–1.
104. W. W. Tarn (1948), *Alexander the Great*, Cambridge: Cambridge University Press, II, p. 181; P. A. Rahe (1981), 'The Annihilation of the Sacred Band at Chaeronea', *American Journal of Archaeology*, 85, pp. 84–7; I. G. Spence (1993), *The Cavalry of Classical Greece: A Social and Military History with Particular Reference to Athens*, Oxford: Oxford University Press, p. 107; J. Buckler and H. Beck (2008), *Central Greece and the Politics of Power in the Fourth Century BC*, Cambridge: Cambridge University

Press, p. 256; P. Sidnell (2006), *Warhorse: Cavalry in Ancient Warfare*, London: Continuum, pp. 97–9; J. Keegan (2004), *The Face of Battle: A Study of Agincourt, Waterloo and the Somme*, London: Pimlico, pp. 95–6 and 154–60; J. Gillingham (1999), 'An Age of Expansion', in M. Keen (ed.), *Medieval Warfare*, Oxford: Oxford University Press, pp. 76–8; M. Bennett (1998), 'The myth of the military supremacy of knightly cavalry', in M. J. Strickland (ed.), *Armies, Chivalry and Warfare: Proceedings of the 1995 Harlaxton Symposium*, Stamford: Paul Watkins, pp. 304–16, especially pp. 310–16; S. Morillo (1999), 'The "Age of Cavalry" Revisited', in D. J. Kagay and L. J. Villalon (eds), *The Circle of War in the Middle Ages: Essays on Medieval Military and Naval History*, Woodbridge: Boydell & Brewer, p. 50; Verbruggen, *Art of Warfare*, pp. 46–9.
105. Sears and Willekes suggested that the wedge formation used by the Macedonians was ideal for exploiting the natural herd instinct of horses, as Alexander led a successful cavalry charge against the Theban Sacred Band at Chaeronea in 338 BC. They highlight an important passage concerning the tactics of the Macedonian heavy cavalry found in the *Tactica* of Arrian, where the author employs the verb διακόπτω (to break through the enemy's line, phalanx or city wall); see M. A. Sears and C. Willekes (2016), 'Alexander's Cavalry Charge at Chaeronea, 338 BCE', *Journal of Military History*, 80, pp. 1017–35; Worley, *Hippeis*, pp. 162–3.
106. McGeer, *Sowing the Dragon's Teeth*, p. 286.
107. *Sylloge Taktikorum*, 46.17–34, pp. 81–6; Nicephorus Uranus, *Taktika*, 64.31, p. 148; *On Skirmishing*, 9, p. 170 (the first mention in the work). Phocas recommends a third line (παράταξις) but does not use the term *saka*; see *Praecepta Militaria*, IV.65–7, p. 42.
108. *Strategikon*, II.1, pp. 23–5; Leo VI, *Taktika*, XVIII.136, pp. 488–90.
109. *Strategikon*, II.1, p. 23.
110. Ibid., II.1, p. 23.
111. *Sylloge Taktikorum*, 46.19, p. 81.

8

Tactical Changes in the Byzantine Armies of the Tenth Century: Investigating the Root Causes

There is a series of questions that emerges from the evidence that has been scrutinised thus far in this study regarding the structural and tactical changes in the Byzantine army of the tenth century. I believe that the answers to these questions would provide us with a clear picture as to whether the Byzantines were, indeed, showing any signs of innovation or tactical adaptation to the strategic situation in the East. Most importantly, however, I wish to track down the catalyst (a battle, an encounter with an enemy nation, etc.) that provided the Byzantines with the impetus to develop many of their tactics in the operational theatres of the East. It is my intention to combine this with the following chapter that will investigate the evidence of adaptation that can be found in the contemporary historical sources about the battles between the Byzantines and the Arabs in the East for the same period – the middle of the tenth century.

Tactical Changes in the Infantry

The questions that have emerged thus far are the following: Why did Nicephorus Phocas change the infantry formation to a διττό (two-fold, double) hollow square, and why did he move the cavalry inside it? Was the double-faced formation of the infantry taxiarchy an answer to the battle tactics applied by his enemies? Why did the c. 930 *Sylloge Taktikorum* refer to the infantry taxiarchy as an αντίστομος (*antistomos*) formation – an oblong formation elongated on the front and back – rather than an αμφίστομος (*amfistomos*) – an oblong formation elongated on the sides, as identified by Phocas a generation later? Is there any correlation between this double-faced formation and the moving of the archers from behind the main infantry force, as described in the c. 600 *Strategikon* and the c. 900 *Taktika*, to the middle of the formation as advised by the authors of the mid-tenth century?

We have already seen the advantages offered to a mixed army marching into battle in a square formation: the enemy cannot attack the formation from all sides without dividing his forces, it is a place of refuge, and it offered far fewer opportunities for a soldier contemplating retreat or desertion. Phocas portrays the infantry square as the base from where the cavalry units would launch their attack against the enemy: 'If it should happen that the enemy hits our cavalry units hard and repels them – God forbid – they must retire inside our heavy infantry units for protection.'[1] Against whom, though, was this infantry square devised as a defensive formation and place of refuge for both infantry and cavalry? The author of the *Praecepta* identifies the enemy in just a few verses:

> If the enemy proceeds in close order with their forces in proper formation, bringing along a vast host of cavalry and infantry, and their forces move in against one side of our units, the *Arabitai* will encircle our four-sided formation in a swarm, as they usually do, confident in their horses. There is no need for the cavalry to head off in pursuit of them because of the speed of their horses, for when pursued they are not overtaken and, aided by the speed of their horses, they quickly counterattack and strike against our men.[2]

The people that Nicephorus is referring to are the Bedouins, and he is using the term *Arabitai* (Αραβίται) to distinguish them from the rest of the Muslim units (Αγαρηνοί, *Agarenoi*). Who were these people and what was their role in the Muslim armies of the tenth century?

The Hamdanids and the Fatimids were using Bedouin tribes as a source of irregular auxiliary troops for many centuries up to the tenth and beyond.[3] The Hamdanids recruited these troops from the northern Syrian tribes of the Banu Kilab and the Banu Numair that dwelt in this region ever since they had migrated from the Arabian Peninsula three centuries ago – some of them succumbing to the process of sedentism, while others retaining their nomadic way of life and spirit of warfare.[4] Although it is impossible to give a precise figure for the total number of Bedouins available for military service, many of the Bedouin tribes that were employed by the Hamdanid and Fatimid governments received yearly stipends by means of *iqta*[c] in return for military service and loyalty.[5] This lightly armed cavalry wore little or no armour and carried a short lance, rather than a bow,[6] while they seem to have used very similar enveloping tactics to the Seljuk Turks.[7] Greater speed and mobility, resulting from their light armour and the superiority of their horses,[8] gave the Bedouins the upper hand in skirmishing tactics. The only thing the Byzantines could do to counter their feigned retreat and encircling tactics was to maintain their formation and refrain from giving chase,[9] while additional cavalry units were kept as a reserve in the intervals on both flanks to counterattack and scatter them.¹⁰

We read in the *Muqaddimah* of Ibn Khaldun (written in 1377) about the methods of waging war employed by various peoples:

> Since the beginning of men's existence, war has been waged in the world in two ways. One is by advance in closed formation. The other is the technique of attack and withdrawal. The advance in closed formation has been the technique of all non-Arabs throughout their entire existence. The technique of attack and withdrawal has been that of the Berbers of the Maghreb.¹¹

The battle tactic of repeated attacks and withdrawals that was accompanied by an attempt to attack the enemy from behind had been central to Arab cavalry tactics since pre-Islamic times, although not using bows but rather lances. It was identified as *karr wa farr*, 'a sudden attack by the army on the enemy, followed by a quick retreat': 'This was repeated during the battle and it often inflicted damage and confusion in the enemy ranks while the Muslim forces remained intact.'¹² In connection with what has been discussed so far about the cavalry falling back to a field fortification for refuge, Ibn Khaldun also notes:

> One of the techniques of the peoples who use the technique of attack and withdrawal, is to set up, behind their enemies, a barricade of solid objects ... to serve as refuge for the cavalry during attack and withdrawal ... The Arabs and most other Bedouin nations that move about and employ the technique of attack and withdrawal, dispose their camels and the pack animals carrying their litters in lines to steady the fighting men. They call it *al-majbudah*.¹³

This was essentially the tactic which was central to the battlefield practices of the Berbers of North Africa and Syria from early Islamic times to the eleventh century and perhaps even later, and of which the author of the *Praecepta* advised his officers to be cautious.¹⁴ Could, however, the inspiration for the Byzantine infantry square, which is based on the imperial army's military encampments, have come from the Muslims as well?

A strong indication seems to be found in the period when it first appeared in the sources, during John Curcuas' campaigns in the East in the 930s–40s. The *Praecepta* refers to the Muslims deploying their troops in square formations (τετράπλευρος παράταξις), but not in the hollow formation indicated for the Byzantine mixed armies.¹⁵ Rather, we should look to the military treatise *On Skirmishing* (written c. 969, but reflecting earlier periods of border warfare in the East) for more clues: 'The enemy are ravaging our country without breaking their military formation [φοσσατικῶς] and not sending their raiding parties out to any great distance, but playing it very safe.'[16] The key term here is the adverb φοσσατικῶς (*fossatikos*) denoting the way in which the Muslims marched through Byzantine territory, namely without breaking their formation. Although the Latin *fossatum* originally denoted a military camp – a synonym for the Greek ἄπλεκτον (*aplekton*) – it acquired several meanings in the military treatises of the tenth century: in *On Skirmishing*, *On Tactics* and in two of Constantine VII's treatises on *Imperial Military Expeditions* it has the general meaning of the army's marching formation on enemy territory.[17]

If we look, though, at Leo VI's *Taktika*, the noun φοσσάτον (*fossaton*) (φοσσάτον τό ἄπλικτον τοῦ ὅλου στρατοῦ καλεῖται – 'entrenchment is the specific term for the camp of the entire army') indicates a term which has the same meaning as in the *Praecepta Militaria*.[18] If we examine these treatises together, could this mean that the Muslim raiding army that 'was playing it safe' was marching in a formation based on their encampment? The answer is possibly, if we consider that Muslim military manuals were keen to stress the importance of being vigilant and keeping a tight formation when marching in enemy territory:

> Should the threat be 'unknown' [i.e. the direction from which it will be launched is not known], he [the commander] should scatter scouting parties and horsemen on all sides of the army, and muster the men according to their [battle] ranks and stations. The commander of the army should be in the middle of the centre [section].[19]

Therefore, we may assume that, as the writing of Nicephorus suggests, the Byzantine armies did adjust their tactics to counter the advantages of the Bedouins in terms of speed, manoeuvrability and surprise. Both the hollow double (διττή) square that kept the cavalry units inside as a point of refuge and regroupment, and the double-faced formation of the taxiarchy, with the heavy infantrymen kept at the back rows for safety in case the enemy broke into the square formation, were meant to deal with the cavalry tactics of the Bedouins. However, as Psellus and Attaleiates indicate in their description of the imperial campaigns in Syria in 1030 and 1068, the battle tactics of the Bedouins continued to be a headache for the Byzantine generals many decades after the compilation of the *Praecepta Militaria*.[20]

Tactical Changes: Investigating the Root Causes

Another theory I want to suggest argues that the inspiration for the hollow infantry square may have come directly from the marching formations described by the ancient tacticians Asclepiodotus and Onasander. We read in the Στρατηγικός (*General*), and more specifically in the chapter 'On Maintaining Military Formation':

> They [the soldiers] must proceed, prepared at the same time for marching and for battle ... A marching formation that is compact and rectangular – not very much longer than its width – is safe and easy to manage for every emergency. The general must place his medical equipment, pack animals, and all his baggage in the centre of his army, not outside. Should he consider that his rear is not quite secure and undisturbed he should form his rearguard of the most vigorous and courageous soldiers.[21]

Therefore, according to Onasander, in a square or rectangular compact formation, the units most vulnerable to attack, such as the medical equipment train, would be placed in the centre, protected from enemy attacks not only from the front but also from the wings and the rear. The author does not mention any regular intervals, but these could perhaps be identified with the gaps created by the different units marching in formation. Asclepiodotus gives us an even better description of a hollow marching square: 'Sometimes the army marches in four parts by divisions, on its guard upon every side against the enemy, and we have a four-sided figure fronting each side, an oblong rectangle or square which fronts on all sides.'[22]

These passages would have served as an inspiration for the authors of the mid-tenth century *taktika*, bearing in mind that classical education was essential for a Byzantine officer, and that recent studies have concluded that the military leaders of the Byzantine army were well aware of the existing military manuals and frequently consulted them.[23] They may not have been as accurate regarding the numbers of units, intervals and the specific places of the different cavalry or infantry *turmae*, but it seems reasonable enough that this formation could have been modified to serve as a battle formation when marching in enemy territory, expecting a sudden attack by the enemy from the front, the rear or the flanks: 'If you are expecting combat, have your army march in formation, whether you are proceeding by *droungoi* or by *tourmai* or by entire battle-lines.'[24]

In relation to what I have discussed so far in terms of the tactical changes in the composition and depth of the infantry taxiarchy in the tenth century, can we say that there is any correlation between the double-faced formation described by the *Sylloge Taktikorum* (4–2–4) and the *Praecepta*

Militaria (2–3–2) and the moving of the archers from behind the main infantry force into the middle of the formation? The *Strategikon* describes the formation of the light infantry as a mixed formation, where the archers are posted 'sometimes to the rear of each file in proportion to the number of men' or 'sometimes within the files, alternating one heavy-armed infantry man with one archer'. This is a tactical deployment, which is also described by Asclepiodotus, and possibly Frontinus, and copied by Leo VI in his *Taktika*.[25] The reason, however, why this formation was impractical and, perhaps, even dangerous for the heavy infantry in the first rows of the formation is explained by the following authors.

We read in the *Strategikos* of Onasander:

> The general will assign his light-armed troops to a position in front of the phalanx, for if placed in the rear they will do more damage to their own army than to the enemy, and if among the heavy-armed, their peculiar skill will be ineffectual . . . but drawn up behind the ranks or in among the heavy-armed they will shoot high, so that the arrows have impetus only for their upward flight, and afterwards, even if they fall on the heads of the enemy, will have spent their force and cause little distress to the foe.[26]

The author of the *Sylloge Taktikorum* also highlights the same danger to the infantry: 'A sixteen-deep phalanx is useless because the archers and the slingers would be blocked by the number of infantrymen directly in front of them.'[27]

What do we know, however, about the Byzantine archers and their equipment in the tenth century and, most importantly, what changes does the *Praecepta Militaria* reveal about their tactical role on the battlefields of the period? The Byzantine archer of the tenth century was equipped with the composite bow introduced during the fourth century AD by the Huns.[28] Composite bows have a long history. They were introduced in Egypt and Assyria sometime in the third millennium BC, as they were in use in Mesopotamia, Anatolia and the northern Asian steppes from about 2400–600 BC. They seem to have permeated into the Muslim world following the conquest of Sassanian Persia in the second quarter of the seventh century, but we have to bear in mind that variations in size, shape and performance are noticeable between regions and civilizations; for example, the Turkish bows tended to be shorter and less powerful than the Tartar ones. This weapon, which was essentially first developed as a cavalry bow – thus being short and light enough to use on horseback – was always of compound construction, consisting of five wooden sections spliced together and having sinews glued to the back of the bow with strips of horn to reinforce its belly. The maximum range of a composite bow could reach some 250 yd (225 m), although the effective range where a 550 g arrow could hit its target was reduced to about 100 yd (90 m). The effectiveness of the English longbow, another medieval 'revolutionary' weapon that has fascinated historians for centuries, was considered to have a range of around 220 yd (200 m), where archers could hit the target with a 2–5 m deviation. Tests conducted at the Royal Armaments Research and Development Establishment by Jones, however, have proved that arrows with long needlepoint bodkin heads, shot from a 70 lb (31.75 kg) bow could not penetrate a 3 mm plate armour at a range of 33 ft (10 m), thus confirming Keegan's view that 'these arrows cannot . . . given their terminal velocity and angle of impact, have done a great deal of harm'.[29]

Hence, the main operational role of the archers would have been to shower the advancing enemy cavalry with clouds of arrows to thin their ranks as much as possible by picking up weaknesses in their armour, and to frustrate and disperse their attack altogether by causing mayhem in their ranks. The factors that were at play in the overall effectiveness of archers against cavalry were (1) the number of volleys released against the enemy before impact, and (2) the speed at which the enemy galloped towards the infantry formation. Regrettably, the lack of contemporary evidence on Byzantine archers does not allow us to form a comprehensive picture about their command structure, tactical role or rate of fire.

Nevertheless, recent studies on similar questions regarding Western European and Muslim armies of the Crusader period can prove helpful. According to Keegan's estimates regarding the frequency of fire of the archers and the average speed of the charge of the French knights at the Battle of Agincourt in 1415, a well-trained and experienced longbow archer would have been able to release one arrow shot every ten seconds, while a well-trained Mamluk *ghulam* in Saladin's army could discharge up to five arrows in just three seconds.[30] The French knights would have probably covered the 250 yd distance between the English and the French armies in about forty seconds or so – managing a speed of between twelve to fifteen miles an hour.[31] This would have given the English longbowmen the chance to shoot their bows three times – four, perhaps, if they were lucky.

If we attempt to translate the operational role of the predominantly Armenian[32] archers in the Byzantine army, based on the evidence of the English longbow and the Mamluk archers, we can say that they would have been able to release three bow shots against the advancing enemy's heavy cavalry at 10–12 second intervals, before the enemy could cover the 250 yd of the bow's maximum range of fire. The fact that Phocas moved the bulk of the archers forward and incorporated them into the main body of the taxiarchy would have increased their range of fire, perhaps by as much as 20 yd or more,[33] as the men in front of the first row of archers would have decreased from twelve to just three. These would have crouched to anchor their spears and *menavlia* to the ground, giving the archers the freedom to shoot at the enemy at close range and achieve their maximum penetrating potential, while this may also have enabled them to shoot a fourth volley of arrows. Can this change, then, be seen as an answer to the superior fighting capabilities of the Arab heavy cavalry? It is possible, but we do not have any definitive answer to that question.

However, the most noteworthy development that we can infer from the writings of the tacticians of the tenth century was the increased significance of the role of archers in the Byzantine army and their incorporation into the main body of the infantry soldiers – the taxiarchy. Not only did their numbers increase significantly to more than half of the total number of foot soldiers in each division (not their overall number, as that remained a third of the total of 4,800 men), but we can also see them fighting in unison, as one body of troops along with the rest of their comrades-in-arms, the heavy infantrymen. This is highlighted by the addition of the last two lines of the double-faced taxiarchy formation, the ουραγός, especially consisting of well-trained and experienced men carrying the rank of tetrarch.

We can better understand the significance of this formation for battlefield tactics if we compare it with a very similar development in Western Europe one and a half centuries later. I am referring to the dismounting of men-at-arms by the Anglo-Norman kings in the early twelfth century. At the engagements at Tinchebrai (1106), Brémule (1119) and Lincoln (1141), units of knights dismounted to fight on foot, although there is no direct evidence for the role of archers except a hint of their presence at Brémule in 1119.[34] At Bourgtheroulde (1124), there are very strong indications that units of archers were deployed either in the front (according to Orderic Vitalis) or on the left wing (according to Robert Torigny) of the men-at-arms to repel the charging French knights, while at Northallerton (1138) the ferocity of the Galwegian attack was met by the English archers 'letting off clouds of arrows'. Significantly, the place of the archers at Northallerton was just behind the spearmen and the men-at-arms in the front ranks, a formation bearing great similarities to the tactical formation of a Byzantine taxiarchy.

However, why did the Anglo-Norman knights fight as infantry? The theory that has been proposed suggests not only the influence of the Anglo-Saxon tradition, which was so forcefully exhibited at Hastings, but most importantly the strength and influence of the central authority.[35] The Anglo-Norman monarchy was one of the richest and most centralised in medieval Europe, thus being capable of raising large numbers of infantry that could mix with the professional standing army of the *familia regis*, which only a commander with enough power and authority could compel to dismount and fight on foot.

A similar development in the Byzantine Empire of the tenth century stimulated a change in the strategic, operational and tactical role of the infantry in the imperial army.[36] In marked contrast with the late sixth-century *Strategikon*, which deals with the infantry as an afterthought, the *Praecepta* dedicates its first two chapters to infantry formations. It is also clear from the figures given that the ratio between infantry and cavalry had changed from 3:1 to 2:1, making the foot soldiers – numerically – the most significant element in the Byzantine army of the time. It was during this period that they acquired their own commander, identified as the *oplitarches* (οπλιτάρχης – the commander of the heavy infantrymen, or *oplitai*).

Byzantinists have praised Constantine VII, Nicephorus Phocas and their successors for this radical change in the organisation, training and equipping of the armed forces, along with the rigorous discipline, operational specialisation and professionalism they enforced, with the idea of τάξις (order, discipline) dominating the military texts of the period.[37] As in Polybius and in Procopius' *Persian Wars*, wherever indiscipline is presented as a problem, the discussion is invariably couched in mention of the failure of leadership.[38] This issue prompted Psellus to comment: 'The decisive factor in the achievement of victory was, in his [Basil II's] opinion, the massing of troops in one coherent body, and for this reason alone he believed the Roman armies to be invincible.'[39]

The tactical innovation that finalised the formation of the infantry phalanx of this period was the introduction of the corps of the *menavlatoi*, a unit that would have been deployed in front of the main army in either a linear or wedge formation, putting their heavy spears (*menavlia*) to the ground and at an angle to attack the horses of the charging enemy *kataphraktoi*. A second step taken by Phocas to bolster the defence of the infantry's formation was the deployment of the ουραγός line of tetrarchs to the front of the formation, thus increasing the depth of the heavy infantry in the front rows to four deep (including the *menavlatoi*). This depth of the phalanx could have been further increased by the manoeuvre described by Nicephorus Uranus, known from ancient times as the 'doubling [of the phalanx] by depth'.

A strong indication that the writings of Phocas came as a result of trial and error from years of fighting experience in the East can be found in the number and place of the unit of the *menavlatoi*. The latter were moved from being a protective shield of just 300 troops to being projected some 55 to 72 metres in front of the main army to the front rank of each taxiarchy, thus also quadrupling their number to 1,200 men. The projection of the corps of the *menavlatoi*, however, failed to produce any results. Why would that be?

Although a unit like the *menavlatoi* could withstand an attack by heavy cavalry if it kept its formation unbroken and if it had adequate support from units of cavalry and archers, its slow speed and limited ability for manoeuvre made any (counter) attack a very precarious undertaking, if not unthinkable, even for well-trained professional soldiers. Furthermore, any unit of heavy infantry that was projected forward from the main army was left with its flanks exposed to enemy attack, thus significantly increasing the chance of being encircled by the enemy cavalry.[40] A characteristic example comes from the Battle of Dyrrachium (October 1081) where the Varangian Guard, some 2,000 heavily armed men, fighting dismounted in their Anglo-Saxon custom in the centre front line of the whole formation but projected a few yards forward, was encircled and annihilated by the Normans when the units of the main Byzantine army of Alexius Comnenus failed to keep up with the advancing Saxons in what, we suspect, would have been an order for an all-out counterattack issued by the emperor.[41]

Finally, we should ask why did the tacticians of the tenth century feel the need to introduce this type of 'anti-*kataphrakt*' soldier? Can we explain this as another reaction to the changing tactics of the Muslim armies in the eastern operational theatres of the empire? Regrettably, primary sources do not provide us with concrete information in order to establish whether there was, indeed, a significant change in the structure and consistency of the Muslim armies fighting against the Byzantines in Cilicia, Syria and Mesopotamia. That said, we know that heavily armed cavalry troops had been serving in Muslim armies for centuries prior to the writing of the *Sylloge Taktikorum* and the *Praecepta* – except for the Fatimids whose army was initially based on the Berber tribes and it was only in the late tenth century (the reign of al-Aziz, 975–96), after the military shortcomings of the Berbers were revealed in their fighting in Syria, that a number of Turkish *ghulam* and Daylami soldiers were purchased.[42]

Rather, I believe that the Byzantine high command introduced this new unit of elite foot soldiers because this was the first time since the late Roman period that the corps of the infantry was asked to stand its ground and defend its position – and the cavalry units inside its formation – against enemy *kataphraktoi*. This new kind of heavily armed infantryman clearly reflects the significant change in the nature of the missions undertaken by the Byzantine infantry units, as we have already seen, in complete contrast to the relatively undisciplined, poorly trained peasant militias of the previous centuries, whose role in warfare was mainly the manning of strategic towns, forts and outposts and a kind of frontier guerrilla warfare (as seen in *On Skirmishing*). As such, they were always overshadowed by the heavy cavalry, a situation that the generals of the tenth century desperately tried to change.[43]

Investigating the Root Cause: Tactical Changes in the Cavalry

We have seen that the depth and composition of the regular cavalry units in the tenth century appear to have been different compared with the recommendations of the authors of the *Strategikon* and the *Taktika*. Phocas describes a double-faced cavalry battalion of 500 men, which was five-men deep and a hundred across with the two middle rows composed of mounted archers. The tacticians of our period also recommended the addition of a third line of cavalry, identified by the author of the *Sylloge Taktikorum*, Nicephorus Uranus and the author of the treatise *On Skirmishing* by its Arabic name (*saqah*). What, however, was the underlining reason behind the introduction of the third cavalry line?

The tactical role of the *saqah* is described by the author of the *Sylloge Taktikorum*:

> The general should make sure that there are intervals in the middle of the second line, so that the third line that comes right after that – which they call *saka*, would be able to send [detachments] through the intervals to fend off [the enemy], for better security.[44]

What we understand from the extract is that the third line afforded more security to the entire cavalry formation in case the enemy managed to defeat the units of the first line and throw them back to the βοηθός (*voethos* – the support line). The *saqah* would be able to send detachments to repulse them in order for the general to keep the four units of the second line intact even longer and await the right moment to use them for any counterattack.

Looking for the source of the influence for the Byzantine addition of the third line of cavalry, we find that *saqah* was the technical term used by Muslim tacticians to denote the rearguard of an army, not in battle but rather in a marching formation. The classic model of a Muslim battle formation included five lines, the first two called the 'fighters', the third and fourth protecting the baggage train with the fifth forming the rearguard.[45] The order of the march, however, described by authors such as al-Harthami, al-Ansari, Ibn Khaldun and the early fourteenth-century Damascene treatise *Nihayat al-su'l wa'l-umniyya fi ta'lim a'mal al-furusiyya*, consists of a vanguard (*muqaddama*), a right flank (*maysara*), a left flank (*maymana*) and a rearguard (*saqah*), along with the main force in the middle of the formation.[46]

Another possible sign of adaptation of the Byzantines to the tactics of their enemies is the slight change that we see in the composition of the triangular *kataphrakt* formation in the intervening years between the compilations of the *Sylloge Taktikorum* (c. 930s) and the *Praecepta Militaria* (c. 969). Essentially, what Phocas did was move the mounted archers to the centre of the formation 'where they [the archers] could be protected by them [the *kataphrakts*]', whereas before they composed the entire middle section between the fifth and the eighth rows. Was this an attempt to secure the flanks of the formation from enemy attacks? The answer could be affirmative if we combine this change with a cavalry manoeuvre that was introduced by our authors in this period, involving the corps of the *prokoursatores*.

The *prokoursatores* were a lightly armed reconnaissance and skirmishing unit that galloped ahead of the main army and were expected to turn the initial skirmishes with the enemy units to their advantage. In case they failed, they were supposed to retire, drawing the enemy onto the main force through the intervals of the cavalry units and line up directly behind them.[47] As the *kataphraktoi* would launch their attack, the *topoteretes* (commander)[48] of the *prokoursatores*

> must dispatch fifty of his men through the two intervals on either side of the kataphraktoi out to the right flank of the kataphraktoi and fifty out to the left to ride beside the kataphraktoi and keep the enemy away from their flanks so that they do not divert or disrupt the kataphraktoi and break up their charge.[49]

It is quite likely that Phocas wished to prevent any Bedouin attacks on the flanks of his *kataphrakt* triangular formation, something which could have had a 'bowling ball' effect on the ranks of his cavalrymen and seriously disrupt their change. As McGeer suggests, this recommendation could well be seen as part of the action–reaction pendulum of the Byzantino-Hamdanid wars, as the latter could have spotted the weak spots in the τρίγωνος παράταξις (*trigonos parataxis*), the place where the mounted archers were, and attacked exactly there. Thus, in order to counter the Hamdanid response and for better protection, Phocas not only put the mounted archers in the centre of the formation but also placed two *banda* of *prokoursatores* on either flank.

Notes

1. *Praecepta Militaria*, II.94–8, p. 28.
2. Ibid., II.101–11, p. 28; cf. *On Tactics*, 10, p. 280.
3. For the employment of Bedouins in the armies of the Hamdanids and the Fatimids, see McGeer, *Sowing the Dragon's Teeth*, pp. 225–46; Lev, 'Infantry in Muslim Armies', pp. 185–206; Lev, 'Fatimids and Egypt', pp. 186–96; Lev, 'Fatimid Army', pp. 165–79; Y. Lev (1991), 'The Evolution of the Tribal Army', in Y. Lev (ed.), *State and Society in Fatimid Egypt*, Leiden: Brill, pp. 81–92; Hamblin, 'Fatimid Army', pp. 57–61, 145–9; Beshir, 'Fatimid Military Organization', pp. 38–9.
4. Cappel, 'Byzantine Response to the Arab', pp. 113–32.
5. Hamblin, 'Fatimid Army', p. 58. For more on the *iqta*ᶜ, see M. Chamberlain (2008), 'The Crusader Era and the Ayyubid Dynasty', in C. F. Petry (ed.), *The Cambridge History of Egypt*, vol. 1: *Islamic Egypt, 640–1517*, Cambridge: Cambridge University Press, pp. 227–9; O. Safi (2006), *The Politics of Knowledge in Premodern Islam: Negotiating Ideology and Religious Inquiry*, Chapel Hill: University of North Carolina Press, pp. 87–90; Kennedy, *Armies of the Caliphs*, pp. 81–2; C. Cahen (1972), 'L'administration financière de l'armée fatimide d'après al-Makhzūmī', *Journal of the Economic and Social History of the Orient*, 15, pp. 163–82; C. Cahen (1953), 'Evolution de l'iqta du IXe au XIIIe siècle', *Annales d'histoire économique et sociale*, 8, pp. 25–62; A. K. S. Lambton (1965), 'Reflections on the *iqṭā*', in G. Makdisi (ed.), *Arabic and Islamic Studies in Honor of Hamilton A. R. Gibb*, Leiden: Brill, pp. 358–76.
6. McGeer, *Sowing the Dragon's Teeth*, p. 239.
7. To give one example, Fulcher of Chartres reports that they were the ones who opened the battle at Ascalon in 1099 by attacking the left flank of the Crusaders in an attempt to encircle them; hence, the Muslims were described as 'a stag lowering his head and extending his horns so as to encircle the aggressor with them'. Fulcherius Carnotensis (1913), *Historia Hierosolymitana*, ed. H. Hagenmeyer, Heidelberg: Carl Winters, 1.31.6, p. 314.
8. Arab horses were smaller, faster and especially bred for the climatic conditions of the Middle East and North Africa: A. Hyland (1996), *The Medieval Warhorse from Byzantium to the Crusades*, Conshohocken, PA: Combined

Books, pp. 40–4, 106–23. Leo VI writes about the Arab φαρία (*pharia*): 'which [the horses] are highly prized and not easily procured'; see *Taktika*, XVIII.129, pp. 484–6. Theophanes and Constantine VII mention a treaty between Constantine IV and the Umayyad caliph Muʿawiya under the terms of which the latter was supposed to pay as annual tribute to the empire, among other things, fifty thoroughbred horses: Theophanes Confessor, *Chronicle*, p. 496; *DAI*, 21.16, p. 86. Horse breeding was a speciality of the region of the Syro-Anatolian *thughūr*, especially amongst the Syrian Christian populations; see Asa Eger, *Islamic–Byzantine Frontier*, p. 259. For the trading of horses in Islamic markets, including a detailed discussion and lists, see Durak, 'Commerce and Networks of Exchange', pp. 96–7 and 246–71.
9. 'It does no good at all to go after them [the Arabitai]': *Praecepta Militaria*, II.110–11, p. 28.
10. Ibid., II.126–7, pp. 28–30.
11. Ibn Khaldun, *Muqaddimah*, II, p. 74. We should note here that the *karr wa farr* method of cavalry warfare was considered somewhat outdated by the late twelfth century, largely owing to the extended use of Mamluk horse archers.
12. Al-Ansari, *Muslim Manual of War*, pp. 72, 92.
13. Ibn Khaldun, *Muqaddimah*, II, pp. 77, 78. We should note that *al-majbudah* is probably a term of North African Berber origin.
14. Nicolle, *Crusader Warfare*, II, pp. 135, 143.
15. *Praecepta Militaria*, I.134, p. 20.
16. *On Skirmishing*, 13.4–6, p. 188.
17. Only a few examples will be given here: *On Skirmishing*, 13.4, p. 188; 18.5, p. 210; *On Tactics*, 9.24, p. 276; Constantine Porphyrogenitus, *Three Treatises*, B.135, 140, p. 90; C.57, C.81, C.86, pp. 96–8; *DAI*, 30.45, p. 142.
18. Leo VI, *Taktika*, XI.1, p. 194; see also, ibid., XI.8, p. 196; *Praecepta Militaria*, V.5.21, V.5.23, V.5.33, V.5.37, V.5.45, V.5.54, pp. 52–4.
19. Al-Ansari, *Muslim Manual of War*, p. 91. We should note that in the fourteenth century, al-Ansari had incorporated in his work an abridged version of the ninth-century Abbasid military scholar al-Harthami; see Shihab al-Sarraf (1996), 'Furusiyya Literature of the Mamluk Period', in D. Alexander (ed.), *Furusiyya*, vol. 1: *The Horse in the Art of the Near East*, Riyadh: King Abdulaziz Public Library, p. 120.
20. Psellus, *Chronographia*, pp. 29–30; Attaleiates, *Historia*, CSHB, 50, pp. 108–9, 116.
21. Onasander, *Strategikos*, VI.5, p. 396. In his chapter on 'The Needlessness of Lengthening the Phalanx in Fear of an Encircling Movement of the Enemy', Onasander notes specifically that 'the general who wishes to guard against an encircling movement of the enemy . . . should make his rear and the flanks of the wings as strong as the front ranks'. The basic idea that all sides of the phalanx should be equally strong corresponds with the fact that all sides of the square are equal; see Onasander, *Strategikos*, XXI.2, p. 452.

Tactical Changes: Investigating the Root Causes

22. Asclepiodotus, *Tactics*, XI.4, p. 318.
23. T. G. Kolias (1996), 'Η πολεμική τακτική των Βυζαντινών: θεωρία και πράξη' [The War Tactic of the Byzantines: Theory and Practice], in K. Tsiknakes (ed.), *Byzantium at War (9th–12th c.)*, Athens: National Research Institute, pp. 158–9. See also G. Theotokis (2014), 'From Ancient Greece to Byzantium: Strategic Innovation or Continuity of Military Thinking?', in B. Kukjalko, I. Rūmniece and O. Lāms (eds), *Antiquitas Viva: Studia Classica*, 4, Riga: University of Latvia Press, pp. 106–18.
24. Leo VI, *Taktika*, IX.5, p. 154.
25. Asclepiodotus, *Tactics*, VI, pp. 272–4; Leo VI, *Taktika*, XIV.60, p. 326. Frontinus describes a battlefield deployment of a mixed formation, where the heavy infantry (*legionarii*) were posted in the first line and in reserve, while the light infantry (*auxilia*) were intermingled between them (*his immiscuerunt auxilia*); see Στρατηγήματα [*Strategemata*], II.iii.21, p. 120.
26. Onasander, *Strategikos*, XVII, pp. 444–6.
27. *Sylloge Taktikorum*, 43.1, p. 66.
28. For a selected bibliography on the composite bows, their history, construction, characteristics and their use, see M. Strickland and R. Hardy (2005), *From Hastings to the Mary Rose: The Great Warbow*, Stroud: Sutton Publishing, pp. 97–112; J. C. N. Coulston (1985), 'Roman Archery Equipment', in M. C. Bishop (ed.), *The Production and Distribution of Roman Military Equipment*, Oxford: Oxford University Press, pp. 220–366; W. F. Paterson (1966), 'The Archers of Islam', *Journal of the Economic and Social History of the Orient*, 9, pp. 69–87; P. E. Klopsteg (1987), *Turkish Archery and the Composite Bow*, London: Butler & Tanner; A. D. H. Bivar (1972), 'Cavalry Equipment and Tactics on the Euphrates Frontier', *Dumbarton Oaks Papers*, 26, pp. 271–91, especially pp. 282–7; W. E. Kaegi (1964), 'The Contribution of Archery to the Turkish Conquest of Anatolia', *Speculum*, 39, pp. 96–108; Kolias, *Waffen*, pp. 220–35; T. May (2007), *The Mongol Art of War: Chinggis Khan and the Mongol Military System*, Barnsley: Pen & Sword Military, pp. 50–3.
29. D. Whetham (2008), 'The English Longbow: A Revolution in Technology?', in L. J. A. Villalon and D. J. Kagay (eds), *The Hundred Years War*, part 2: *Different Vistas*, Leiden: Brill, pp. 213–30; Keegan, *Face of Battle*, pp. 94–8. See also C. J. Rogers (2008), 'The Battle of Agincourt', in L. J. A. Villalon and D. J. Kagay (eds), *The Hundred Years War*, part 2: *Different Vistas*, Leiden: Brill, pp. 37–131; K. DeVries (1997), 'Catapults are Not Atomic Bombs: Towards a Redefinition of "Effectiveness" in Premodern Military Technology', *War in History*, 4, pp. 454–70.
30. Keegan, *Face of Battle*, p. 70; Paterson, 'Archers of Islam', pp. 69–87, especially pp. 82–6; Strickland and Hardy, *From Hastings to the Mary Rose*, p. 100. See the very useful source on Muslim archery by a late fourteenth-century Muslim writer of a military treatise (*furusiyya*): Taybugha al-Baklamishi al-Yunani (1970), *Saracen Archery*, trans. J. D. Latham and W. F. Paterson, London: Holland Press.

31. Keegan notes at this point that the French horses were probably a big hunter type. These were certainly bigger and heavier than the Arab horses and probably slower as well; Keegan, *Face of Battle*, p. 70. I have not come across a study, however, that would calculate, with some margin for error, the average speed of Muslim horses in combat.
32. Beshir, 'Fatimid Military Organization', p. 43; Hamblin, 'Fatimid Army', pp. 26–7, 151–2.
33. *Sylloge Taktikorum*, 43, p. 67.
34. Strickland and Hardy, *From Hastings to the Mary Rose*, pp. 70–1. On the role of archers at Tinchebrai, Brémule, Bourgtheroulde, Northallerton and Lincoln, see J. Bradbury (1985), *The Medieval Archer*, Woodbridge: Boydell & Brewer, pp. 41–57.
35. Morillo, '"Age of Cavalry" Revisited', pp. 54–6.
36. Haldon, *Warfare, State and Society*, pp. 218–19.
37. McGeer, *Sowing the Dragon's Teeth*, pp. 202–11; Haldon, *Warfare, State and Society*, pp. 218–19; Toynbee, *Constantine Porphyrogenitus*, pp. 282–322; Treadgold, *Byzantine State and Society*, pp. 487–94.
38. Whately, *Battles and Generals*, pp. 89–96. See also R. L. Moore (2002), 'The Art of Command: The Roman Army General and His Troops, 135 BC–AD 138', PhD dissertation, University of Michigan.
39. *Sylloge Taktikorum*, 45.8, p. 72; Psellus, *Chronographia*, I.33, p. 17.
40. The example of the Battle of Pydna (22 June 168 BC) clearly illustrates the dangers of operating even a highly professional heavy infantry like the Macedonian phalanx without adequate support from other arms, such as archers and cavalry. The gaps that were created between the units of the Macedonian *phalangites*, after the first line lost its cohesion during the attack because of the uneven terrain, prompted Lucius Aemilius Paullus to order his legions to strike at exactly that weak spot, attacking the *phalangites* on their exposed flanks; see A. Goldsworthy (2000), *Roman Warfare*, London: Cassell, p. 55; J. D. Montagu (2006), *Greek and Roman Warfare: Battles, Tactics, and Trickery*, London: Greenhill, pp. 220–8. At Courtrai, the front and flanks of the Flemings were protected by ditches. At Mons-en-Pévèle, one wing was supported by the village hedge and the other was up against a brook. At Falkirk, the Scots had a marsh in front of their battle order, and archers were stationed on the wings to cover the flanks. At Bannockburn, their front was protected by artificial pits, while the main battle order was drawn up in a wood; see Verbruggen, *Art of War*, p. 182.
41. Theotokis, *Norman Campaigns in the Balkans*, pp. 161–2.
42. Lev, 'Evolution of the Tribal Army', pp. 81–93; Hamblin, 'Fatimid Army', p. 32.
43. Haldon, *Warfare, State and Society*, pp. 197–200.
44. *Sylloge Taktikorum*, 46.17, p. 81.
45. Al-Ansari, *Muslim Manual of War*, pp. 100–2; Ibn Khaldun, *Muqaddimah*, p. 75. A translation of the battle array described by Fakhruddin Mubarak

Shah, a fourteenth-century ruler of an independent Bengali kingdom, can be found in Nicolle, *Crusader Warfare*, II, pp. 127–9.

46. Al-Ansari, *Muslim Manual of War*, pp. 84, 90; Ibn Khaldun, *Muqaddimah*, p. 75; G. Tantum (1979), 'Muslim Warfare: A Study of a Medieval Muslim Treatise on the Art of War', in R. Elgood (ed.), *Islamic Arms and Armour*, London: Scolar Press, p. 198; Nicolle, *Crusader Warfare*, II, p. 136.
47. *Sylloge Taktikorum*, 46, p. 79; *Praecepta Militaria*, IV.93–106, p. 44; McGeer, *Sowing the Dragon's Teeth*, p. 303, especially diagram 22 on p. 304.
48. This is the first time that the unit of the *prokoursatores* acquire their own commander, reflecting once again the high degree of specialisation in the units of the Byzantine army of the period, *Praecepta Militaria*, IV.13, p. 38.
49. Ibid., IV.127–32, p. 46.

9

Byzantine–Arab Battles of the Tenth Century: Evidence of Innovation and Adaptation in the Chronicler Sources

The objective of this chapter is to examine the most important primary sources for the period of the Byzantine expansion in the tenth century. These include two Byzantine sources, namely Leo the Deacon and John Skylitzes, whose accounts of the Byzantine wars in the Balkans are considered by several modern historians as the best and most detailed in hand; a local Syriac source, Yahya ibn Said al-Antaki from Antioch; and three Muslim sources, al-Mutanabbi, Abu Firas and Ibn Zafir, who provide us with invaluable information about the Byzantine–Arab conflicts of the 940s–60s in Cilicia and Syria. I will focus my analysis on the chroniclers' social, religious and educational backgrounds, the dates and places of the compilation of their work, their own sources and the way they gleaned information from them, and their biases and sympathies, which shed light on their level of impartiality as historians. This section will be followed by a comparative analysis of the sources strictly from a military perspective, reaching significant conclusions regarding their value as 'military historians'.

Leo the Deacon

The work of Leo the Deacon is considered one of the best histories of the so-called period of the 'Reconquest' in Byzantium and a much-valued source of tenth-century Byzantine warfare.[1] The few facts we know about Leo and his life come primarily from scarce references in his *History*, as this was the trend amongst Byzantine classicising historians like Procopius and his successors who followed in the footsteps of Herodotus and Thucydides, neither of whom provided much personal information in their own works.[2] Leo was born around the year 950 in the small town of Kaloë in western Anatolia, a bishopric dependent on Ephesus, just south-west of Philadelphia. He was already in Constantinople as a youth around the year 966 pursuing his secondary education (εγκύκλιος παίδευσις), as he tells us in his fourth book.³ The language in his work reveals a writer with a classical education, who was well versed in the ancient authors – especially Homer, as attested by the frequent quotations in his work, but also Herodotus, Diodorus, Dionysius of Halicarnassus, Arrian, Dio Cassius, Herodian, Procopius, Agathias and Theophylact – whose books he had probably found in the imperial library.⁴ In fact, Leo generally uses archaic and literary vocabulary for military units, equipment and ships, although he is not always consistent. On a number of occasions, he includes a contemporary term for further explanation – for example, his use of the term *triremes*, which 'the Romans call *dromones*'.⁵ The classical education of the period would typically have included grammar, rhetoric and philosophy,⁶ with the study of the former focusing on Homer, and especially the *Iliad* given Leo's special interest in warfare.⁷ He was probably ordained as a deacon in 975, as the minimum age for the diaconate was 25, and he immediately became a member of the palace clergy after Basil II's ascent to the throne in 976.⁸ He was evidently in Constantinople in 985, as he informs us of the downfall of Basil the Nothos, while he also reported on the Battle of Abydos in 989 and an earthquake that damaged Hagia Sophia in the same year.⁹ The *terminus post quem* for the composition of his *History* is placed in the year 995, when Leo delivered an oration in praise of Basil II. There is, however, a theory according to which Leo became Bishop of Caria and lived well beyond the year 1000,¹⁰ although we cannot find any clues in his *History* that he survived that terrible (for medieval clerics) year.

Leo's work has been described as being midway between a world chronicle and the humanistic memorial of an emperor.¹¹ The *History* is divided into ten books based on the reigns of two emperors, Nicephorus Phocas (963–9, Books I–V) and John Tzimiskes (969–76, Books VI–X), along with digressions into the reigns of Romanus II (959–63) and Basil II (976–1025). Indeed, Leo writes that he intended to cover Basil's reign up to 995 as well, but he obviously failed to do so.¹² Perhaps he thought he would benefit by publishing his first ten books to begin with, and then carry on with the reign of Basil, possibly with his 995 oration playing the role of some sort of preview.

Leo's narrative is based on the deeds of the aforementioned emperors, as the four dates and the two sets of regnal years covered in his work clearly suggest. He is only peripherally interested in the year of occurrence of major events. It is worth noting that the continuous narrative and the strict chronological order of events used by chroniclers up to this period was gradually replaced in the tenth century by this new and innovative historiographical method, which puts the analysis of specific individuals like the emperors at the core of the narrative.[13] Leo's dating differs significantly from that of other chronographers and his irregular indications of the change of seasons are primarily related to military campaigns, which were after all his primary focus (spring and autumn signalled the beginning and end of expeditions respectively).[14]

Leo refers to the sources of his narrative at the beginning of his work:

> But I will now set down in writing subsequent events, both those that I saw with my own eyes (if indeed *eyes are more trustworthy than ears*, as Herodotus said),[15] and those that I verified from the evidence of eyewitnesses.[16]

Although Leo stipulates that he was an eyewitness of a number of the events he describes in his work (primarily for the period 959–76), it is very difficult to determine exactly what events he witnessed and what are based on evidence from oral accounts. We know that he was a young student in Constantinople in 968 at the time of an eclipse (22 December 968). Thus, his relative youth would not have allowed him to be an eyewitness to the events prior to the 970s, but he was old enough to attend to Emperor Basil II as a deacon in his Bulgarian campaign of 986.[17] Leo uses λέγεται (*legete*, 'it is said') thirty-seven times and φασί(ν) (fasi(n), 'they say') fifteen, probably wishing to indicate to his readers that he had not examined personally the information he received from his sources.[18]

Unfortunately, Leo does not identify a single one of these as being either written or oral. Treadgold has suggested that Leo would have personally known Symeon Logothetes and the author of the chronicle of Pseudo-Symeon, both covering the period between 842 and 948/963 in relative detail, and that he would have largely relied on their work for the period before 959.[19] It has also been proposed that both Leo and Skylitzes used a common source – now lost – for the years 969 to 971, a history that was vehemently hostile to the Macedonian dynasty and Nicephorus II – although it did praise John I – that extended to the year 971 and was probably, although not conclusively, composed by Nicephorus 'the Phrygian', a well-educated deacon in the imperial palace.[20] In fact, Nicephorus must have been well informed about events and gossip at court, as he provides considerable information about warfare and diplomacy up to the campaign against the Rus' in 971, and he may even have been an eyewitness.[21] Cheynet has suggested that Leo may have taken some information from members of the Parsakoutenoi clan of the Phocas family, while Moravcsik has raised the possibility of a participant in Tzimiskes' campaign against the Rus', as a first-hand account.[22] Finally, Morris and Treadgold have also suggested that Leo may have known secretaries in the imperial chancery, who could have given him access to an 'official panegyric' of John Tzimiskes and to other state documents.[23] Leo's omission or lack of knowledge of several major events that characterised Phocas' reign, however, like his relationship with Athanasius of Lavra and his patronage of that great religious house, his controversial legislation regarding church and monastic property, his devaluation of the currency, his foreign relations with Otto II, and many others should make us view Leo's writings with some caution and examine them in parallel with other contemporary or later sources.

John Skylitzes

John Skylitzes' life is rather obscure but we know that he was a high-ranking judge in the capital, who lived in the second half of the eleventh century and held the titles of *protovestiarius*, *kouropalates* and *proedrus*, and the offices of prefect of Constantinople and *drungarie* of the Watch.[24] Called a 'Thracesian' by Zonaras and Cedrenus, he would probably have been born in western Anatolia into a family important enough to acquire a surname – Skylitzes, or 'Little Dog'. The only other people with that particular surname that we know of are probably the historian's descendants, who distinguished themselves as members of the clergy later in the twelfth century.[25]

The *Synopsis Historion* was written almost certainly during the early years of the reign of Alexius Comnenus (1081–118) and, as the title suggests, it is a comprehensive digest of historical works already in existence written in a simple, unaffected language and terminology, that is to say a narrative that could be clearly understood even by the masses. The author makes no claim to be dealing for the first time with neglected material but that he rather wishes to rewrite past histories: 'All of this [existing knowledge of past authors] I put together in summary form and this [my work] I now bequeath to future generations as an easily digestible nourishment, "finely ground-up" as the proverb has it.'[26] The first edition of the *Synopsis*, written most likely in 1092–4, covers the period from the year 811 and the death of Nicephorus I to the year 1057 and the abdication of Michael VI, and was later used by Cedrenus. Skylitzes chose to expand his first edition up to 1079, with this supplemented edition later used by Zonaras.[27]

The disparate lengths of text devoted to each reign, ranging from just a few pages for Michael I or Romanus II to as much as forty pages for the reigns of Basil II and Constantine IX, make the *Synopsis* an invaluable source for the history of the empire from the mid-tenth century onwards. As I have already mentioned, it is theme rather than chronology that is the dominant organising principle in the *Synopsis*. Like Leo's *History*, the work of Skylitzes is structured around the reigns of the ruling emperors and all natural events that occur (famines, earthquakes, eclipses, the birth of conjoined twins, etc.) are mentioned in relation to the corresponding reign and even interpreted as signs of divine approval or censure of the emperor's deeds and government. Along with Leo, he is described as a 'biographer-encomiast ... concerned to extol [his] subjects' martial prowess and, to a lesser degree, their actions in daily life'.[28]

The events that Skylitzes mentions in his work can be divided into two categories: (1) those that take place in Constantinople, encompassing a whole range of things such as the appointment of a patriarch, the foundation of a church or court gossip, and (2) foreign affairs, meaning war, either civil or foreign, with the theatre of operations shifting around the borders of the empire, sometimes to give a description of a battle, siege or naval campaign and sometimes to simply list a number of places conquered from the enemy.[29] However, our author's dating of events, namely the setting down of the 'year of creation' (*anno mundi*), the 'year of the incarnation' (*anno domini*), the *indiction*, the regnal years of the emperor and/or the Sassanian ruler, is rather inconsistent and bears no resemblance to the systematic chronological system that can be found in the works of Theophanes Continuatus or Georgios Sygkellos, two authors whom Skylitzes revered.

Skylitzes names ten historians who wrote after Theophanes in the foreword of his *Synopsis*: Theodore Daphnopates, Genesius, Nicephorus the Deacon, Theodore of Side, Theodore of Sebastea, John the Monk, Leo the Deacon, Nicetas the Paphlagonian, Manuel the Protospatharius and Demetrius of Cyzicus. Nevertheless, this does not mean that he had read all of them, since we know that he made use of the latter three from references by other historians.[30] We do know, however, that in the sections of the *Synopsis* that deal with the ninth and the first half of the tenth centuries the sources are still extant, including the Περί Βασιλείων (*Peri Vasileion*) of Joseph Genesios,[31] who wrote under the command of Constantine VII, and Theophanes Continuatus, whose fifth book (*Vita Basilii*) Skylitzes used intensively at the end of his Book VI.[32]

The sources Skylitzes relied on to write about the reigns of the emperors of the tenth century are very difficult to identify, causing much speculation and several different theories to emerge among modern scholars. I have already mentioned one of the suggestions, first presented by Siuziumov and Kazhdan, according to which Skylitzes and Leo used the *History to 971* by Nicephorus 'the Phrygian', a well-informed deacon at the imperial palace who apparently had considerable information on war and diplomacy up to the campaign against the Rus' in 971, so much so that it has been argued he may have been an eyewitness to many of the events.[33] His history was used by Skylitzes for the years between 944 and 971, and by Leo the Deacon for the years 969 to 971. The 'schizophrenic' description of the reign of Phocas, however, which is in some respects favourable and in others extremely hostile to the emperor, can be explained by the potential use of an alternative, anti-Phocas source written sometime before 1000.[34] Moravcsik and Shepard have also suggested the use of a 'war diary' for Tzimiskes' reign, especially for his wars in Bulgaria, but it is difficult to know whether this was a direct or indirect source.[35] Skylitzes would probably have known the work of Leo 'the Asian' – most definitely to be identified with Leo the Deacon – although it is difficult to pinpoint exactly how far he made use of Leo's history due to the lack of a sufficient number of textual similarities between the two works.[36]

For Basil II's reign, none of Skylitzes' underlying source materials survive, although piecemeal evidence has led a number of historians to believe that, for the years 976–1025, Skylitzes most likely drew on a lost history of Theodore of Sebasteia, who continued the history of his uncle – Theodore, Bishop of Side – that went back to 811.[37] Nevertheless, Holmes has suggested that the minimalistic treatment of Phocas and the extensive coverage of Bardas Skleros in Skylitzes' appraisal of the period of revolt of 976–89 points to the fact that our author may have been more interested in source materials that focused on the activities and ambitions of Skleros than Phocas or Tzimiskes (for example, in his narrative on the Battle of Arkadiopolis in 970, Skylitzes highlights the role of Skleros in defeating the Patzinaks).[38] For the so-called period of the *Epigonoi* that followed the death of Basil II until 1057, Skylitzes made direct use of the history of John the Monk who, in turn, had paraphrased and revised the work of Demetrius of Cyzicus. It has been suggested that the latter had direct access to Constantinopolitan annals that constituted an extraordinary source of information on the empire's wars between 1025 and 1043.[39]

We can argue conclusively that, as a whole, the *Synopsis* draws on quite a limited number of sources which Skylitzes had to hand, and it does not seem wrong to assume that in the part of his work where he depends on Theophanes Continuatus he just paraphrases or embellishes the single source that he is using. Flusin and Holmes, however, have argued that this harsh criticism should not be applied to the work in its entirety and that there are parts in the *Synopsis* where Skylitzes' editing process has been thorough, especially where the work relies on sources – apart from Psellus – which have not survived. Indeed, some scholars even find the idea of the author personally collecting evidence in support of the information he received from his sources quite attractive.[40] Thus, Skylitzes seems to have kept the promise set out in the preface of his work:

> I read the histories of the abovementioned writers with great care. I conjured away from them all comments of a subjective or fanciful nature. I left aside the writers' differences and contradictions. I excised whatever I found there which tended toward fantasy; but I garnered whatever seemed likely and not beyond the bounds of credibility.[41]

Yahya ibn Said al-Antaki

The historical chronicle of Yahya ibn Said al-Antaki is the most prominent literary evidence of the flourishing and diverse relations that existed between medieval Islam and Byzantium, especially during the reign of al-Hakim (996–1021).[42] Yahya is an important historical figure because he is the only Greek Christian author of an extant Arabic chronicle, which constitutes the basic source on the Byzantine–Arab interaction in Syria and Mesopotamia in the second half of the tenth and the first half of the eleventh centuries. Other Arab sources on the subject add only fragmentary evidence, or what has survived has been abridged and incorporated into the works of subsequent authors. Yahya is also a font of rare information on the local east Anatolian and Syrian environment and geography, and an informed and exceptionally factual and objective commentator on the political and religious conflicts in the eastern Mediterranean basin. His narrative remains focused on the facts, and he rarely and reluctantly engages with the anecdotal stories of events found among contemporary Greek sources such as Leo the Deacon.[43]

Yahya was born an Orthodox Christian of the Melkite rite in Fatimid Egypt around 980. He left Egypt in 1014–5 for Byzantine Antioch in order to take refuge from the religious persecution of the Fatimid caliph, al-Hakim, and remained there until his death in 1066.

A well-educated man and, according to one theory, a physician,[44] his work is a continuation of the *Nazm al-jawhar* by the Greek patriarch Eytychius of Alexandria (d. 940), to whom he was related (probably his nephew) and with whose output he was dissatisfied. He wrote the first two drafts of his work in Egypt, completing a third one after he settled in Antioch, although we know that he kept revising it up to the year 1034 (the year of the death of Emperor Romanus III) to bring it up to date with events taking place in the Middle East, such as the Byzantine conquest of Edessa in 1031.[45]

Yahya's history is not a chronicle in the strict sense of the term. His style of historical writing is distinctly individual. He neither follows the strict annalistic pattern of other major Islamic chronicle traditions, where events are placed rigidly under the year they took place (as in *The History of al-Tabari* or *The Chronicle of Ibn al-Athir*), nor does he imitate the Egyptian local tradition for the biographies of prominent individuals. Rather, he organises his work based, in some parts, on the reigns of caliphs (first the Abbasid and then the Fatimid) and in others on geographical areas, although he seems to have been interested solely in the region of the eastern Mediterranean (Egypt, Syria and Byzantium's eastern provinces).[46]

Yahya's sources can be divided into three categories: Greek, Syrian-Christian and Muslim.[47] For the history up until the reign of Constantine VII, it has been argued that Yahya used a number of Byzantine sources, such as Theophanes Continuatus and Symeon Logothetes, simply on account of the number of instances where the narration and even some expressions are similar. We do not know, however, whether he had access to the Greek texts or to Arabic translations.[48] For the later period dealing with the reigns of Romanus II, Nicephorus Phocas, John Tzimiskes and Basil II (especially regarding Bardas Skleros' revolt), Yahya probably had access to Greek sources, although the common references become rarer and in some cases his account contradicts the Greek. According to Vasiliev, it is quite possible that he may have had access to the ecclesiastical sources of the Patriarchates of Alexandria and Antioch, due to his family connections.[49] There are also four Muslim sources that our author would have had at his disposal: Thabit ibn Sinan, Ibn Zulaq, Ali ibn Muhammad al-Shimshati and al-Musabbihi. The first three were amongst the most famous historical writers in Arabic of the tenth and eleventh centuries. Some scholars also suspect that Yahya may have been familiar with the work of his contemporary Ibn Miskawaih.[50]

Abu at-Tayyib Ahmad ibn al-Husayn al-Mutanabbi

Perhaps one of the greatest poets in the Arabic language of all time, Abu at-Tayyib Ahmad ibn al-Husayn al-Mutanabbi (915–65) was an Iraqi-born son of a humble family, raised in the Mesopotamian city of Kufa and educated in Damascus.[51] His nickname, Mutanabbi, means 'the one who wants to become a prophet' and tells us a lot about his political ambitions of becoming a *wali* (a holy man or a Muslim prophet). It was during his revolutionary activities in the early 930s, a period when he was imprisoned, that he began writing his poems. Having sought riches and personal glory in the courts of several Muslim lords, he was introduced in 948 to Sayf ad-Dawla by Abu'l-Asahir, the Hamdanid governor of Antioch. In the person of Sayf ad-Dawla, Mutanabbi found a powerful and influential patron, an educated man who was interested in poetry and philosophy, a champion of Islam who was an Arab of pure blood rather than a Turk or an Iranian Daylami (bearing in mind the increasingly dominant role of this group in the courts of the Abbasids during our period), to whom he was devoted and whose wars against the Byzantines he was more than happy to immortalise.

From 948 to 956 (the year Mutanabbi fell out with Sayf and departed for Egypt), he accompanied Sayf ad-Dawla in his campaigns in Syria, Cilicia and Mesopotamia, and was inspired by the latter's military achievements into writing some of his finest pieces of poetry. Mutanabbi's poems, along with those of Abu Firas, whom I discuss below, belong to a specific category of lyric poetry that originated in pre-Islamic Arabia, known as *qasida* (Arabic for 'intention').[52] The classic form of *qasida* maintains a single elaborate metre throughout the poem and every line rhymes. It typically runs for more than fifty lines, and some times for more than a hundred. Since it very often takes the form of a panegyric, written in praise of a king or a nobleman – a genre known as *madih* (praise) – the theme most popular already since pre-Islamic times is the relationship between the poet, the patron and the poem. These three are interdependent, since the poet depends on the patron's favour for sustenance, while the patron sees his glory enhanced by the poem.[53]

The ideal poet–patron match is celebrated in Mutanabbi's works, where he idealises Sayf ad-Dawla as the romanticised embodiment of Arab-Muslim chivalry. Yahya fought side by side with the Hamdanid prince on several military campaigns and his panegyric seems to be sincere. The objectivity of Mutanabbi's material, however, has been questioned regarding his writings about Sayf ad-Dawla and the Ikshidid caliph Abu Kafur (in whose court he sought refuge after his departure from Aleppo).

Considering the obvious contrast between the two patrons as they appear through his poems, the anecdotal literary material surrounding Mutanabbi appears to be largely ideologically motivated and, although possessing considerable cultural literary value, it should not be taken as objectively historical.[54]

Abu Firas

Born to a Greek mother, Abu Firas (932–68) was a high-ranking member of the Hamdanid family and a cousin of Nasir and Sayf ad-Dawla.[55] At an early age, he was made governor of Menbij (Hierapolis), a strategically located city in the Taurus frontier and a focal point for the Byzantine–Arab wars of the 950s–60s, and later of Harran where, despite his youth, he distinguished himself in the conflicts with the Nizari tribes of Diyar Mudar in the western part of the Jazira. Completely devoted to Sayf ad-Dawla, whom he revered as a father, he took part in almost all of Sayf's expeditions and was taken prisoner by the Byzantines in 962, spending four years in captivity in Constantinople.

A collection of his poems, titled *al-Rumiyat*, is one of our most important sources on the life of Sayf ad-Dawla and the Byzantine–Arab wars of the period. Four sections of his work are of particular interest: a historical piece narrating the role of the Hamdanids in the Islamic world, especially against the Byzantines, two poems for the 950 and 958 campaigns respectively, and a poem composed during his captivity in the capital, where he met with Nicephorus Phocas. He was, of course, an eyewitness to many of the events he described in his poems, but he must also have had access to other contemporary witnesses, especially at the Hamdanid court, which included his cousins, and perhaps some sort of archival material in Aleppo.

Ibn Zafir

Although not a contemporary to the events he describes, Ibn Zafir (1171–226) is another very important source for the history of the Byzantine–Arab conflicts of the 950s–60s, whose history contains information not found in any other work, even though his sources remain unknown.[56] He was an important, educated man, and for some time a professor at the al-Kamiliya madrassa in Cairo where he succeeded his father. A few years before his death, he was appointed vizier to the Egyptian sultan. His *Book of Histories of Lost Dynasties* (*Kitab al-duwal al-munqatiʾa*) contains a comprehensive history of the Hamdanid, Tulunid, Ikshidid, Fatimid and

Abbasid dynasties. His information is unique and extremely valuable, especially for the campaigns of Sayf ad-Dawla.

A Comparison of the Strengths and Weaknesses of the Principal Narrative Sources, and their Respective Value as 'Military Historians'

The aim of this section is to examine the primary sources of our period strictly from a military perspective and attempt to reach some conclusions regarding their value for the history of tenth-century warfare in the region of the Balkans and Asia Minor. The major questions that will be raised are: To what extent are the figures they provide for army size reliable, both in absolute numbers and in the ratios given between cavalry and infantry? What was our chroniclers' knowledge of the local geography where the military operations took place, and to what extent – if at all – were they familiar with the terrain of the battles, sieges or army campaign routes they describe? How accurate or detailed are their descriptions of castles and fortifications and how far do their narratives permit the accurate reconstruction of a chain of events, especially regarding the battlefield manoeuvres of armies in action?

Another important question that should be raised at this stage relates to the dangers modern historians face in using chronicler material for their interpretation of events – especially in terms of battle tactics, strategies and military campaigns. This topic was first raised by Verbruggen in the mid-1950s and has been picked up since by Keegan and, among others, DeVries, Morillo and Abels.[57] Verbruggen criticised followers of the so-called 'old school' of military historians, namely Delpech and the Prussian general Kohler, for producing works that lack the critical faculty which is indispensable to the study of the art of medieval warfare, and contrasted their work with other historians like Oman and Delbruck. His main argument is that it is necessary to check the military value of each chronicler's account for possible inventions and legends, which can be spotted solely through the comparison of many sources, contemporary or not.

What exactly are the limitations of medieval sources and what danger do they pose for a modern researcher? Since my study includes people from many levels of society, a palace cleric (Leo the Deacon), an army officer and court official (John Skylitzes), an educated physician (Yahya ibn Said), a poet (Mutanabbi), a Muslim prince (Abu Firas) and a madrassa professor (Ibn Zafir), a brief presentation of these limitations should be made at this point. In many cases, several of the clerical sources give an incomplete narrative of battles, sieges or entire campaigns, simply because reporting these events in a manner similar to a modern war correspondent was not their objective. It has to be stressed, however, that it is undoubtedly an oversimplification to argue that all clerics were ignorant of military affairs, keeping in mind that Orderic Vitalis and William of Poitiers are two of the foremost sources for Anglo-Norman military history, while Suger, Abbot of St Denis, gives a vigorous account of King Louis VI at war. Similar to their Muslim counterparts, they were decidedly non-professional military historians and we should not expect specialised military insight from their work. Furthermore, the authors of medieval Islamic historical sources were more often than not religious scholars who had devoted their lives to the in-depth study of the Quran before becoming interested in the writing of history, which they saw through the prism of faith.[58]

The chroniclers' works are also affected by invention and/or exaggeration depending on their biases and sympathies. The most characteristic example is Mutanabbi's *qasida* poetry, which glorifies Sayf ad-Dawla and his exploits against the Byzantines. Other times, their narrative is embellished with religious elements owing to their religious convictions, as very often they tend to ascribe victory to a miracle from God, such as Skylitzes' narrative for the battle at Dorystolon – 'The Romans are said to have benefited from the enhanced supernatural aid at that time, for a storm arose in the south'[59] – or Usama's declaration: 'Victory in warfare is from Allah (blessed and exalted is he!) and is not due to organization and planning, nor to the number of troops and supporters.'[60] Indeed, what a striking contrast to Procopius' dictum, placed into the mouth of Leo Phocas' by Leo the Deacon: 'For wars are usually won not so much by a pitched battle as by cautious planning',[61] or Leo VI's conviction that: 'it is not true, as some inexperienced persons may hold, that wars are decided by a multitude of men and courage, but by the favour of God and by generalship and discipline',[62] and Vegetius' argument that 'in every battle it is not numbers and untaught bravery so much as skill and training that generally produce the victory'.[63] For medieval warriors, God was essentially the Old Testament God of Battles, who was the ultimate arbiter and whose aid was vital in ensuring both personal safety and corporate victory. Essentially, men regarded battle as a judicial duel on a grand scale. Therefore, besides the importance of tactics and morale, it is worth emphasising the religious context within which many late antique and medieval historians wrote, and which heavily influenced them in allowing a role for the divine in historical causation.[64]

Another issue concerns the problems raised by using inaccurate terminology to translate a term that was used by secular or ecclesiastical sources at a particular place and time. In his monumental *The Art of Warfare in Western Europe During the Middle Ages*, Verbruggen argues: 'Among the best of the most reliable sources are those written in the vernacular ... They provide a clear and distinct terminology.'[65] He then goes on to back up his argument by listing several German terms that were diffused into the *lingua romana* between the sixth and the ninth centuries, to conclude with the following comment: 'Even though chroniclers do not always indicate the difference between these units, the terminology is clear and what is being discussed is unmistakable.'[66]

Verbruggen's position on this subject has come under attack by Abels and especially Morillo, who have directed attention to the fact that 'we [modern historians] must constantly be on our guard in examining the histories of those individual [historical] places so as not to read back our own views of the world into their times'.[67] Morillo urges caution when it comes to translating individual historical terms, as it is paramount to place the term firstly in its appropriate sociological, economic and administrative context before drawing any conclusions. Indeed, how we see history depends on what terms we use to describe the past, and the variety of cultures and languages only serves to make the historian's task even more challenging.

In order to give an example of how pertinent it is to show caution when it comes to translating primary sources and subsequently basing historical interpretations on those translations, I will refer to the use of the term κατάφρακτος (*kataphraktos*) by the Byzantine sources and the controversy that for many years surrounded the 'reintroduction' of the unit into the operational theatres of the East in the mid-tenth century. I deliberately chose to put the term *reintroduction* in inverted commas above because I believe that the views of several Byzantinists who deal with the organisation of the army of this period are misleading.

The established opinion has the heavily armed cavalry unit of the *kataphraktoi* introduced into the ranks of the Byzantine army during the reign of Emperor Nicephorus II after centuries of absence.[68] Nicephorus' legislation was responsible for their recruitment, armament and training, while their disappearance from the sources during the reign of Basil II has led some to assume that they were disbanded after Byzantium's victories in the East and the Balkans. Although I support the view that the triangular formation of the *kataphrakts* is, indeed, an innovation of this period – perhaps it may date to the campaigns of John Curcuas in the 930s–40s[69] – I am more inclined to agree with Kolias' view on this issue, whose main argument for the reintroduction of the heavy cavalry unit is based on three major points: the use of the term *kataphraktos* by the primary sources,

Phocas' legislation regarding the land of the cavalry *stratiotai* (soldiers), and the empire's strategy in the operational theatre in the East.[70]

The last recorded use of the term *kataphraktos*, before its reappearance in the military manuals of the tenth century, comes from the anonymous military treatise *On Strategy*.[71] Until about a decade or so ago, this treatise was believed to have been compiled in the sixth century, which gave us an intervening period of four centuries until the term's next appearance in the period of the Byzantine 'Reconquest'.[72] It is now accepted, however, that the work belongs to the compendium of Syrianus Magister, along with two other works, the *Rhetorica Militaris* and the *Naumachia*, and can be firmly placed chronologically in the reign of Emperor Theophilus in the middle of the ninth century.[73] However, simply because the term *kataphraktoi* does not appear again for a hundred years until the mid-tenth century, this does not mean that the Byzantine army did not have any heavy cavalry of this type in the intervening century.

The major point to consider here relates to the tendency of medieval historians to use archaic terms to describe battle formations, units, individual combatants and their weaponry, and whether this reflects actual continuity with tradition or points to the classicising tendencies of the author. Morillo and Abels have underlined the widespread propensity of medieval chroniclers to demonstrate their familiarity with classical terms then 'in vogue' and their general knowledge of classical authors.[74] They believe that we should not confuse the historians' 'showing off' of their classical education with actual reality, although we should be very careful before we associate any historian's agenda with personal aggrandisement or, simply, vanity.

If we take, for example, the classicising terminology of Orderic Vitalis, one of the foremost sources on Anglo-Norman military history of the eleventh and twelfth centuries, we find that it is misleading because the author uses the term *pedites* to describe both foot soldiers and dismounted knights.[75] Comparing this example with Leo the Deacon's repeated use of 'πανσίδηροι ιππότες' (*pansidiroi ippotes*, ironclad knights) and his generally archaic vocabulary and lack of consistency in the use of military terms,[76] we can clearly see why the information we get from primary sources – especially ecclesiastical ones – should be dealt with caution and undergo careful scrutiny. In fact, Kolias believes that the first recorded use of the term *kataphraktos* by the author of the *Sylloge* (c. 930) – and subsequently by Phocas – is simply incidental and there are plenty of other terms that could have been used instead, such as καβαλλάριος, καβαλλαρικός, κλιβανοφόρος or επιλωρικοφόρος (*kaballarios, kaballarikos, klibanoforos, epilorikoforos*).[77] Indeed, some are repeatedly used by Leo

VI in his *Taktika* and by the authors of *On Skirmishing* (c. 969) and *On Tactics* (c. 991–5).[78] Consequently, as there is no historiographical gap in the use of the term *kataphraktos* that would support the absence of that elite unit of heavy cavalry from the battlefields of the East, the study of the primary sources could easily become a distorting lens rather than an open window into the past.

Finally, we should bear in mind that certain of our authors would not have been experienced in military affairs, a fact which bears the risk of inaccurate or erroneous reports of events.[79] A good example of this is the report of Emperor Tzimiskes' feigned retreat during the final engagement with the Rus' outside Dorystolon (July 971): Skylitzes (an experienced army officer) reports this manoeuvre in every detail, while Leo the Deacon merely mentions that the Byzantine troops retreated at head-long speed.[80]

Therefore, contemporary military historians should refrain from labelling medieval sources as 'dull,' generally devoid of any interest in battles and/or sieges, dependent on rhetorical devices or having the tendency to reduce battles to a series of conventional images. We should, rather, bear in mind that

> what is described in a battle description depends on unconscious cultural and conscious intellectual decisions about what is important to describe ... the way ancient authors describe the details of battle can tell us about the mental rigging of the societies in which they lived.[81]

Hence, in order to properly evaluate our sources as military historians we must know the background, life and specific context in which they wrote. We must also become cognisant of how they understood battle and what the literary models that underscored their writing were, certainly not whether their descriptions fit our understanding.

Returning to the main topic of my research, the reliability of the numbers provided by our sources and to what degree modern historians can take such estimates at face value, we should bear in mind that there were several factors that affected the numbers and troop estimates that medieval historians included in their works. Both the Byzantine army, as a continuator of the Roman army, and the Abbasid armies before the second quarter of the tenth century had a sophisticated system of recruitment based on the enrolment of soldiers into military registers (kept in the *logothesio* and the diwan respectively). They were thus able to compile a detailed census of the number of soldiers that should have been available to serve at any given time by calling for annual musters.[82] The point here, however, is not whether the governments had accurate and up-to-date accounts of their armies – we have to take that as a given, although certainly not all military rolls would have reflected all the deaths, retirements, desertions or the number of soldiers who would have been unable to serve for financial reasons. The issue is whether the people who wrote the histories available to modern historians would have been able to consult these records.

By way of examples, we know that Procopius was the secretary of Belisarius and accompanied him as the latter campaigned across much of the sixth-century Mediterranean world;[83] Agathias of Myrina (530–82/594), a poet, lawyer and continuator of Procopius' *Wars*, employed several oral sources for his work – including a friend and interpreter working in the Persian royal annals – and it has been argued that he may have had access to military dispatches or diaries of Narses' entourage;[84] Menander the Guardsman (writing in the mid-sixth century), a military officer at the court of Emperor Maurice (582–602), had access to the state archives for the proceedings, negotiations, correspondence and reports of envoys and embassies.[85] According to Vasiliev, Yahya of Antioch quite possibly had access to the ecclesiastical archives of the Patriarchates of Alexandria and Antioch because of his family connections.[86] The continuator of Trajan the Patrician's *Chronicle*, identified as the *protoasecretis* in charge of state archives and future patriarch Tarasius,[87] demonstrates an unusual tendency to provide a variety of statistics that must have come from official archives. These include the numbers and origins of the workmen employed to restore the Aqueduct of Valens in 766/7 and a rare figure for the official establishment of the Byzantine army in 772/3.[88] Treadgold argues that Tarasius would have been one of a few historians who drew on systematic archival research along with his subordinate secretaries; his, now lost, history would form one of the main sources of Theophanes. The author of the *Life of Basil*, probably Theodore Daphnopates, also appears to have made use of state archives in the capital as patrician and *protoasecretis*.[89] In fact, Daphnopates was a high-ranking official during the reigns of Romanus I and Constantine VII and drew up official correspondence in many important matters, such as the negotiations for a peace treaty with Symeon of Bulgaria in 925.[90] For the period between 1025 and 1043, Skylitzes made use – through the history of John the Monk – of the lost history of Demetrius, Bishop of Cyzicus, which in turn would have incorporated important records from state annals probably kept in the patriarchate. These annals would have provided the bulk of Skylitzes' information on the wars in the East for the aforementioned period.[91] Anna Comnena's *Alexiad* contains extracts of imperial *chrysobulls* from diplomatic correspondence and documents of important treaties otherwise unknown; therefore, we can assume that the imperial archives and library would have been open to her and her husband, Nicephorus Bryennius.[92]

Even if chroniclers did consult official documents about army numbers, however, whether directly or through friends and/or relatives, we should bear in mind that the social circle of these educated people would have been narrow and elitist, and there arises the additional problem of how they interpreted and processed their data.[93] Furthermore, the account of a chronicler who reported on the size of medieval armies would have been affected by their sympathies (for example, by reporting an inflated number for the enemy troops in order to enhance the victory of their patron in the eyes of their readers, or deflating the size of their own army if the battle was lost in order to minimise criticism over the defeat),[94] their inherent tendency to exaggerate, their reliance on oral testimonies, which always bears the risk of inflation and/or miscalculation, the period in which the chronicler was writing their work and their level of experience in military matters (for example, a dismounted knight may be counted as infantry by an inexperienced chronicler).[95]

Embarking on my analysis of the Byzantine sources, the first significant point that I want to make concerns the frequency and accuracy of the numbers provided by the chroniclers in question. Both Leo the Deacon and Skylitzes report more troop estimates (five and six times more respectively) for the period after the 950s, focusing on the operational theatre of the Balkans rather than Syria. In the first nine books dealing with the reigns of Phocas and Tzimiskes, Leo the Deacon provides us with a total of ten troop estimates. From these, however, only two can be found in the first two books that examine the early period of the reign of Romanus II and the crucial expedition against Crete (960/1), and even these seem to be overinflated. Thus, we are informed that 40,000 Muslims – apparently just a part of the garrison – were surprised by a detachment of men during a night raid led by Phocas himself in the first few months of the Siege of Chandax,[96] while Leo also gives us the impossible number of 400,000 men for Phocas' expeditionary force against Tarsus in 965.[97] Skylitzes also seems not to have had accurate information about the Byzantine campaigns prior to chapter 14, which deals with the reign of Phocas. He is familiar with the alleged massacre of 5,000 Muslim horsemen by Tzimiskes in the latter's campaign against Adana (964),[98] while he also gives us the rather doubtful number of 100,000 men for the army sent by the Fatimids to save Antioch in 970.[99]

Both Skylitzes' and Leo the Deacon's accounts become more detailed and accurate regarding army size estimates and numbers for battle casualties when they examine Tzimiskes' campaigns in Bulgaria. In Books VI, VII and VIII, Leo the Deacon provides another eight troop estimates of both the Byzantine and enemy armies (for example the 30,000 'Scythians' and 10,000 Byzantines at Arkadiopolis in 970), including the number of casualties after the ensuing battle (20,000 'Scythians' and only 55 Byzantines).[100] Although Leo seems to exhibit a good understanding of the Byzantine army, Treadgold has expressed his concern about the disparities in the numbers of casualties between the Byzantines and their enemies, as in this case the near impossibility of the former losing only 55 men compared with the staggering casualties inflicted on the Rus'.[101] Leo would, almost definitely, have had access to official counts of the battle, but the Rus' numbers must have been simply exaggerated guesses to enhance Tzimiskes' victory. Also very useful are Leo's estimates for Tzimiskes' army numbers and consistency in the battle outside Preslav (971), where the emperor allegedly had 15,000 heavy armed infantry and 13,000 cavalry (including the Immortals), both of which are reasonable numbers for the scale of the campaign that had been undertaken.[102]

In complete contrast with has gone before, Skylitzes' narrative in chapter 15 contains twelve estimates of troop numbers for both the Byzantines and their enemies at Arkadiopolis and Dorystolon. In the first case, the mixed Rus', Patzinak and Magyar force had 308,000 men, which is surely an exaggerated figure aimed at enhancing the Byzantine victory, while the imperial army had deployed the much more plausible number of 12,000 men, although the fact that it is recorded to have sustained only 25 casualties seems unlikely.[103] Turning to the battle at Dorystolon, useful figures provided by Skylitzes include the numbers and consistency of the army's vanguard commanded by the emperor himself (5,000 infantry and 4,000 cavalry), the Bulgar army that was found training outside Preslav (8,500 men) and the garrison installed in the city's citadel (8,000 men), the size of Svyatoslav's army (330,000 men) and the number of Bulgar prisoners captured by the latter after the fall of Preslav (20,000 men).[104] With the exception of the size of Svyatoslav's force at Dorystolon and perhaps the number of Bulgar prisoners, the rest of the figures seem relatively reliable and accurate, all the more so since they escape the large, round multiples of ten thousand.[105] However, despite both of our chroniclers' detailed accounts of Tzimiskes' campaigns in the Balkans, and the suggestion that both had used a common source written probably between 969 and 971 and an official record of Tzimiskes' campaigns in Bulgaria, none of the numbers provided by our chroniclers seem to coincide.

It is not surprising that our Arab sources were much better informed about the Byzantine–Arab wars in Cilicia and Syria in the 950s–60s than any other available source. Two of our best sources for this period regarding numbers and troop estimates for armies and battle casualties are Yahya ibn Said al-Antaki and Ibn Zafir. Accordingly, each of them provides us with nine figures of army sizes and battle casualties focusing on the period between 938 and 964. Their information is crucial to reconstructing the series of campaigns that dominated the Byzantine–Arab conflict of the 930s–60s. Ibn Zafir, in particular, gives us significant information that, as I mentioned earlier, cannot be found elsewhere, thus making his *Book of Histories* an indispensable source for the study of the military campaigns of the 940s–50s.

For Sayf ad-Dawla's campaign against Anzitene and Hisn-Ziyad (938), Ibn Zafir informs us about the number of Byzantine forces under Curcuas which arrived to intercept them, some 200,000 men, which is surely an exaggeration. Our chronicler also reports a unit of 20,000 Byzantine 'patricians' – no doubt a *tagmatic* unit from the capital – that was attacked during the battle by Sayf's *ghulam* corps, capturing 70 of them.[106] Given the total of about 4,000 men for each *tagma*, the total of 20,000 for the Byzantine 'patricians' does not seem far-fetched.[107] The size of Curcuas' army is reported by Abu Firas as well, although his figure of 80,000 may also be an exaggeration for a Byzantine expeditionary force before 955.[108] One could argue that these inflated numbers for Curcuas' armies were reported either to enhance the victory of the Arabs over the Byzantines, or they come as a direct result of that practice from one of our authors' sources. Finally, another reasonable number (40,000 men) is given for the size of the Byzantine army that was sent against Sayf ad-Dawla in the summer of 951 and, although no battle was fought, the campaign of the Arabs against Tzamandos and Charsianon was deemed a success.[109]

Ibn Zafir is, once again, our main source for the numbers engaged in the two most crucial battles of the period that resulted in the defeat of the Byzantine army at Marash in the summer of 953 and Hadath in the following year. For Marash, he mentions only an elite unit of 600 cavalry, which routed Phocas' army of 'considerable' numbers. This sounds once again like an attempt to magnify Sayf's victory to the ears of his audience, especially since the Hamdanid prince managed to take Constantine Phocas (the domestic's son) as prisoner.[110] It seems more likely that Ibn Zafir and Abu Firas drew their information from a common source, as the unit of 600 elite cavalry is also reported by the latter along with the capture of Constantine Phocas and Leo Maleinos (the son of the *strategos* of Cappadocia).[111] For the Battle of Hadath, we are informed that Bardas

Phocas mustered 50,000 troops for his campaign, both infantry and cavalry that also contained units of foreigners (Rus', Bulgars and Armenians) and suffered 3,000 casualties.[112] Ibn Zafir also reports Sayf ad-Dawla's tactical move of personally leading a unit of 500 of his *ghulams* against the centre of the enemy formation. All these numbers seem reasonable enough, bearing in mind that the Byzantines would have dispatched large numbers of soldiers to prevent the capture of the strategic frontier fortress of Hadath.

Yahya provides us with four numbers for Tzimiskes' campaign at Amida, Arzen and Mayyafariquin (June 958), and the latter's victory over Sayf ad-Dawla's lieutenant Nadja al-Kasaki. Al-Kasaki had brought with him some 10,000 troops, 5,000 of which he lost, while the Byzantines captured 3,000 prisoners.[113] This victory was followed by another in the autumn, this time against the emir himself at Ra°ban, where the Byzantines took 1,700 Muslim prisoners, although we have no clue as to the numbers involved in the battle. These estimates look reasonable enough and to them we can also add Abu Firas' information that Basil Parakoimomenos, the eunuch who had arrived from the capital to bring reinforcements to Tzimiskes after his initial success, had brought 12,000 men to the east.[114] Yahya also reports the number of Muslim troops mustered by Sayf ad-Dawla for his famous expedition in November 960 (30,000 men), when he was ambushed on his return journey by Leo Phocas and Constantine Maleinos at a mountain pass close to Marash, although the size of the Byzantine force is unknown.[115] Two years later, during the imperial expedition to capture Anazarbus and the ensuing battle with an army from Tarsus, Yahya reports 5,000 Muslims dead and 4,000 taken prisoner, while he is also aware of the massacre of the cavalry troops from Tarsus by Tzimiskes in his campaign against Adana (964), putting the number at 4,000 men. This event is reported by Skylitzes as well, although the latter writes that 'about 5,000' were killed.[116]

Another crucial point is the geographical knowledge of our chroniclers and the degree of their familiarity with the areas where the events they describe took place. Do they provide us with detailed and accurate enough geographical information so that we can track the route of each army? Do we get any descriptions of the battlefields, the fortifications of major towns and cities, or the topography of their surrounding areas? If we begin our analysis with the Byzantine sources, it is evident that neither Leo the Deacon nor Skylitzes was familiar with the geography of the relevant regions. When it came to place names, rivers, plains and, most importantly, the topography and fortifications of major town and cities

(Chandax, Tarsus, Preslav and so on), they relied on eyewitness information passed on to them.

In Leo the Deacon, there is a very good description of the walls of the Cretan capital of Chandax and its surroundings. In fact, our chronicle is careful to point out the significance of carefully reconnoitring the surrounding area and the walls of a city before imposing a siege on it 'so that he [the emperor] might attack wherever it was vulnerable'.[117] This advice follows the recommendations found in the *Praecepta Militaria* and Vegetius' *De Rei Militari* on the precautions taken by an army general before a pitched battle or a siege operation.[118] Leo seems to have been well informed not solely about the topography of the city – 'for on the one side it had the sea as a sure defence, and on the other side it was set on a nearly flat and level rock, on which the walls were laid' – but also about the size of its walls, which were 'wide enough so that two wagons could easily make a circuit on top of the ramparts and pass each other ... and in addition two extremely wide and deep moats were dug around it'.[119] We are also provided with a very basic description of the fortifications of the city of Tarsus at the end of Book III, where we are told that Tarsus had a double circuit wall of 'extraordinary height', encircled by a moat 'of very great depth ... terminating in battlements'.[120] Two further significant details reported by Leo are the cutting of the walls by the River Kydnos, 'which had a strong current from its sources' and was crossed by three bridges, and the plain outside of the city that 'was suitable for cavalry'.[121]

Regrettably, Leo's geographical and topographical information for the imperial expeditions in Syria in the 960s is limited and vague, leaving one unable to track the route of an army in much detail. For example, if we examine Phocas' campaign to capture the strategic Cilician city of Tarsus (964/5), we are informed that he departed from the capital and 'after making an encampment in Cappadocia ... marched towards Tarsus',[122] with no more details about the route of the army. In a similar fashion, 'the Emperor Nicephoros ... took the Roman forces, and hastened to Syrian Antioch, where he set up camp', referring to the first and unsuccessful siege of the city in 968.[123] Leo's narrative is equally disappointing in Book VI, where he reports the battle at Arkadiopolis (970); he does not provide any geographical or topographical information or any description of the battlefield. In fact, not even the name of the place is mentioned, with vague comments like 'the Scythians were encamped not far away but nearby [the imperial camp]' or 'they [the Byzantine ambushing party] were ordered to lie in wait in the thickets on either side [of the pass]' being prominent in the text.[124]

For Tzimiskes' expedition in Bulgaria (971), however, Leo does seem to have had better informed sources for the campaign, as he provides us with some mythological facts about the River Danube and appears familiar with the geography of the northern Black Sea coast.[125] He is also aware of the first stop of the imperial forces at Adrianople, although this is not surprising as it was the practice of imperial armies to camp at Adrianople, situated at the crossroads between the road systems leading west (Thessaloniki) and north (Sardica or Varna), before any major campaign in the Balkans.[126] He also refers to the 'difficult and narrow paths [*kleisourai*] leading to Mysia [Bulgaria]', which should take us back to his description of the geography of Bulgaria in Book IV.[127] Apart from his brief comment about the plain outside Preslav, which was 'suitable for cavalry', and the identification of the city of Dorystolon as the assembly point for Svyatoslav's army, he gives no significant topographical details for the battlefield or its surrounding area.[128]

Skylitzes' knowledge of the geography of the Balkans and, especially, Cilicia and Syria is a significant weak point in his work, as he provides us with even less information than Leo the Deacon about the topography of the military operations he describes. For John Curcuas' campaigns in Mesopotamia and Cilicia in the 920s–40s, our chronicler has the tendency to compress the chronology of events:

> The magister John Kourkouas, domestic of the Scholae, was ravaging Syria [the entire area under Arab domination] and sweeping aside all resistance. He took possession of many fortresses, strongholds and cities of the barbarians and then came to the renowned Melitene which he besieged.[129]

On the Magyar invasion of imperial territories in the Balkans in 934, Skylitzes reports: 'In the month of April, seventh year of the indiction, the Turks invaded Roman territory and overran all the west right up to the city.'[130]

On Nicephorus Phocas' Cilician campaigns as emperor, Skylitzes reports only the one that took place in the second year of his reign (Spring 964), for which he informs us of the emperor's stop at the fortress of Drizion – probably in Cappadocia – before entering Cilicia and 'destroying the cities of Anazarbos, Rhossos [a port south of Alexandria] and Adana in addition to no small number of fortresses'.[131] Compared to our chronicler's previous reports regarding imperial expeditions against the Arabs in the East, this time Skylitzes seems to be much better informed about the course of events because he appears to be better acquainted with the major cities upon which the emperor laid siege. This includes, for example, a small fort where he left his wife (Theophano) and children before proceeding into enemy territory. Both Skylitzes and Leo the Deacon report this expedition, along with the siege of Mopsuestia and Tarsus as already mentioned. It is unlikely, however, that they used a common source, as Skylitzes reports that Mopsuestia succumbed due to famine, while Leo states that the city's walls were undermined by fire.

Crossing over to the two most crucial battles on the Balkan front, at Arkadiopolis and Dorystolon, Skylitzes fails to give us any specific details about the geography of the region or the topography of the battlefields. For Skleros' ambush of the Patzinak force at Arkadiopolis, he simply reports that 'he [Skleros] set up ambushes and traps by night in some suitable places',[132] leaving no clue to posterity as to the exact nature of the place where Skleros lured the Patzinaks into his trap. Concerning Tzimiskes' expedition against the Rus' in 971, Skylitzes certainly seems to be better informed about the campaign and he does mention a number of locations, namely the place where the emperor met with a supposed Bulgar embassy (Raidestos, on the European banks of the Sea of Marmara), and the first strategic objective of his campaign ('Great Preslav, where the palace of the Bulgar kings lay'). He imparts no information, however, on the exact route of the army, on the topography of the area of Preslav, which was the scene of a battle between units of the Byzantine and Bulgar armies, or on the fortifications of the city and its citadel, apart from a very general observation that the latter was 'very well fortified and impregnable'.[133]

The *Synopsis* once again lets us down for the siege of Dorystolon and the ensuing battles outside the city between the imperial forces and Svyatoslav's troops. We can infer from Skylitzes' account, however, that there was a plain outside the city's main gate where 'the Emperor concentrated all of his forces', and that Dorystolon did not have a moat, as he reports that Svyatoslav had to dig a deep trench around the city during the night to prevent the Byzantines from bringing in their siege machines.[134] Skylitzes' reliance on oral sources is apparent because, despite the mere handful of place names he mentions in this part of his *Synopsis*, strangely enough he seems to know the name of the estate where the admirals Leo and Nicephorus resided during the siege.[135] Thus, although the *Synopsis* may feature one of the best accounts of Tzimiskes' campaigns in Bulgaria, it provides little consistent information about the topography of the fields of battle, a factor which played a vital role in their outcome.

It is not surprising that our Arab sources were much better informed about the geography of the areas where the military operations they describe took place. Their accounts contain valuable information on the topography of these regions – place names, rivers, small towns and forts – that allows us to reconstruct the route of the Arab and Byzantine armies in their annual campaigns over the Taurus in the 940s–50s. We are also fortunate to have two eyewitnesses of the Hamdanid campaigns of the period up to 956, Mutanabbi and Abu Firas, with their versions being supplemented by the equally valuable information provided by Yahya – who drew from contemporary Greek, Syrian and Muslim sources, as already seen – and Ibn Zafir. Mutanabbi accompanied Sayf ad-Dawla on all his expeditions into Byzantine territories from 337/948 to 345/956, dedicating at least one poem for each campaign to glorify the exploits of his patron. In this section, I will focus on three of Sayf ad-Dawla's most celebrated campaigns and examine what geographical and topographical information historians can deduct from the Muslim sources.

For Sayf's grand expedition in 339/950, Mutanabbi identifies as the main invasion target the themes of Cappadocia and Charsianum in central Asia Minor. Upon crossing into Byzantine territory through al-Safsaf and Hisn al-Uyun,[136] they proceeded to Samandu (Tzamandos) in the region south of Caesarea in Cappadocia. Travelling north towards Harsana (Charsianon castle) through Halys (Kilizil Irmak), Sayf's army defeated Bardas Phocas at the valley of the Luqan (Batu al-Luqan, probably the Lykos-Kelkid Irmak).[137] On his return journey in October, however, he was ambushed by Leo Phocas at a defile which is identified only by Yahya, as the Darb al-Kankarun.[138]

Three years later, Sayf launched one of his most remarkable expeditions against the Byzantines and Mutanabbi dedicated two of his poems to it. With the invasion route taking the Arabs through the Taurus via Harran, Duluk, Lake Sanga and the defile of al-Qulla, they ravaged the area between Melitene and Arqa.[139] Mutanabbi adds that on their return journey they found the defile of Mauzar in the Euphrates occupied by enemy forces. Eventually, the ensuing battle forced Sayf to divert his army north through Melitene, the River Qubaqib, the fortress al-Minsar on the Euphrates, Hisn al-Ran and Samosata, where he was informed of an enemy incursion into northern Syria. He then marched along the Euphrates to Duluk, wanting to catch the Byzantines while on retreat, and inflicted a significant defeat on them at Gayhan, not far from Marash.

Finally, the 345/956 invasion, as glorified by Mutanabbi, is the most detailed account of Sayf's expeditions.[140] Entering Byzantine territory through Harran, Hisn ar-Ran and Hisn al-Hamma, Sayf marched towards Anzitene on the *thema* of Mesopotamia (John Tzimiskes was *strategos* of

Mesopotamia at the time). He ravaged a number of communities in the regions of Tell-Bitriq and Arsanas, and during his journey back to Syria through Diyar-Bakr, defeated an army under the leadership of Tzimiskes that blocked his way at the defile of Darb Baqasaya (close to the Arghana-Su, a tributary of the Tigris).[141]

Mutanabbi's poems are indispensable for tracking the route of the Arab armies on a modern-day map, as they contain rare geographical information about the Arab–Byzantine frontier regions not found in any other contemporary source.[142] The historian Canard lists a number of these unidentified places, such as Darb al-Qulla and Darb al-Mawzar in the region south of Melitene, Hisn al-Ran on the left side of the River Euphrates, Sumnin close to a small lake called Goldjik (south-west of Harput) and others.[143] However, Mutanabbi's topographical information about the battlefields of the period is rather vague. For example, if we examine his narrative on the 950 expedition, he identifies the River Halys and the valley of the River Lykos for the first engagement in September, while a defile was the place for the second engagement later that autumn against Leo Phocas' troops. He carries on in a similar fashion for the rest of Sayf ad-Dawla's campaigns. It would have been extremely useful to modern historians if Mutanabbi had provided his readers with some more specific information on the nature of the battlefields, such as the length and breadth of the terrain on the opposite sides of the river, whether it was flat or suitable for cavalry (soft ground, rocky or exhibiting any other obstacles). Let me repeat, however, that answering these questions in the fashion of a modern war correspondent was not the aim of any of the Muslim historians of the period; thus, we should not expect of them to provide us with any military insight in their works.

Although Yahya's history is much less detailed regarding the route of the Hamdanid armies compared with Mutanabbi's, it does contain most of the major locations and cities that were besieged by the Arabs and the Byzantines, and identifies their main strategic objectives. For example, we know that Sayf ad-Dawla besieged the fortress of Barzuya in 336/948 while Leo Phocas was besieging Hadath, and that the strategic city of Qaliqala (Theodosiopolis) was the emir's strategic objective for the next year's expedition.[144] Harsana (Charsianon) and the region of Hadath are mentioned by Yahya for the 339/950 expedition, although he simply writes that 'several Byzantine fortresses were captured [by Sayf ad-Dawla]', falling short of listing their names or location.[145]

For the crucial campaigns of 342/953 and 343/954, Yahya mentions only the defiles of the Merwan and Samosata, and Hadath respectively.[146]

Thankfully, his narrative becomes a bit more detailed for the 345/956 campaign, where he identifies the Hamdanid emir's strategic objectives: Batn Hinzit, the River Arsanas, where the emir pitched camp, and Tell Bitriq, where Tzimiskes had pitched his, along with the defile of Tailleurs (Darb Baqasaya), where a battle took place between the two.[147] In the rest of his work, Yahya's narrative contains a number of geographical snippets of information that can be used to corroborate the accounts of Mutanabbi and Abu Firas. Although he gives some rare details, such as the fortress 'called al-Yamani' that was besieged by Tzimiskes in the summer of 347/958,[148] Yahya's geographical and topographical knowledge of the operational theatres in question is relatively limited.

Since a large part of Leo the Deacon's *History* covers the military exploits of the Phocades, he narrates in detail Nicephorus Phocas' and his brother Leo's victories in Asia Minor, Crete and the Balkans, making his *History* one of the most important primary sources available that examine the Byzantine military expansion of the tenth century. For the Cretan expedition to capture Chandax in 960, Leo writes that the emperor 'drew up his army in three battles, studded it thickly with shields and spears and . . . launched a frontal assault against the barbarians'.[149] With the exception of the aforementioned tripartite formation and a frontal assault – probably by elite heavy infantry units judging by the nature of the terrain (a beachhead was established opposite the city walls) – we know nothing about the course of the battle, the manoeuvres of each army, their consistency or their officers in charge (although Leo does mention a certain Pastilas, *strategos* of the Thrakesion theme, in the following paragraph).

Leo describes the Byzantine formation again in Book II, this time for the spring offensive of 961, as 'deep and oblong' with no further details on the numbers of men involved (lines, rows, etc.) or their equipment (infantrymen, archers, *menavlatoi*, etc.).[150] On the crucial battle in the plain outside the city of Tarsus in 965, Leo's narrative once again lacks the necessary details regarding battlefield manoeuvres, although there is a comment about 'the Roman divisions moving into action with incredible precision' – obviously wishing to highlight the effect of years of intense drilling and training introduced by Phocas since becoming Domestic of the Sholae. We are, however, informed about the Byzantine formation, which included units of ironclad horsemen (πανσίδηροι ἱππόται) placed at the front of the army, while archers and slingers would follow from behind. The emperor is mentioned as the commander of the right wing of the regular cavalry, placing Tzimiskes in command of the left with units of regular cavalry.[151]

Moving on to the campaigns of Tzimiskes in Bulgaria, Leo notes that for the first confrontation outside Preslav the Byzantines were organised into a deep formation (φάλαγγα, *phalanga*), and that 'the approach of a disciplined army' caused the 'Tauroscythians' to be struck 'with panic and terror'.[152] For the battle itself, Leo reported that the Bulgars had drawn their army in a 'strong and close' infantry formation to receive the enemy attack, but we have no more details about the encounter, with the sole exception of a crucial manoeuvre by the Byzantine cavalry (Immortals) which, according to Leo, decided the outcome: 'When the battle was evenly balanced on both sides, at this point the emperor ordered the Immortals to attack the left wing of the Scythians with a charge.'[153]

The siege of the city of Dorystolon and the series of battles that decided its outcome are, by far, the most detailed part of Leo's work. Clearly influenced by ancient historians, the author adds liveliness to his battle descriptions by providing several character sketches, while keen to portray Tzimiskes' strategic and fighting abilities. Leo identifies five battles that took place on the outskirts of Dorystolon, describing all of them in some detail.[154] As we will see in the following section, however, Leo's description of battle manoeuvres is brief and in many places quite vague, while his narrative – following the Homeric model – is also dominated by several individual confrontations that overshadow the battle itself. For example, following the arrival of the Byzantine squadron in the Danube and the ensuing battle between units of both armies, Leo narrates the encounter in just a few verses, simply mentioning the heavy equipment worn by both armies. He states that 'both sides fought valiantly, and it was unclear who would be victorious, as both sides pushed each other back in turn', and then goes on to write about the individual achievements of Sphendoslavos (third in rank in the Rus' expeditionary army) and Theodore Lalakon, a technique which he repeats a few paragraphs later with the famous battle between Anemas and Ikmor – evidently, having been exposed to the heroic tradition of the period, either oral or written.[155]

His work contains two descriptions of the Byzantine battlefield formation, the first given before the initial battle between the two armies, where we are offered an important piece of information regarding the place of the *kataphraktoi* in the mixed formation: 'After the emperor deployed the Romans in the van and placed ironclad horsemen on both wings, and assigned the archers and slingers to the rear and ordered them to keep up steady fire...' Second, before the final encounter and the surrender of Dorystolon to the Byzantines, Leo notes an encircling manoeuvre conducted by a unit under the command of Bardas Skleros, which seems to follow the recommendations of the contemporary military manuals, as I will explain further below.[156]

If we turn to the *Synopsis* and its accounts of the major battles of the tenth century, Skylitzes' narrative lacks several essential details, such as names and units (although some *tagmatic* units are named for the Bulgarian expedition of 917),[157] along with the description of the battle tactics employed by the opposing armies. Thus, for the Byzantine defeat at Achelous in 917, Skylitzes mentions only that 'the Bulgars were thoroughly routed and many of them slaughtered', even though he would have been aware of the anecdotal episode with the domestic and his runaway horse causing panic amongst those of his troops pursuing the Bulgars.[158] For Leo Phocas' ambush of Sayf ad-Dawla in the mountain passes of the Taurus in 960, Skylitzes gives us only some basic details of the chain of events:

> Once Chamdan was there [a defile named Adrassos, located probably in the theme of Lykaonia, although its exact location is still uncertain][159] and had advanced well into the narrow passage, he was surrounded by the forces lying in ambush. Men concealed for this purpose rose up from their concealed positions, rolling great stones down to them and shooting all kinds of missiles at them.[160]

Unfortunately, Skylitzes gives no battle details for any of the imperial army's expeditions in Syria in the 960s, including the one close to Amida against a Tarsion army in 964, the one in the Tarsus plain the following year, and Nicephorus' battles against Khorasanian troops at Antioch in 968.

Skylitzes' description of Tzimiskes' operations in the Balkans in 970–1 is one of the best and most detailed in his work and can only be compared with the one written by Leo. Both seem to have used a common source – probably eyewitness accounts and official reports of the campaigns – as mentioned before. For the ambush of the Patzinak force at Arkadiopolis (970), our author is well informed about the events and the battle manoeuvres that decided the outcome of the engagement. He mentions the name of the commander in charge of the unit set to lure the Patzinaks into the ambush, the division of the enemy army and the basic stages of the engagement, including the pretended fleeing of Alakasseus' unit ('a leisurely retreat'), the counterattack by the Romans from the flanks and the front, and the eventual encirclement of the Patzinak army.[161] Regrettably, since Skylitzes seems to show little interest in the strategic backdrop of long campaigns and in the precise details of the engagements,[162] the battles outside Preslav and Dorystolon are presented in a brief but simple manner. At Preslav, Skylitzes notes in just a few lines that 'they [the Byzantines] surprised eight thousand five hundred fully armed men [Bulgars] . . . these resisted for a time but then, overcome, turned and fled. Some were ingloriously slain, some reached safety inside the city'.[163] No details of the opposing formations or battlefield manoeuvres are given, although it is quite likely that our author would have known more about this engagement since he gives us the number of the Bulgars (8,500), a number that seems quite reasonable.

For the siege of Dorystolon, the *Synopsis* provides us with an account of five confrontations between the imperial army and the Rus', and includes the famous single battles between Theodore of Mistheia and Ikmor, which were also included in Leo's *History*.[164] Skylitzes' narrative of the actual battles, however, is rather disappointing. He gives no indication of the division of the armies, the units that were engaged in combat, their consistency, numbers or their leaders. Most importantly, he provides us with only the most basic details of the armies' manoeuvres in the battlefield. The only place where Skylitzes' narrative becomes a bit richer is in his report about the first engagement between the two armies: 'For some time the battle was equally matched but when it drew on towards evening on that day the Romans rallied each other . . . Then they charged the Scyths' left wing and put down many of them by the irresistible nature of this manoeuvre.'[165] Indeed, this rare comment on a battlefield manoeuvre of the Byzantines, most likely by the heavy cavalry, is what makes his account vital for our study of the battle.

Although Skylitzes' and Leo the Deacon's accounts agree on the main points of the Battle of Dorystolon, it is evident that the *Synopsis* lacks the description of personalities characterising Leo's *History*.[166] As McGrath has pointed out in her study of the rhetoric of the battle, Leo the Deacon wanted to communicate the psychological background of the conflict and the fear that dominated the minds of the simple soldiers, adding liveliness to the battle descriptions with a number of psychological insights, while his observations were often directed by his patriotism and personal views. Skylitzes, rather, pays more attention to the heroic exploits of individual protagonists in contrast to the 'brutal economy' applied to the strategy, geography and economy of so many of the raids and sieges mentioned in his *Synopsis*. These differences in the presentation of basic events, along with the period of writing, make both accounts indispensable for the study of tenth-century warfare for a modern historian.

Although our Muslim sources for the period are more detailed when it comes to place names, rivers, unidentified castles and numbers for the armies campaigning in the region, their accounts lack the necessary details regarding the course of the battles fought between Byzantine and Arab armies, battlefield manoeuvres and the consistency of the units taking part. The only case where we have some information about the course of a battle is regarding Hadath in 954, even though this is utterly inadequate to reconstruct the actual battle in any level of detail. First, we have Mutanabbi's amazement about the multi-ethnic composition of the Byzantine army 'with only the interpreters being able to understand all these [different] languages', and his reporting of the *kataphraktoi* cavalry corps (for the first time in the tenth century).[167] Mutanabbi, Abu Firas and Ibn Zafir also note the crucial manoeuvre that won the battle for Sayf ad-Dawla, namely the emir's charge against the enemy with his retinue of 500 *ghulam* cavalry that was probably directed against the centre of the Byzantine formation, where they might have been able to make out Phocas' banner.[168] Finally, Yahya tells us that the battle lasted until the 'moment of ᶜ*asr*', which is the time for the Muslim afternoon prayer.[169]

In conclusion, it should not come as a surprise that the Byzantine sources focus mainly on the military campaigns of the emperors in the Balkans, while the Muslim chroniclers are much better informed about the Byzantine–Arab conflicts in Cilicia and Syria. Leo the Deacon and Skylitzes report several troop estimates – most of them quite reasonable – in the chapters that deal with Phocas' and Tzimiskes' campaigns in Bulgaria, while in the handful of cases where they provide us with any numbers for the imperial expeditions in the East, these are exaggerated and, thus, unreliable. Conversely, despite both being eyewitnesses of the wars in the East, Mutanabbi and Abu Firas rarely provide us with any figures for the Hamdanid invading armies and when they do, their figures are inflated, which may have served to enhance the victory of the Arabs over the infidels in the eyes of their readers. For this, we must rely on the careful comparison of the accounts of Yahya of Antioch and Ibn Zafir.

Mutanabbi has included in his poems several rare geographical and topographical snippets of information that cannot be found in any other primary source of the period. His accounts, however, like the other Muslim sources, lack the necessary details to reconstruct the topography of the battlefields. Although they identify a key geographical characteristic of the place where a battle was fought, they do not give us anything about the nature of the ground – a factor which played a vital role in the outcome of any battle. As regards the Byzantine sources, their descriptions of cities and battlegrounds are generally vague, with few places being described adequately.

Finally, when it comes to the description of the major battles of this period, the Byzantine sources are, once again, focused on the Balkan theatre of operations and, except for the siege of Chandax and Tarsus, both our chroniclers focus on the operations against Preslav and Dorystolon in 970/1. Although both their histories are indispensable for the reconstruction of these campaigns, yet none of them provides us with names and units, along with any analysis of the battle tactics employed by the opposing armies. Rather, they choose to focus on the achievements of individual protagonists. Muslim sources are even more disappointing in this respect, due to their almost complete lack of interest in reporting the battles of the period.

Notes

1. Leonis diaconi (1828), *De Velitatione bellica Nicephori augusti*, ed. C. B. Hass, Corpus Scriptorum Historiae Byzantinae, 4, Bonn: Weber; *History of Leo the Deacon*; Treadgold, *Middle Byzantine Historians*, pp. 236–46; N. M. Panagiotakes (1965), *Leon o Diakonos*, Athens: Association of Byzantine Studies; Karpozilos, *Βυζαντινοί Ιστορικοί και Χρονογράφοι*, pp. 475–525; A. Markopoulos (2003), 'Byzantine History Writing at the end of the First Millennium', in P. Magdalino (ed.), *Byzantium in the Year 1000*, Leiden: Brill, pp. 183–97. For a translation into modern Greek, see B. Karales (2000), *Λέων Διάκονος. Ιστορία* [Leo the Deacon: History], Athens: Kanaki.
2. G. Greatrex (1996), 'Stephanus, the Father of Procopius of Caesarea?', *Medieval Prosopography*, 17, pp. 125–45; J. Ljubarskij (1991), 'Writer's Intrusion in Early Byzantine Literature', in I. Ševčenko, G. G. Litavrin and W. K. Hanak (eds), *XVIIIth International Congress of Byzantine Studies: Selected Papers, Main and Communications*, Moscow: Byzantine Studies Press, pp. 433–56.
3. *History of Leo the Deacon*, IV.11, p. 123.
4. Treadgold, *Middle Byzantine Historians*, p. 236.
5. *History of Leo the Deacon*, I.3, p. 60.
6. A. Markopoulos (2008), 'Education', in Jeffreys et al. (eds), *Oxford Handbook of Byzantine Studies*, pp. 785–95. See also Moffatt, A. (1979), 'Early Byzantine School Curricula and a Liberal Education', in I. Dujcev (ed.), *Byzance et les slaves: mélanges Ivan Dujcev*, Paris: Association des amis des études archéologiques, pp. 275–88; R. Browning (1978), 'Literacy in the Byzantine World', *Byzantine and Modern Greek Studies*, 4, pp. 39–54; P. Lemerle (1971), *Le premier humanisme byzantin*, Paris: Les Presses Universitaires de France, especially ch. 4, pp. 75–108.

7. Markopoulos has suggested that Leo imitates Homer's account of events – especially single combats and sieges – at the siege of Troy: A. Markopoulos (2000), 'Ζητήματα κοινωνικού φύλου στον Λέοντα τον Διάκονο' [Issues of Gender in Leo the Deacon], in S. Kaklamanes and A. Markopoulos (eds), *Ενθύμησις Νικολάου Μ. Παναγιωτάκη* [Studies in Memory of Nikolaos M. Panagiotakes], Heraklion: Crete University Press, p. 488. See also G. Buckler (1929), *Anna Comnena: A Study*, London: Oxford University Press, pp. 178–87.
8. Panagiotakes, *Leon o Diakonos*, pp. 8–9; Treadgold, *Middle Byzantine Historians*, pp. 236–7.
9. *History of Leo the Deacon*, X.9–11, pp. 215–21.
10. Panagiotakes, *Leon o Diakonos*, pp. 16–29. Treadgold believes that Leo died soon after becoming bishop in 996 or 997; see *Middle Byzantine Historians*, p. 238.
11. Cheynet, 'Les Phocas', p. 303, n. 43.
12. *History of Leo the Deacon*, X.x, p. 218.
13. Markopoulos, 'Byzantine History Writing', pp. 183–97.
14. *History of Leo the Deacon*, p. 23; Treadgold, *Middle Byzantine Historians*, p. 246.
15. Herodotus, *Histories*, I.8 (Candaules to Gyges), followed by a popular Latin saying 'Auribus oculi fideliores sunt'.
16. *History of Leo the Deacon*, I.i, p. 58.
17. *History of Leo the Deacon*, IV.xi, p. 122; X.viii, pp. 213–14.
18. See the detailed footnote by Talbot in *History of Leo the Deacon*, p. 14, nn. 39, 40. For the use of oral tradition in historiographical works, see M. Richter (1994), *The Oral Tradition in the Early Middle Ages*, Turnhout: Stroud. I was also delighted to read and discuss some very useful points in S. John (2012), 'The Use of Oral Evidence in the Twelfth-Century Historical Writing of the First Crusade', *The Crusades and the Latin East Seminar Series*, London: Institute of Historical Research (12 March).
19. Treadgold, *Middle Byzantine Historians*, pp. 236 and 238.
20. Treadgold, *Middle Byzantine Historians*, pp. 226–36; M. I. A. Siuziumov (1916), 'Ob istochnikakh L'va D'iakona', *Vizantiiskoe obozrenie*, 2, pp. 106–66; A. Kazhdan (1961), 'Iz istorii vizantiiskoi khronografii X v. 2. Istochniki L'va D'iakona i Skilitsy dlia istorii tretei chtverti X stoletiia', *VizVrem*, 20, pp. 106–28; R. Morris (1994), 'Succession and Usurpation: Politics and Rhetoric in the Late Tenth Century', in P. Magdalino (ed.), *New Constantines: The Rhythm of Imperial Renewal in Byzantium, 4th–13th Centuries*, Aldershot: Ashgate, pp. 208–9.
21. Treadgold, *Middle Byzantine Historians*, p. 233.
22. Cheynet, 'Les Phocas', p. 303, n. 43; Treadgold, *Middle Byzantine Historians*, p. 240.
23. Morris, 'Succession and Usurpation', p. 209; Treadgold, *Middle Byzantine Historians*, p. 246.

24. *Oxford Dictionary of Byzantium*, III, p. 1914. On Skylitzes and his work, see E. S. Kiapidou (2010), *Η Σύνοψη Ιστοριών του Ιωάννη Σκυλίτζη και οι Πηγές της (811–1057): Συμβολή στη Βυζαντινή Ιστοριογραφία κατα τον 11ο αιώνα* [The *Synopsis Historion* of Ioannes Skylitzes and its Sources (811–1057): Contribution to the Byzantine Historiography of the Eleventh Century], Athens: Kanaki. This is an expanded version of the author's doctoral dissertation, to which I had access: 'Η Σύνοψη Ιστοριών του Ιωάννη Σκυλίτζη και οι Πηγές της (811–1057)' [The *Synopsis Historion* of Ioannes Skylitzes and its Sources (811–1057)], PhD thesis, University of Ioannina. See also Treadgold, *Middle Byzantine Historians*, pp. 329–39; Karpozilos, *Βυζαντινοί Ιστορικοί και Χρονογράφοι*, III, pp. 239–91; H. Hunger (1978), *Die hochsprachliche profane Literatur der Byzantiner*, 2 vols, Munich: Beck, I, pp. 389–93; C. Holmes (2005), *Basil II and the Governance of the Empire (975–1025)*, Oxford: Oxford University Press, pp. 66–240. For an edition of the Greek text, see Joannes Scylitzes (1973), *Ioannis Scylitzae Synopsis historiarum*, ed. I. Thurn, Corpus Fontium Historiae Byzantinae, 5, Berlin: De Gruyter; for an English translation of the text with a critical introduction, see John Skylitzes (2010), *A Synopsis of Byzantine History, 811–1057*, trans. J. Wortley, Cambridge: Cambridge University Press.
25. Treadgold, *Middle Byzantine Historians*, p. 329.
26. John Skylitzes, *Synopsis of Byzantine History*, p. 3.
27. On the dating of the two editions of Skylitzes' work, see Treadgold, *Middle Byzantine Historians*, pp. 330–1.
28. Markopoulos, 'Byzantine History Writing', p. 195.
29. John Skylitzes, *Synopsis of Byzantine History*, pp. xxvi–xxvii.
30. Treadgold, *Middle Byzantine Historians*, pp. 334–5; Kiapidou, *Synopsis Historion*, pp. 135–42.
31. Treadgold, *Middle Byzantine Historians*, pp. 181 and 333.
32. On the theory that Theodore Daphnopates is the author of Theophanes Continuatus' work, see Treadgold, *Middle Byzantine Historians*, pp. 188–96; A. Markopoulos (1985), 'Theodore Daphnopates et la continuation de Theophane', *Jahrbuch der Österreichischen Byzantinistik*, 35, pp. 171–82; John Skylitzes, *Synopsis of Byzantine History*, p. xix; Holmes, *Basil II*, p. 93.
33. Treadgold, *Middle Byzantine Historians*, p. 233. See also Kiapidou, *Synopsis Historion*, pp. 238–44.
34. Holmes, *Basil II*, pp. 92–3, n. 62; John Skylitzes, *Synopsis of Byzantine History*, pp. xx–xxi.
35. Moravcsik, *Byzantinoturcica*, I, pp. 398–9; Holmes, *Basil II*, pp. 110–11.
36. Kiapidou, *Synopsis Historion*, pp. 244–7.
37. Treadgold, *Middle Byzantine Historians*, pp. 247–58 and 334; Holmes, *Basil II*, pp. 96, 120–70; N. M. Panagiotakes (1996), 'Fragments of a Lost Eleventh-Century Byzantine Historical Work', in C. Constantinides et al. (eds), *Philhellen: Studies in Honour of Robert Browning*, Venice: Istituto ellenico di studi bizantini e postbizantini di Venezia, pp. 321–57.

38. Holmes, *Basil II*, pp. 268–78; Treadgold, *Middle Byzantine Historians*, pp. 254–5.
39. Treadgold, *Middle Byzantine Historians*, pp. 258–70.
40. John Skylitzes, *Synopsis of Byzantine History*, p. xxii; Holmes, *Basil II*, p. 130; D. I. Polemis (1965), 'Some Cases of Erroneous Identification in the Chronicle of Skylitzes', *Byzantinoslavica*, 26, pp. 74–81.
41. John Skylitzes, *Synopsis of Byzantine History*, p. 3.
42. Yahya ibn Said al-Antaki (1932–57), *Histoire de Yahya-ibn-Sa'īd d'Antioche, continuateur de Sa'īd-ibn-Bitriq*, ed. I. Krachkovskii and A. A. Vasiliev, *Patrologia Orientalis,* Paris: Firmin-Didot, vol. 18, fasc. 5; for a recent edition in Italian, see *Yahya al-Antaki Cronache dell'Egitto Fatimide e dell' Impero Bizantino 937–1033* (1998), trans. B. Pirone, Milan: Jaca Book. The best work on the chronicle of Yahya is J. H. Forsyth (1977), 'The Byzantine–Arab Chronicle (938–1034) of Yahya B. Said Al-Antaki', 2 vols, PhD dissertation, University of Michigan; see also Vasiliev, *Byzance et les Arabes*, 2.II, pp. 80–91.
43. Forsyth, *Byzantine–Arab Chronicle*, pp. 27–8.
44. Forsyth, *Byzantine–Arab Chronicle*, p. 11; Vasiliev, *Byzance et les Arabes*, 2.II, p. 82.
45. Vasiliev, *Byzance et les Arabes*, 2.II, p. 83.
46. Forsyth, *Byzantine–Arab Chronicle*, p. 3; Vasiliev, *Byzance et les Arabes*, 2.II, p. 84.
47. Vasiliev, *Byzance et les Arabes,* 2.II, p. 85; see also Forsyth, *Byzantine–Arab Chronicle*, pp. 34–100.
48. Vasiliev, *Byzance et les Arabes*, 2.II, p. 86.
49. Vasiliev, *Byzance et les Arabes*, 2.II, p. 87.
50. Ibid., pp. 89–90; Forsyth, *Byzantine–Arab Chronicle*, pp. 35–6, 83–4.
51. Larkin, *Al-Mutanabbi*.
52. R. Jacobi (1996), 'The Origins of the Qasida Form', in S. Sperl and C. Shackle (eds), *Qasida Poetry in Islamic Asia and Africa*, Leiden: Brill, pp. 21–35; in the same edition, S. P. Stenkevych, 'Abbasid Panegyric and the Poetics of Political Allegiance: Two Poems of Al-Mutanabbī on Kafur', pp. 35–63.
53. S. K. Jayyusi (1996), 'The persistence of the qasida form', in S. Sperl and C. Shackle (eds), *Qasida Poetry in Islamic Asia and Africa*, p. 18. See also Latham, 'Towards a Better Understanding of al-Mutanabbī's Poem on the Battle of al-Hadath', *Journal of Arabic Literature*, 10, pp. 1–22.
54. Stenkevych, 'Abbasid Panegyric', pp. 42–4.
55. Vasiliev, *Byzance et les Arabes*, 2.II, pp. 349–52; see also S. Dahan (1944), *Le Diwan d'Abu Firas al-Hamdani (poète arabe du IVe siècle de l'hégire)*, Beirut: Institut français de Damas; Patoura, Αιχμάλωτοι, pp. 95–7.
56. Vasiliev, *Byzance et les Arabes*, 2.II, pp. 118–20.
57. Verbruggen, *Art of Warfare*, pp. 10–18; Keegan, *Face of Battle*, pp. 27–36; R. Abels and S. Morillo (2005), 'A Lying Legacy? A Preliminary Discussion

of Images of Antiquity and Altered Reality in Medieval Military History', *Journal of Medieval Military History*, 3, pp. 1–13; K. DeVries (2004), 'The Use of Chroniclers in Recreating Medieval Military History', *Journal of Medieval Military History*, 2, pp. 1–17; S. Morillo (2001), 'Milites, Knights, and Samurai: Medieval Military Terminology and the Problem of Translation', in R. P. Abels and B. S. Bachrach (eds), *The Normans and Their Adversaries at War: Essays in Memory of C. Warren Hollister*, Woodbridge: Boydell & Brewer, pp. 167–84.
58. C. Hillenbrand (1999), *The Crusades: Islamic Perspectives*, Edinburgh: Edinburgh University Press, pp. 31–3.
59. John Skylitzes, *Synopsis of Byzantine History*, p. 292. The ascription of an 'act of God' to the course of a battle that usually manifests in the form of a storm or a wind is a common occurrence in late antique and medieval primary sources.
60. Usamah ibn Munqidh (1929), *Memoirs of an Arab-Syrian Gentleman. Or, An Arab Knight in the Crusades: Memoirs of Usāmah ibn-Munqidh (Kitāb al-i'tibār)*, trans. P. Khuri Hitti, New York: Columbia University Press, p. 177.
61. *History of Leo the Deacon*, II.3, p. 73. Whately has pointed out that 'just as the elements of a siege are different from those of a pitched battle, the "why" of a Procopian siege is often different from that of a Procopian pitched battle', referring to Procopius' morality and his frequent ascription of a city's survival to divine assistance; see Whately, *Battles and Generals*, pp. 101–5 and 146. See also K. DeVries (1999), 'God and Defeat in Medieval Warfare: Some Preliminary Thoughts', in D. J. Kagay and L. J. A. Villalon (eds), *The Circle of War in the Middle Ages: Essays on Medieval Military and Naval History*, Woodbridge: Boydell & Brewer, pp. 87–97.
62. Leo VI, *Taktika*, prologue, p. 8.
63. Vegetius, *Epitome of Military Science*, I.1, p. 2.
64. Strickland, *War and Chivalry*, ch. 3, pp. 55–97.
65. Verbruggen, *Art of Warfare*, pp. 13–14.
66. Ibid.
67. Morillo, 'Milites, Knights, and Samurai', p. 183.
68. McGeer, *Sowing the Dragon's Teeth*, pp. 214–15; Ahrweiler, 'Recherches sur l'administration', p. 16; Lemerle, *Agrarian History*, pp. 130–1; Treadgold, *Byzantium and its Army*, pp. 173–4; Haldon, *Warfare, State and Society*, pp. 220–2, although he considers the possibility that this unit might simply have come into greater prominence rather than being 'reintroduced'.
69. Haldon, *Warfare, State and Society*, pp. 221–2.
70. T. G. Kolias (1993), *Νικηφόρος Β΄ Φωκάς (963–969). Ο Στρατηγός Αυτοκράτωρ και το Μεταρρυθμιστικό του Έργο* [Nikephoros Phokas the Second (963–969): General Imperator and his Reforms], Athens: Vasilopoulos.
71. In his description, the author uses the term for a mixed heavy infantry and cavalry battle formation 'τους δε πεζούς παρ' εκάτερα τους καταφράκτους έχοντας' (the infantry with the *kataphraktoi* on either side); *On Strategy*, 35.20, p. 108.

72. This treatise on diverse aspects of warfare has been edited twice: by Herman Kochly and Wilhelm Rustow in 1855, and by George Dennis in 1986. On both occasions it was published as an independent monograph: H. Kochly and W. Rustow (1853–5), *Griechische Kriegsschriftsteller*, Leipzig: Engelmann; G. T. Dennis (1985), *Three Byzantine Military Treatises*, Corpus Fontium Historiae Byzantinae, 25, Washington, DC: Dumbarton Oaks.
73. P. Rance (2007), 'The Date of the Military Compendium of Syrianus Magister (Formerly the Sixth-Century Anonymus Byzantinus)', *Byzantinische Zeitschrift*, 100, pp. 701–37; S. Cosentino (2000), 'The Syrianos' *Strategikon* – A Ninth-Century Source?', *Bizantinistica: Rivista di studi bizantini e slavi*, 2, pp. 243–80; A. D. Lee and J. Shepard (1991), 'A Double Life: Placing the Peri Presbeon', *Byzantinoslavica*, 52, pp. 15–39, esp. 25–30.
74. Abels and Morillo, 'A Lying Legacy?', pp. 1–13; Morillo, 'Milites, Knights, and Samurai', pp. 167–84. One should highlight, however, the revisionist views of Bachrach on the subject; B. S. Bachrach (2007), '"A Lying Legacy" Revisited: The Abels-Morillo Defense of Discontinuity', *Journal of Medieval Military History*, 5, pp. 153–93.
75. *The Ecclesiastical History of Orderic Vitalis* (1969–80), ed. M. Chibnall, 6 vols, Oxford: Oxford University Press, VI, pp. xxi–xxv.
76. *History of Leo the Deacon*, pp. 37–51.
77. Kolias, Νικηφόρος Β΄ Φωκάς, p. 24; M. Wojnowski (2012), 'Periodic Revival or Continuation of the Ancient Military Tradition? Another Look at the Question of the *Katafraktoi* in the Byzantine Army', *Studia Ceranea*, 2, pp. 195–220.
78. A long list of references can be found in Leo VI, *Taktika*, p. 660 (καβαλλαρικός); Dennis, *Three Byzantine Military Treatises*, p. 355 (καβαλλάριος, καβαλλαρικός).
79. On historians misunderstanding a battle manoeuvre because of lack of experience, see Verbruggen, *Art of Warfare*, p. 13.
80. John Skylitzes, *Synopsis of Byzantine History*, p. 291; *History of Leo the Deacon*, IX.viii, p. 196. See also S. McGrath (1995), 'The Battles of Dorostolon (971): Rhetoric and Reality', in T. S. Miller and J. Nesbitt (eds), *Peace and War in Byzantium*, Washington, DC: Catholic University of America Press, pp. 152–64.
81. Lendon, 'Rhetoric of Combat', pp. 273–6. For a recent monograph evaluating Procopius as a military historian, emphasising the author's classicising manner, cultural adherents and approaches to battle, see Whately, *Battles and Generals*. See also Kagan's monograph focusing on the battle narratives of both Ammianus Marcellinus and Caesar, examining the narrative techniques they employed and their means of expressing causality: K. Kagan (2006), *The Eye of Command*, Ann Arbor: University of Michigan Press.
82. Haldon, *Warfare, State and Society*, ch. 4; Haldon, *Byzantium in the Seventh Century*, ch. 6; C. E. Bosworth (1975), 'Recruitment, Muster and Review in Medieval Islamic Armies', in V. J. Parry and M. E. Yapp (eds), *War, Technology and Society in the Middle East*, London:

Oxford University Press, *pp. 59–77*; Kennedy, *Armies of the Caliphs*, pp. 155–6; Tantum, 'Muslim Warfare', pp. 190–1, who translates parts of the *Nihayat al-Su'l*; Constantine Porphyrogenitus, *Three Treatises*, B.34–150, pp. 84–92; *On Tactics*, ch. 29.

83. The latest study on Procopius' *History* of the Justinianic wars is Whately, *Battles and Generals*.
84. See the discussion in Cameron, *Agathias*, pp. 39–43.
85. *The History of Menander the Guardsman* (1985), ed. R. C. Blockley, Cambridge: Cambridge University Press, pp. 18–19.
86. See note xlix.
87. Treadgold, *Middle Byzantine Historians*, pp. 19–20; D. Afinogenov (2002), 'A Lost 8th-Century Pamphlet against Leo III and Constantine V?', *Eranos*, 100, pp. 15–17.
88. Theophanes Confessor, *Chronicle*, pp. 608 (765/66) and 616–17 (772/73); Treadgold, *Middle Byzantine Historians*, p. 21; Treadgold, *Byzantium and its Army*, p. 64.
89. *Vita Basilii Imperatoris*, pp. 11*–13* (asterisks used by the editor of the translation to indicate the introductory pages).
90. Treadgold, *Middle Byzantine Historians*, p. 179.
91. Ibid., p. 261.
92. Ibid., pp. 344–86.
93. This problem is examined in: W. Treadgold (2005), 'Standardized Numbers in the Byzantine Army', *War in History*, 12, pp. 1–14; Treadgold, *Byzantium and its Army*, pp. 43–4; W. Treadgold (1980), 'Notes on the Numbers and Organization of the Ninth-Century Byzantine Army', *Greek, Roman and Byzantine Studies*, 21, pp. 269–77; Treadgold, 'Remarks', pp. 205–12. See also J. Haldon (2000), 'Theory and Practice in Tenth-Century Military Administration: Chapters II, 44 and 45 of the *Book of Ceremonies*', *Travaux et mémoires*, 13, pp. 330–3.
94. There is a remarkable passage in Ibn Khaldun's *Muqaddimah* where the author condemns the uncritical use of sources as it produces gross exaggerations because of 'the common desire for sensationalism'; see Ibn Khaldun, *Muqaddimah*, pp. 11–13. See also H. Delbruck (1913), *Numbers in History*, London: Wentworth, pp. 11–23. Delbruck's methods have come under criticism by B. S. Bachrach (1999), 'Early medieval military demography: some observations on the methods of Hans Delbruck', in Kagay and Villalon (eds), *Circle of War in the Middle Ages*, pp. 3–20. See also J. Flori (1993), 'Un problème de méthodologie: la valeur des nombres chez les chroniquers du Moyen Age, à propos des effectifs de la première croisade', *Le Moyen Age*, 119, pp. 399–422; J.-C. Cheynet (1995), 'Les effectifs de l'armée byzantine aux Xe–XIIe siècles', *Cahiers de civilisation médiéval*, 38e année, 152, pp. 319–20. For the use of Byzantine numbers by modern historians, see W. Treadgold (1989), 'On the Value of Inexact Numbers', *Byzantinoslavica*, 50, pp. 57–61.

95. Regarding the estimates of enemy numbers by an inexperienced observer, I quote a passage from the *Strategikon*: 'The arrangement of cavalry and infantry formations and the disposition of other units cause great differences in their apparent strength. An inexperienced person casually looking at them may be very far off in his estimates. Assume a cavalry formation of six hundred men across and five hundred deep ... If they march in scattered groups, we must admit that they will occupy a much greater space and to the observer will appear more numerous than if they were in regular formation ... Most people are incapable of forming a good estimate if an army numbers more than twenty or thirty thousand ... More accurate information about an army's numerical strength may be obtained from deserters, from prisoners, from the passage of narrow defiles, from camps when all the enemy forces make camp together' (*Strategikon*, IX, pp. 102–3). On all of these difficulties in reporting the figures of armies in the Middle Ages, see Verbruggen, *Art of Warfare*, pp. 5–9; J. France (1999), *Victory in the East: A Military History of the First Crusade*, Cambridge: Cambridge University Press, pp. 122–42; Haldon, *Warfare, State and Society*, pp. 99–106.
96. *History of Leo the Deacon*, I.vii, p. 66.
97. Ibid., IV.i, p. 105.
98. John Skylitzes, *Synopsis of Byzantine History*, p. 257.
99. Ibid., p. 274.
100. *History of Leo the Deacon*, VI.xii, pp. 159–60.
101. Treadgold, *Middle Byzantine Historians*, pp. 244–5.
102. *History of Leo the Deacon*, VIII.iv, pp. 179–80.
103. John Skylitzes, *Synopsis of Byzantine History*, pp. 275–6.
104. Ibid., pp. 282–5.
105. Despite Treadgold's position that 'numbers for official establishments can be round without being either rounded or estimated' and that 'the overall principle of standardization in the Roman army is indisputable', I feel uneasy when faced with nice, round multiples of ten for the total of an expeditionary army; see Treadgold, 'Standardized Numbers', p. 3.
106. Ibn Zafir, pp. 121–2.
107. Cheynet, 'Les effectifs', pp. 322–6.
108. Abu Firas, p. 358.
109. Ibn Zafir, p. 125.
110. Ibid., p. 126.
111. Abu Firas, p. 362.
112. Ibn Zafir, p. 125
113. *Histoire de Yahya*, pp. 774–5; Kamal al-Din, p. 183.
114. Abu Firas, pp. 369–70.
115. *Histoire de Yahya*, p. 781.
116. Ibid., pp. 783, 793; John Skylitzes, *Synopsis of Byzantine History*, p. 257.
117. *History of Leo the Deacon*, I.v, p. 64.

118. *Praecepta Militaria*, IV.192–208, p. 50; Vegetius, *Epitome of Military Science*, III.9, p. 84–5.
119. *History of Leo the Deacon*, I.v, pp. 64–5.
120. Ibid., III.x, p. 101.
121. Ibid., III.x, pp. 101–2; IV.i, p. 105.
122. Ibid., III.x, p. 101.
123. Ibid., IV.x, p. 119.
124. Ibid., VI.xii, p. 159.
125. Ibid., VIII.i, pp. 176–7.
126. Ibid., VIII.ii, p. 177. See also T. Loungis (1997), 'Επιθεώρηση ενόπλων δυνάμεων πριν από εκστρατεία' [Inspection of Armed Forces before a Campaign], in K. Tsiknakes (ed.), *Byzantium at War (9th–12th c.)*, Athens: National Research Institute, pp. 93–110, especially pp. 106–10.
127. *History of Leo the Deacon*, IV.v, pp. 110–11.
128. Ibid., VIII.v, pp. 180–1.
129. John Skylitzes, *Synopsis of Byzantine History*, p. 216.
130. Ibid., p. 220.
131. Ibid., p. 257.
132. Ibid., p. 276.
133. Ibid., p. 284.
134. Ibid., pp. 286–8.
135. Ibid., p. 288.
136. These two invasion points are only mentioned by Abu Firas, p. 359.
137. Mutanabbi, *Poem on Hadath*, pp. 309–14; Ibn Zafir, p. 124; Kemal al-Din, p. 181.
138. Yaḥyā, p. 769.
139. Mutanabbi, *Poem on Hadath*, pp. 322–6, 327–8. The ravaging of Melitene and Arqa, along with the battle at Marash, are reported by Ibn Zafir, p. 126. Abu Firas, a contemporary eyewitness, also mentions the stopping points at Diyar-Bakr, Kaᶜb, and the defile of Mauzar; see Abu Firas, pp. 361–3.
140. Canard, *Hamdanides*, p. 788.
141. Mutanabbi, *Poem on Hadath*, pp. 340–4; neither Abu Firas nor Ibn Zafir reports this expedition.
142. M. Canard (1936), 'Mutanabbi et la guerre byzantino–arab: intérêt historique de ses poésies', in *Al Mutanabbi: recueil publiee à l'occasion de son millènaire*, Beirut: Imprimerie catholique, pp. 99–114.
143. Ibid., p. 102.
144. *Histoire de Yahya*, pp. 767–8.
145. Ibid., p. 768.
146. Ibid., pp. 771–2.
147. Ibid., pp. 774–5.
148. Ibid., p. 775.
149. *History of Leo the Deacon*, I.iii, pp. 61–2.
150. Ibid., II.vi, p. 76.

151. Ibid., IV.iii, pp. 106–8.
152. Ibid., VIII.iv, pp. 179–80.
153. Ibid., VIII.iv, p. 180.
154. Ibid., VIII.x, pp. 185–6; IX.i, pp. 186–7; IX.ii, pp. 188–90; IX.vi, pp. 192–3; IX.viii, pp. 196–7.
155. Ibid., IX.iii, pp. 188–9; IX.vi, pp. 192–3.
156. Ibid., VIII.ix, p. 185.
157. John Skylitzes, *Synopsis of Byzantine History*, p. 197.
158. Ibid., pp. 197–8.
159. Canard, *Hamdanides*, p. 804.
160. John Skylitzes, *Synopsis of Byzantine History*, pp. 234–5.
161. Ibid., pp. 275–8.
162. Holmes, *Basil II*, pp. 162–8.
163. John Skylitzes, *Synopsis of Byzantine History*, p. 282.
164. Ibid., pp. 285, 286, 287, 289, 291.
165. Ibid., p. 285.
166. McGrath, 'Battles of Dorostolon', pp. 152–64; Holmes, *Basil II*, pp. 162–8.
167. Mutanabbi, *Poem on Hadath*, p. 333.
168. Ibid., p. 333; Abu Firas, p. 364; the number 500 is mentioned only by Ibn Zafir, p. 125.
169. *Histoire de Yahya*, p. 772.

10

Tactical Innovation and Adaptation in the Byzantine Army of the Tenth Century: The Study of the Battles

This chapter will focus on the tactical changes that took place in the imperial army in the tenth century, through the study of the major battles of the period. The most useful primary sources for identifying these changes are the military treatises of the time, which can furnish a great amount of significant details on how armies should – in theory – have been organised and deployed on the battlefield up to the period when they were compiled. I have already discussed their recommendations about the marching and battle formations, the armament and the battlefield tactics of the Byzantine army units, and I have provided a number of arguments and thoughts as to whether these changes reflect any kind of innovation or tactical adaptation to the strategic situation in the East.

In order to determine whether theory translated into practice, I will examine – albeit briefly – the most important pitched battles of this period at Hadath (954), Tarsus (965), Arkadiopolis (970), Dorystolon (971), Alexandretta (971), Orontes (994) and Apamea (998) through the accounts of contemporary lay and ecclesiastic sources, in regard to a number of questions, such as how successful the Byzantines were at adapting to the changing military threats posed by their enemies in the East; how far we can see the Byzantines responding to the tactical and strategic threats of enemies in ways not anticipated by the manuals; and what these reveal about the place of literacy in the Byzantine command structure, the training of the officer class, and the question of professionalism. The conclusions drawn by the study of these campaigns will shed some light on the fighting tactics, training, morale and esprit de corps of the – predominantly Eastern – armies that participated in these campaigns.

On the morning of 29 Jumada II 343/30 October 954, one of the most famous battles of the Byzantine–Arab wars of the period unfolded, ending in disaster for the army of Bardas Phocas, and expediting his replacement by his son Nicephorus the following year. Close to the mountain of al-Uhaydib, near the town of al-Hadath (Adata), the army of the Domestic of the Scholae, some 50,000-strong if we believe the figures supplied by Ibn Zafir, met with the expeditionary force of Sayf ad-Dawla that had arrived in the area on 17 Jumada II/18 October to restore the fortifications of the town.

First, we should point out Mutanabbi's amazement at the multi-ethnic composition of the Byzantine army, which comprised 'men of every language and every nation, and only the interpreters could understand them'.[1] Mutanabbi writes about Byzantine and Rus' troops, to which Abu Firas adds Armenians and Slavs, while Ibn Zafir notes Rus', Bulgar and Armenian infantry and cavalry soldiers. The important thing here is not the multi-ethnic origin of Phocas' army in itself, but rather that all of the aforementioned Muslim sources underlined the presence of Rus' and Armenian troops, which might indicate that their contingents would have been significant enough to be noticed by two eyewitnesses (Mutanabbi and Abu Firas).[2]

There could be a correlation between the comments made by our Muslim sources and what the author of the *Praecepta Militaria* notes about nationalities and the qualities necessary for a foot soldier serving in the imperial army. Although the large number of Armenian and Rus' soldiers does not necessarily reveal any change in the tactics and organisation of the Byzantine armed forces of the period that could be understood as a tactical adaptation to their enemies in the East, this is certainly another indication of the ample supply of high-quality professional soldiers the Byzantine generals had at their disposal in this period.

Since Sayf ad-Dawla was already near the town of Hadath before Phocas' arrival, it may be assumed that it was him who commanded the higher ground. We know nothing of the composition of the opposing armies or the ratio between infantry and cavalry, except for Mutanabbi's mention of the presence of *kataphraktoi*.[3] Mutanabbi, Abu Firas and Ibn Zafir simply note the crucial manoeuvre that won the battle for Sayf ad-Dawla, namely the latter's charge against the enemy with his retinue of 500 *ghulam* cavalry – possibly directing it against the centre of the Byzantine formation where they might have been able to make out Phocas' banner.[4] In all likelihood, Bardas would have carried his banner in the centre just behind the unit of the *kataphrakts* that would have dominated the first line, in accordance with the recommendations of the military manuals.[5]

Historians have been unable to provide any conclusive answer as to how the *ghulam* cavalry managed to break through the line of the Byzantine army. Did they break the first line of the Byzantine cavalry, with the rest of the units just melting away? Or did the disorganised units of the *kataphraktoi* and thematic cavalry seek refuge inside some sort of infantry square, thus probably forcing Sayf to attack them as well? The number of 3,000 dead, both infantry and cavalry, reported by Ibn Zafir points to the latter. This is, however, just speculation as we cannot be certain what exactly happened that day.[6]

Leo the Deacon, our main source on the Byzantine campaign to conquer the strategically important Cilician city of Tarsus, does not provide us with any detailed analysis of the battle that took place outside the city. We read that Nicephorus Phocas had brought with him large numbers of heavy cavalry, hence his decision to clear the fields and meadows outside the city from any kind of vegetation that could hamper the movement of his cavalry units or conceal an ambush from his enemies. Even though we know nothing about the size, composition and battlefield formation of the Tarsiot army, Leo reports the deployment of the imperial army before the battle as follows:

> [The emperor] arranged the divisions on the battlefield, deploying the iron-clad horsemen (πανσιδήρους ιππότας) in the front ranks (κατά μέτωπον), and ordering the archers and slingers (σφενδονητάς) to shoot at the enemy from behind. He [Phocas] took his position on the right wing, bringing with him a vast squadron of [regular] cavalrymen (μυρίανδρων ιππέων ἴλη), while John who had the sobriquet Tzimiskes . . . fought on the left.[7]

Leo's account confirms the recommendations of the manuals to have three cavalry units on the first line, with two units of regular cavalry surrounding the *kataphrakts* in the middle, and archers following behind to provide cover in case of an enemy attack. Their range, however, was insufficient to threaten the enemy in the event of an advance.[8] The position of the infantry is not mentioned here, but it would probably have followed immediately behind the cavalry in the centre.

As we saw at the Battle of Hadath, the customary place of the commander-in-chief of the army was in the middle of the second line (between two divisions of regular cavalry), to keep him away from danger. Nicephorus is mentioned to have taken his place on the right wing of the army that attacked the Tarsiots, possibly wishing to perform a manoeuvre against the left flank of the enemy to win the field – although Leo the Deacon mentions nothing of the sort. Finally, we understand from Leo's description of the advance of the Byzantine cavalry units, moving forward with 'incredible precision (αμήχανω κόσμο, *amihano kosmo*), as the entire plain sparkled with the gleam of their armour' – a product no doubt of years of intensive training and drilling – that it did have the desirable psychological effect on the Tarsiots, who 'immediately turned to flight . . . overwhelmed by a terrible cowardice'.[9] Regrettably, the sources offer no more details about the tactics employed that day.

The Byzantine–Fatimid confrontation in the Cilician port of Alexandretta in 971 was the culmination of the imperial capture of the city of Antioch two years earlier. With the Byzantine expansion in northern Syria coinciding with the Fatimid takeover of Ikhshidid Egypt and its lands in the Middle East, northern Syria inevitably became a battleground where two of the superpowers of the age eventually clashed. Already since the autumn of 970, Jafar ibn Falah, the Fatimid governor of Syria, began preparations for an expedition to overthrow the Byzantines from Antioch. In the campaigning period that followed, he dispatched a large army against the Syrian capital.

The best sources for what is probably the most detailed battle of the period of the 'Reconquest' are late Muslim chronicler accounts from the fourteenth and fifteenth centuries.[10] The Byzantine armies of the East, under the leadership of the Duke of Mesopotamia, defeated and routed the invading Muslim army in a single and decisive battle, thus managing to save Antioch and northern Syria from immediate occupation. As usual, our sources provide us with no details of the battle tactics or any evidence regarding the units of the opposing armies. Skylitzes' figure of 100,000 men for the Fatimid force is surely pure fiction, although we should not doubt that the Muslim army would have been quite numerous. He shares, however, an important piece of information for the composition of the Fatimid army: 'Egyptians [Fatimids], Persians, Arabs [Berbers], Elamites [Daylamites], together with the inhabitants of Arabia Felix [Yemen] and Saba [Ethiopia]'[11] campaigned to avenge the 'affront' of the capture of Antioch.

According to a fifteenth-century Yemeni named Idris ibn al-Hasan, the Fatimid commander Futuh had brought with him some 20,000 troops, a much more plausible figure than Skylitzes' hundred thousand, who after establishing themselves and gaining control of the region around Antioch withdrew due to the arrival of Byzantine reinforcements. This source, however, does not specify the reason for the Fatimid retreat, for it seems highly unlikely for an army of some 20,000 to withdraw immediately upon the arrival of the enemy without any apparent motive.

Al-Maqrizi, in his monumental work on the Fatimid Caliphate written in the beginning of the fifteenth century, points to a side expedition conducted by a large corps of some 4,000 men send by Futuh to control the key port of Alexandretta with its passing caravans that supplied Antioch. Although Maqrizi mentions a battle between the Fatimids and the Byzantine garrison that ended up in defeat for the former, the key information to understand what really took place comes from another Egyptian (fourteenth century) called Abu Bakr ibn al-Dawadari. The latter writes about the corps of 4,000 men – thus pointing to a common source with Maqrizi – marching west to Alexandretta, but the stratagem that won the Byzantines the battle can be found only in the following account:

> They [the Fatimids] proceeded until they were on the point of gaining the camp of the Byzantines. Thereupon they saw pavilions of the Byzantines in the field ... and they were rapidly diverted to plundering. The *Turbazi* [the Byzantine commander] meanwhile, alerted to their presence, took the warriors of his army and withdrew from the valley. Thus when the Berbers entered the tents bent on plunder, the *Turbazi* was able to attack them and they were defeated as the sword caught them from all sides.[12]

It was this victory that proved instrumental for the deliverance of Antioch from the Fatimid siege and put northern Syria firmly within the Byzantine sphere of influence, at least for the next twenty years.

Crossing the continents into Thrace, Bardas Skleros led some 12,000 men – predominantly eastern units according to Skylitzes – to drive off the Rus' army that was ravaging the region, allegedly numbering some 308,000 men. This is definitely a gross exaggeration, although we can be fairly confident that the Rus' greatly outnumbered the Byzantine forces. This discrepancy prompted the Byzantine commander to avoid any confrontation with the Rus' on open ground, reverting rather to defeat his enemy by other means, 'to get the better of the enemy by military cunning; to gain the upper hand over so great a number by skill and dexterity'.[13]

Bardas' next move was to gather as much intelligence as possible about his enemy: 'When he [Bardas] had carefully studied the matter of how the enemy might best be attacked ... he dispatched the patrician John Alakasseus with a small detachment whose orders were to advance and reconnoitre the enemy.'[14] Gathering intelligence about the enemy regarding its numbers, composition, equipment, leaders and morale was one of the most important tasks of a commander prior to an engagement, and this was usually assigned to the *prokoursatores*, also known as *trapezitai* or *tasinarioi*.[15]

Tactical Innovation and Adaptation

The most crucial stage of this engagement, however, was the luring of the unit of Patzinaks into an ambush set by Bardas in a suitable location close to the town of Arkadiopolis. The commander's orders are described by Skylitzes as follows:

> When he encountered the enemy, he was to give battle, but as soon as blows were struck he was to turn his back and give the impression of running away. He was not to flee at full tilt ... but gently and without breaking rank. Then, wherever it was possible, [his men] were to turn about and set upon the enemy again. Their orders were to keep on repeating the operation until [the enemy] was well within the ambushes and traps.[16]

This feigned flight described in the *Synopsis* agrees in its outline with battle tactics found in chapters 11 and 17 of *On Skirmishing*. Indeed, we read in the treatise:

> Let him [the general] search for a suitable and very secure location, if possible, with a fortress nearby ... Units of them [the infantry] should be concealed in ambuscades on both sides of the road. Let the general take position close behind the infantry ... and with him the cavalry units. Up to a hundred selected men should be dispatched by the general to prepare ambushes.[17]

As for the luring of the enemy unit into the defile occupied by friendly forces, we read:

> At times, he [the officer in command] charges into them, at times he begins to run away, and he provokes them into pursuing. If they pursue up to that place in which the infantry is concealed and some of the enemy pass right by them, then our men should charge out of their hiding places and check the pursuing enemy.[18]

Soon after the feigned flight by Alakasseus' troops, the trapped Patzinak horsemen realised what was about to befall them. Halting their pursuit, they stood their ground at some distance from the trap set up by Bardas, thus prompting an attack by the general's troops and the rest of the units in hiding. They attacked from both sides in an organised manner, rank by rank as recommended by both *On Skirmishing* and *Praecepta Militaria*, specifically for the case 'when the enemy remains at a distance and in a disorganised mass'.[19] With the annihilation of the first party of Patzinaks came the engagement with the main army of the enemy who, although shaken by the death of their comrades, nevertheless 'attacked on the Romans, the cavalry leading the charge, the infantry following behind'. The ensuing melee is not described in any detail by Skylitzes or by Leo the Deacon and we can only imagine a struggle of endurance between the two armies fighting in a confined space and at close quarters. The retreat of the Patzinaks was followed by the pursuit of the defeated force, and although none of our sources mentions specifically the unit who undertook this mission, once again we read in our manual: 'When they [the enemy] do turn to flight, it is not the kataphraktoi who should undertake the pursuit but their two accompanying units trailing behind them [the regular cavalry].'[20]

The action won by the imperial forces at Arkadiopolis provided Tzimiskes with excellent intelligence regarding the composition, fighting tactics and morale of his enemy, which he put to good use the following year in his campaigns in Bulgaria, which culminated in the siege of the city of Dorystolon. In April 971, the emperor invaded Bulgaria at the head of an army of around 30,000 men, marching through the mountain and forest passes that had been left undefended, probably due to the Rus' being engaged with suppressing a Bulgar rebellion.

It would have been very interesting to have had the chance to see how this army would have reacted in case of an ambush by Bulgar or Rus' forces, since the Haimos and Rhodope Mountains provide excellent opportunities for ambushes, as Leo the Deacon recalls over Nicephorus' expedition of 966:

> When the Emperor [Nicephorus] saw this [frontier region], he did not think he should lead the Roman force through dangerous regions with its ranks broken, as if he were providing sheep to be slaughtered by the Mysians; for it is said that on several previous occasions the Romans came to grief in the rough terrain of Mysia, and were completely destroyed.[21]

Leo was, no doubt, referring to the defeat of Nicephorus I's army at Pliska in 811 and Krum's tactic of attempting to block the imperial army's access to his lands by constructing wooden palisades in several important passes along the Byzantines' invasion route. This is another sign of the Byzantines learning from experience.

The first engagement between the Byzantine and Rus' armies took place right after the arrival of the imperial army opposite the main gate of the city of Dorystolon. Leo the Deacon reports in significant detail the battle formation of the two armies, and most importantly the place of the *kataphraktoi* in the mixed formation of the Byzantines: 'After the emperor deployed the Romans in the van (κατά μέτωπον, *kata metopon*) and placed the ironclad horsemen (πανσιδήρους ιππότας, *pansidirous ippotas*) on both wings (κατά θάτερον κέρας παριστησάμενους, *kata thateron keras paristisamenous*), and assigned the archers and slingers to the rear and ordered them to keep up steady fire.'[22] In other words, the only line of three divisions consisted of the main infantry forces ordered in the centre of the formation and the heavy cavalry of the *kataphrakts* on both wings to cover the flanks.[23] In order to reinforce this line, Tzimiskes deployed archers and slingers immediately behind the main infantry phalanx to support them as long as they were within missile range. At the same time, his crucial tactical move was to put two extra units of *kataphrakts* as a reserve in the wings. Since the Rus' did not have any cavalry units and were not accustomed to fighting on horseback, as Leo the Deacon insists in his narrative,[24] Tzimiskes would probably have planned to make a decisive attack against the wings of the enemy infantry phalanx, the most vulnerable point of its formation.

What we know about the battle is that the Rus' were the first to attack with 'their habitual ferocity and passion', but the Byzantines halted their advance and being able to break through their formation at a couple of points made them retreat and regroup their shield wall. It is not entirely clear as to which units of the imperial army were involved in this first melee that lasted well into the evening.[25] With his infantry and cavalry failing to break through the dense Rus' lines, Tzimiskes decided to throw into battle his reserve heavy cavalry in the wings in a final counterattack intended to break the enemy's morale: 'So they [the cavalry] pressed forward with an extraordinary assault and the trumpeters sounded the call to battle, and the shout arose from the Romans in a body.'[26] With the Rus' retreating back to the city of Dorystolon, it seemed clear that the Byzantines would have to make preparations for establishing a siege.

Regrettably, the two following engagements between the Rus' and the Byzantines outside the city are poorly documented.[27] The fourth engagement, however, came after the apparent success of a sortie detachment against a Byzantine siege machine. Eager to take advantage of his troops' high morale, Svyatoslav drew up their battle lines and led them forward, while the Byzantines formed a deep phalanx to meet them. Although nothing more can be discerned from the sources regarding the Byzantines' battle division, we may presume that they were deployed in the same formation as the first engagement, namely infantry taking up its position at the centre while two *kataphrakt* units covered its flanks, with reserve cavalry, archers and slingers being deployed immediately behind them.

On 24 July, Svyatoslav ordered the Rus' to throw open the gates of the city and draw their forces, not to the plain opposite the city but rather close to the walls where the area must have been much narrower. Skylitzes provides a short description of the topography of the battlefield:

> The emperor noticed how narrow the place was, and that it was due to this factor that the enemy's resistance was possible: the Romans had so little elbow room they were unable to display the kind of performance which was appropriate to their valour.[28]

Skylitzes recognised Svyatoslav's tactic to counter the Russian lack of heavy cavalry by drawing the Byzantines to a narrow location with woodland on one flank and marshes on the other, where they could not manoeuvre easily, and then attack from the sides as on previous occasions. Leo the Deacon comments that 'the emperor organised the Romans and led them out of camp',[29] presumably in the same three-division formation that we saw in the first engagement, with *kataphrakt* cavalry again reinforcing both wings. The Rus' were the ones who opened the battle, attacking the centre of the Byzantine formation, while their javeliners caused many casualties and much confusion to the cavalry in an effort to prevent any encircling manoeuvres by the latter. As the battle took place in the height of summer, however, and the men suffered due to lack of water, the emperor gave orders for water skins to be brought forward so that the soldiers would refresh themselves and their horses and carry on with the battle unhindered. According to the *Praecepta*, this was the job of the bowmen and slingers who were guarding the intervals of the infantry square.[30]

Skylitzes reports the battle manoeuvre that proved critical for the course of the battle. Phocas soon realised how narrow the place was and that he could not make full use of the capabilities of his heavy cavalry units:

> He [Phocas] ordered the commanders to retreat towards the plain, withdrawing from the city, thus giving the impression of running away. They were not, however, to be in a hurry, but to take their time and retreat only little by little. Then, when they had drawn their pursuers some distance from the city, they were suddenly to turn about, give their horses their heads and attack those men.[31]

This feigned retreat worked perfectly as the Rus' took the bait and went after the retreating Byzantine units, thinking they were running away. Indeed, even Leo the Deacon reported that this action was a retreat, rather than a feigned flight to lure the Rus' out to the plain, which can be explained either by the army's professional execution of the manoeuvre or Leo's poor knowledge of military affairs (contrary to Skylitzes who, although he wrote almost a century after this event, was able to understand this tactic due to his great experience in military matters).[32]

What tipped the balance of battle in favour of the Byzantines was the emperor's decision to commit to action his elite *kataphrakt* cavalry units that were stationed as a reserve in the wings of the army.[33] Skylitzes mentions a number of high-ranking officers who took the lead against the Rus', but Leo reports that the emperor personally commanded the charge – probably taking his own personal guard as well – in order to raise the morale of his soldiers. This cavalry attack, along with a push from the centre by the infantry and a simultaneous attack from a cavalry unit on one of the wings (the sources do not specify which one) under Bardas Skleros, resulted in the envelopment of the Rus' army and their eventual retreat inside the city. This encircling manoeuvre as a method of deciding a battle, coupled with a push from the centre, falls into the category of tactics recommended by the contemporary military manuals examined.[34]

Four years after the Byzantine victory over a Fatimid force of some 4,000 at Alexandretta in 971, Emperor John I Tzimiskes took Apamea on the Orontes before invading southern Syria and Galilee. While the war in the East was a sideshow compared with Basil II's Bulgarian wars that lasted for the better part of three decades (991–1018), the Battles of the Orontes (994) and Apamea (998) were the climax of the struggle for dominance in Syria between the Byzantine and Fatimid Empires.

Our predominantly Muslim sources do not provide adequate information about the Battle of the Orontes. Yahya, a Syriac source, reports only that Emperor Basil II ordered the Duke of Antioch, Michael Bourtzes, to mobilise his forces against the advancing army of the Fatimid general (of Turkic origin) Manjutakin (d. 1007).[35] Bar Hebraeus, a thirteenth-century Syriac bishop (of Jewish origin), indicates a number of approximately 50,000 men for the imperial army, 'some horsemen and some foot men'.[36] What is certain, however, is that a large part of Bourtzes' army would have been Armenians since they played a fundamental role in repopulating the regions of Melitene, Tarsus, Adana and Antioch in the previous decades.[37] The Fatimid army seems to have been quite numerous as well, perhaps even 30,000-strong.[38] Although the sources do not identify exactly the composition of Manjutakin's forces, these would probably have consisted of a multi-racial army of Turks, elite Iranian Daylami infantry, former Ikhshidid and Kafurid troops and Bedouins from North Africa and Syria.[39]

Al-Maqrizi reports that Manjutakin departed from Damascus in Rabi^c I 394 (15 April–14 May 994) against Antioch, where he attempted a show of force by ravaging the environs of the city before marching east to Aleppo.[40] Departing from Antioch, the Byzantine commanders met with the Hamdanid troops sent by Lu'lu' al-Kabir, guardian of the Hamdanid emir Sa^{c}id al-Dawla and nominal ruler of the emirate since 991. Manjutakin, who was besieging Aleppo, upon news of the Byzantine–Hamdanid march south turned west to meet his opponents with the two armies encamping on the Orontes.[41]

We have little information about what precisely happened next. The Byzantine–Hamdanid army occupied the west ford of the Orontes, with Yahya presenting Bourtzes as reluctant to cross it and fight the Fatimids head on. Manjutakin retained the strategic initiative by sending a part of his army – the Bedouins and a section of his Daylami or Turkish troops – to reconnoitre the positions of the Hamdanids, who were guarding one part of the ford, while he would attack the Byzantine force with the rest of his troops. Our sources report that the Hamdanids panicked and fled, thus allowing the Bedouin attacking party to pillage their camp in their usual unruly behaviour. Realising the desperate situation, the Byzantines melted away leaving their commanders in the field along with their baggage train and 5,000 dead.

There are several questions that remain unanswered for this battle. First, would the Byzantines have had time to form their units into a double-ribbed hollow square to face their opponents who were crossing the Orontes? The sources tell us that they had posted sentries to guard the ford, so it would seem unlikely that they would have been entirely taken by surprise. After all, Bourtzes is presented by Yahya as a reluctant – or, rather, very cautious – commander, keeping in line with the recommendations of all military treatises since antiquity, which advised extreme caution and order when crossing a river where the enemy holds the opposite ford.[42] Second, had the Byzantines pitched their camp in the order they were to deploy in battle, in accordance with this recommendation in the *Praecepta Militaria* of Nicephorus Phocas (c. 969): 'They [the soldiers] must keep their places in the camp exactly as they set to deploy in battle formation, so that, in the event of a sudden report of the enemy, they will be found ready as though in battle formation'?[43]

Regrettably, the sources are not clear: both al-Qalanisi and Bar Hebraeus highlight an attack against the centre of the enemy formation by an elite corps of 'Egyptian' infantry that broke through the enemy lines to win the day for Manjutakin. We may presume that this elite unit would most likely have consisted of Daylamis; after all, these hardy infantrymen with their battle-axes, spears (*zupins* or *mizraq*), short swords, bows and shields were renowned for their attacks in close-quarter phalanx formations. It would have been very interesting if we had any clues as to how the Byzantine generals and commanders would have responded to these troops marching against them: would the Armenians have formed a solid 8-man deep phalanx against the Daylamis and the Turkish cavalry? Would the heavy cavalry units have been able to be properly deployed and used in battle? Was the Daylami infantry attack preceded by an attack from the Turkish *ghulams*, probably in coordination with the archers, which we know that Manjutakin had in his army? Sadly, the narrative of the sources raises more questions than answers.

The same operational theatre and, probably, the same location would be the focal point of another disaster for the Duke of Antioch. The sources report that Damianos Dalassinos, the officer who replaced Michael Bourtzes in 995, attempted to take advantage of the political uncertainty – if not to use the term civil war – in the Fatimid Caliphate over the previous year and march against Apamea, a strategic base for expeditions against Aleppo to its north-east. On Jumada II 388 (30 May–27 June 998), the Fatimid (Kutama Berber) general Jaysh ibn al-Samsana marched east towards Apamea to relieve the city with the help of the Fatimid fleet.[44] The setting of the battle was like the one in 994, with the two armies facing each other on the opposing sides of the Orontes. The twelfth-century Damascene politician and historian Ibn al-Qalanisi writes about the battlefield: 'The battle took place in a large meadow surrounded by a mountain called al-Mudiq [Qalʿat al-Mudiq] on which we cannot ride but one-by-one and on the side of which is the lake of Apamea and the river called al-Maqlub [Orontes].'[45]

Although we have no numbers for the Duke of Antioch's forces, an indication of its size can be discerned from the large number of dead after the battle – some 6,000 men according to Yahya.[46] Qalanisi reports some 10,000 men in total for the Fatimid expeditionary force, of whom 1,000 were Bedouin cavalry.[47] This force also included a contingent of Daylami infantry, reportedly receiving the main Byzantine attack on the day, while our chronicler also mentions a unit of 500 *ghulam* cavalry under a certain Bishara the Ikhshidite.

This time the strategic initiative lay with Damianos Dalassinos, who ordered his troops to cross the Orontes and attack the Fatimid force, with the main attack directed against the centre of their formation and the Daylami infantry. The attackers managed to break through the enemy formation and force them to flee, with both the left wing of the Fatimid army – led by a certain Mansur the Slav – and the right wing under the command of the general, al-Samsana, following suit. The Byzantines reportedly pursued and killed some 2,000 of the Fatimids, when the only force that stood its ground and offered stout resistance was the unit of the 500 *ghulam* cavalry.

What happened next is a typical example of the importance of the role of the commander of the army for the morale of his soldiers: while Dalassinos was protected by only a bodyguard of some ten *ghulams* and his sons, he felt confident enough to expose himself to enemy missiles. A Kurdish *ghulam* would have spotted the imperial banner and, in the spirit of the heroic atmosphere seen very often in both Christian and Muslim chronicler accounts, charged against the duke inflicting two deadly blows to his head and torso.

The sources leave many questions unanswered, first and foremost about the battlefield. This was a location which is very likely to have hampered the movements of the cavalry on both sides, while also making any encircling manoeuvres relatively difficult to undertake. Unfortunately, we get absolutely no information regarding the battlefield formation of the opposing armies. The fact that the general of the Fatimid army, al-Samsana, was in command of the right wing instead of the usual centre could mean, according to al-Ansari, that the right wing would have been in a more elevated position, thus offering a better vantage point for the commander-in-chief.[48] This could also simply mean that he had put his Daylami infantry at the centre of the formation, perhaps projected a bit forward than the rest of the army to act as a shield, while he chose to remain in the flanks in command of a cavalry unit.

As regards the order in which the units of the imperial army would have crossed the Orontes, there is a recommended order of 'fighting march' for the crossing of a river, bridge or narrow pass, although all tacticians strongly discourage it: 'The *tagmata* cross first: first the *Scholae*, second the *Excoubitae*, third the *Arithmos*, fourth the *Hikanatoi*; and likewise for the *themata*.'[49] It appears, therefore, that the army crossed precarious spots during the march based on order of precedence. Would this order, however, have been followed during the forced crossing of a bridge, most likely under enemy fire? Would the crack troops of the imperial army – the *tagmata* – have crossed first or would an elite unit of infantry have preceded them – namely the Armenians – to establish a bridgehead and make their crossing less confusing and dangerous? In trying to tackle these questions, one should keep in mind the examples of Stirling Bridge and Bannockburn, and the significant difficulties in coordination and manoeuvring encountered by the heavy cavalry of the English kings when attempting to cross a river and be deployed for battle while the enemy army of the Scots held the opposite ford.[50] At the battle on the outskirts of Antioch in 1098 between the Latins of the First Crusade and the army of the governor of Mosul, Kerbogha, Bohemund of Taranto was afraid that the Seljuks would allow one or two of their divisions across the river and fall upon them, while the rest of the army crossed the bridge over the Orontes. Despite Bohemund's fears, however, the Latins were left free to be deployed as they wished.[51]

If the Byzantine forces had time to cross the Orontes relatively unencumbered, other questions are raised: Was the topography of the battlefield suitable for them to be deployed in their regular infantry hollow-square?[52] What unit of the imperial army attacked – and eventually managed to break – the formation of the Daylami infantry deployed in the centre of the Fatimid force? Which units undertook the pursuit of the fleeing Fatimid forces?

For the latter point we should bear in mind once again the recommendations found in the mid-tenth century Byzantine manuals on the pursuit of a defeated foe: 'Until the enemy is in general flight, our units must not break ranks but should follow up in proper formation in the manner discussed'; 'When they [the enemy] do turn to flight, it is not the kataphraktoi who should undertake the pursuit but their two accompanying units trailing behind them [the regular thematic cavalry].'[53] This should be viewed in comparison with Qalanisi's comment about Dalassinos 'standing near his standard ... in order to contemplate the victory of his army and to come to the possession of the booty captured'. If we take Qalanisi's comment at face value, was the pursuit of the defeated Fatimid forces a disorganised race to seize booty and pillage the enemy camp instead of following the recommendations of the tacticians of the period? Once again, an examination of the sources provides us with more questions than it answers and it would be careless to make any sort of assumptions based on whatever we can derive from our authors.

Conclusions

An important conclusion to contemplate from the study of the battles of the period is the key role of the commander in maintaining the morale of the soldiers and in the eventual outcome of the battle in general. Sayf ad-Dawla won the day at Hadath by launching an attack against the unit of the enemy commander, Bardas Phocas. Although we are not informed of the precise location where the elderly Byzantine commander would have placed his banner, it is likely that this would have been in the second line of cavalry, behind the elite triangular unit of the *kataphrakts*. Sixteen years later, and in a different operational theatre outside Dorystolon, one of the emperor's bodyguards called Anemas, probably in command of the elite *tagma* of the Immortals, charged against the middle of the Rus' infantry formation exactly on the spot where he could see Svyatoslav.[54] Finally, at Apamea in 998 the Byzantine general Damianos Dalassinos was surprised and killed by a Kurdish horseman right when the Muslims had been routed, thus leading to a counterattack and the eventual retreat of the demoralised imperial troops.[55] In this light, why was the killing of the enemy commander so important?

The answer lies in the fact that the overall command and the imposition of order and discipline (described as τάξις in the Byzantine manuals) in a campaigning army stemmed from its commander-in-chief. Against this background, it is easy to imagine that once the person at the top of the pyramid of command was killed, the organisational and psychological consequences would be devastating. Interestingly, the sole example that springs to mind of a high-ranking officer assuming the command of an army after the general's death during battle is Beyazid's taking over from his father, Murad I, after Kosovo (1389). Therefore, superior officers were more often than not kept out of harm's way, commanding the reserve of the army, deploying their units into battle only when this was necessary to decide its outcome.[56]

Tacticians like Frontinus, Onasander and Leo VI underline the importance of the spreading of false rumours about the enemy commander's death in battle.[57] The recent devastating defeat of the Byzantine arms at Achelous River in 917 might have prompted the author of the *Praecepta* to stress the importance of attacking the enemy leader.[58] Twice the author is careful to note that the triangular *kataphrakt* formation should be directed 'right at the spot where the commander of the enemy army is standing'.[59] Even though such a recommendation is not given in the *Sylloge Taktikorum*, this does not mean that Nicephorus' advice comes as a direct result of experience gained on the battlefields of the East. This should be seen more as an attempt to instil the desirable τάξις, this 'incredible precision' (αμήχανω κόσμο) that Leo the Deacon reports during the charge of the *kataphrakts* at Tarsus in 965.[60]

There is a clue regarding the Byzantine cavalry formation at the battle outside the city of Tarsus that I wish to expand upon. We read about the emperor's arraying of his troops: 'He [Phocas] took his position on the right wing, bringing with him a vast squadron of cavalrymen (μυρίανδρων ιππέων ύλη, *miriandron ippeon ili*), while John who had the sobriquet Tzimiskes . . . fought on the left.'[61] Here is a striking contradiction with the military manuals of the period: according to the *Sylloge* and the *Praecepta* the place of the commander-in-chief of the army was to be in the second line, right between the two middle units of regular cavalry.[62] Therefore, does this say anything about the emperor's decision to take over the command of the right flank, while putting his second-in-command and one of his ablest officers in charge of the left?

The *kataphraktoi* were the *force de frappe* of the Byzantine army charged with smashing into enemy lines, right where they could see the enemy commander, while the regular cavalry seems to have had more of a supporting role in the crucial stages of the engagement, trailing the *kataphrakts* but without falling far behind them or breaking their formation.[63] Accordingly, being in charge of one of the regular cavalry units in the flanks would not only have given Nicephorus a clear view of the *kataphrakts'* charge in the centre, but he would have been able to take impromptu decisions more quickly, such as detaching troops from his unit to support the *kataphrakts*, and without exposing himself to real danger.

Why is there such a difference between Nicephorus' place in battle in 965, and the references in the *Praecepta* just a few years later? We find no specific recommendations in the *Praecepta* regarding the exact desired position of the general, but rather three references in three different parts of the work; however, other treatises, such as the *Taktika*, the *Sylloge* or the *Strategikon*, are much more specific.[64] Is the reason that Phocas wanted to leave it up to his readers to decide which place the commander should take in an engagement? Has a part where he potentially wrote about the position of the commander-in-chief now become lost? Or is it simply because he could not trust his units of regular cavalry to perform their duty as planned based on previous experience that he took personal command of the unit? Whatever the case, the only certainty is that more questions have been raised than can be answered due to lack of sufficient evidence from our primary sources.

Another question that emerges is why did Tzimiskes – the second-in-command in 965 – deviate from the tactical developments of the period just six years later against the Rus'? I am referring here to John's decision to place *kataphrakts* on either of his army's wings at Dorystolon in 971. The emperor would have been fully aware of the tactical weakness of his enemy's battle formation – after all, the empire had already been hiring Rus' soldiers individually and in groups since the 940s. The Rus' army comprised almost exclusively of heavy infantry forming a tight phalanx, and had a significant tactical vulnerability, its exposed flanks. Therefore, Svyatoslav resorted to challenging the Byzantines in battle at a narrow point at some distance from the city walls, with woodland on the one flank and the marshy regions stretching back from the river on the other.

It must have been clear to Tzimiskes that the recommended double-ribbed infantry square would not have been of much practical use, as the Rus' did not have any cavalry units capable of encircling them or attacking its rear like the *Arabitai* did. In my view, as there was no danger of any enemy cavalry attacking the flanks of the Byzantine infantry, it would have made more sense to have had the latter receive the 'ferocious charge' of the Rus' in the hope that their formation would hold fast, while the *kataphrakts* would then counterattack the enemy's centre and flanks. Finally, Tzimiskes had placed a number of javeliners and archers to maintain a constant barrage of missile fire against the Rus', falling in line with the *Strategikon*'s recommendation that '[if] the foe is superior in infantry; entice him into the open, not too close, but from a safe distance hit him with javelins'.[65]

Therefore, to return to the main question posed previously regarding the Byzantine battle formations and tactics at Tarsus and Dorystolon, the main reason the Byzantine army was deployed in such a different way in those battles depended not only on the battlefield but, more importantly, on the composition, deployment and fighting tactics of its enemies. Before the invasion of the Patzinaks in the eleventh century, Byzantine emperors and generals had mostly faced armies in the Balkans who based their tactics on fighting on foot (with the exception of Boulgarofygon in 896). In the East, however, they were up against armies with full complements of well-disciplined and trained heavy and light cavalry, and infantry fighting with swords, bows and long lances.

Unfortunately, there are no hints in the sources as to whether the Muslim army that defended Tarsus in 965 was a mixture of urban militia (*aḥdāth*) and *ghazis* (volunteers or 'fighters for the faith') or whether the city had received reinforcements from other parts of the Muslim world.[66] Presumably, the Tarsiots could muster an army of 4,000 effectives only up to a decade before their capitulation.[67] Whatever the case, the defenders would have been unable to face the imperial army in open battle, a fact which Nicephorus' previous experience of warfare in the region would, no doubt, have taught him: hence, his decision to deploy his elite cavalry in the centre of his battle formation, flanked by two units of regular cavalry, while it is also likely that the Byzantine infantry would have been deployed in a hollow square immediately behind the cavalry formation, to

Tactical Innovation and Adaptation

guard against any encircling enemy manoeuvres. Therefore, the unfolding and outcome of the battles at Tarsus and Dorystolon is a characteristic example of the way in which the Byzantine commanders of the period had learned to use their experience in different operational theatres – and against different enemies – to their advantage, adjusting their strategies and tactics accordingly.

Finally, the battles at Arkadiopolis and Alexandretta are two very good examples of how to lure an enemy into a trap before attacking from all sides to win the field. In the case of Alexandretta, our Muslim sources tell us that the Byzantine commander had pulled his forces out of the city and into the surrounding heights, leaving the camp seemingly unprotected to the hands of his enemy. Two things require further explanation in this case: first, was the Byzantine commander aware that the 4,000-strong Muslim force dispatched to besiege Alexandretta was largely comprised of Berbers? Al-Dawadari is the only source naming one of the two leaders of this detachment as a Berber chieftain called Aras. He also implies that the *Turbazi* was, in fact, alerted of the course of the Fatimid force against them; therefore, it is likely that he would have had some sort of intelligence about its size and composition. Second, did the Byzantine commander making his withdrawal from the camp seem like a panicked retreat, waiting until the last moment so that he would be seen leaving the valley by the enemy? Or had he already placed his troops on higher ground waiting for the Berbers to enter the camp and indulge themselves in looting it? Regrettably, the sources leave us in the dark.

What I intend to point out here by raising the aforementioned questions is the degree of adaptability shown by the Byzantine commanders in the field and whether they proved themselves capable of giving their enemies a taste of their own medicine. A professional army attempting to lure an enemy force into a trap was a well-known tactic since antiquity.[68] It is not simply the application of past knowledge that we should highlight here, however, but rather the combination of good intelligence and experience that led the Byzantine commander at Alexandretta to use this stratagem against a known enemy, the Berbers. In the East, the Byzantines had been in constant contact with Berber tribes in Syria, even before the first Arab conquests of the seventh century, and although the process of their sedentarisation had been progressing steadily until the early tenth century, numerous tribes simply chose to retain their traditional way of life, raiding Byzantine territories and trying to convert agricultural lands to pasturelands.[69] Their 'Achilles' heel', however, was their indiscipline and their greed for booty, weaknesses which the Byzantines knew very well. Hence, Leo the Deacon reports the application of another stratagem known since ancient times, the spreading of golden and silver coins to cover an army's retreat, a tactic which could potentially break the enemy's discipline by enticing them to halt their pursuit in search for the coins.[70]

The Byzantine military manuals – the *Praecepta Militaria* in particular – make repeated references to the significance placed upon the regulation of the pursuit of an apparently beaten foe.[71] The *Praecepta*, however, is the only treatise that offers a plan which outlines in detail the stages of the pursuit and the considerations which governed the commander's commitment of his reserves once the enemy had been put to flight, all in the spirit of the rigorous discipline and professionalism (τάξις) instilled in the ranks of the imperial armies of the period. The degree of discipline and training with which the – predominantly Eastern – troops at Arkadiopolis performed their careful withdrawal into the defile while outnumbered and under constant fire, and the timed attack by the rest of the infantry units from all sides – not en masse but rather rank by rank – with everything unfolding in accordance with the writings of the military manuals, leaves little doubt that the Byzantine commanders had learned a good deal from their experience in the Taurus passes.

Notes

1. Mutanabbi, *Poem on Hadath*, v.19, p. 333.
2. Mutanabbi, *Poem on Hadath*, v.14 p. 333; Abu Firas, p. 364. See also Ibn Zafir, p. 125.
3. Mutanabbi, *Poem on Hadath*, p. 333.
4. Mutanabbi, *Poem on Hadath*, p. 333; Abu Firas, p. 364; the number of 500 is only mentioned by Ibn Zafir, p. 125.
5. *Praecepta Militaria*, II.31–6, 63–6, pp. 14–16; *Sylloge Taktikorum*, 46.17, p. 74.
6. Ibn Zafir, p. 126.
7. *History of Leo the Deacon*, IV.iii, p. 107.
8. *Praecepta Militaria*, IV.39–51, pp. 40–2; *Sylloge Taktikorum*, 46.12, pp. 79–80.
9. *History of Leo the Deacon*, IV.iii, p. 108.
10. The reports of the three Muslim sources for the battle have been gathered in P. E. Walker (1972), 'A Byzantine Victory over the Fatimids at Alexandretta (971)', *Byzantion*, 42, pp. 431–40.
11. John Skylitzes, *Synopsis of Byzantine History*, p. 274.
12. Walker, 'Alexandretta', p. 438.
13. John Skylitzes, *Synopsis of Byzantine History*, p. 276.
14. Ibid.

15. *On Skirmishing*, 2.28–31, p. 152; *Praecepta Militaria*, IV.192–5, p. 50.
16. John Skylitzes, *Synopsis of Byzantine History*, pp. 276–7.
17. *On Skirmishing*, 11, p. 182. See also 17, pp. 204–10.
18. Ibid., 11, pp. 182–4.
19. *Praecepta Militaria*, II.55–62, pp. 24–6; *On Skirmishing*, 4.10–13, p. 156.
20. *Praecepta Militaria*, II.124–6, p. 28.
21. Haldon, *Byzantine Wars*, pp. 71–6.
22. *History of Leo the Deacon*, VIII.ix, p. 185.
23. *Strategikon*, XII.B.13, p. 144; Leo VI, *Taktika*, XIV.61, p. 326; Vegetius, *Epitome of Military Science*, II.15, p. 47; III.16, p. 98.
24. *History of Leo the Deacon*, VIII.ix, p. 185.
25. Ibid., VIII.x, p. 186.
26. Ibid.
27. *History of Leo the Deacon*, IX.i–ii, pp. 187–9; John Skylitzes, *Synopsis of Byzantine History*, pp. 286–7.
28. John Skylitzes, *Synopsis of Byzantine History*, p. 291.
29. *History of Leo the Deacon*, IX.viii, p. 196.
30. *Praecepta Militaria*, I.146–8, p. 20.
31. John Skylitzes, *Synopsis of Byzantine History*, p. 291.
32. *History of Leo the Deacon*, IX.viii, p. 196.
33. John Skylitzes, *Synopsis of Byzantine History*, p. 292; *History of Leo the Deacon*, IX.ix, pp. 196–7.
34. *Praecepta Militaria*, IV.28–30, 158–60, pp. 40, 48; *Sylloge Taktikorum*, 46.10, p. 79.
35. *Histoire de Yahya*, 23, p. 440.
36. Bar Hebraeus, I, p. 179.
37. Michael the Syrian, p. 187; Asolik of Taron (1917), *Histoire Universelle*, trans. F. Macler, Paris: Leroux, pp. 199–200.
38. Bar Hebraeus, I, p. 179.
39. Lev, 'Fatimid Army', p. 174; Ibn al-Qalanisi, pp. 41–2; Maqrizi, vol. I, pp. 274–6; Yahya makes the distinction between the unruly Arab troops (Bedouins) and 'propres troupes' (Daylami infantry and Turkish cavalry): *Histoire de Yahya*, 23, p. 441.
40. Maqrizi, I, p. 271.
41. Forsyth, 'Yahya ibn-Said al-Antaki', p. 491.
42. Polyaenus, *Stratagems*, 2.4.2, p. 71; Frontinus, Στρατηγήματα [*Strategemata*], I.vi, pp. 53–5; Vegetius, *Epitome of Military Science*, III.1, p. 64; *Strategikon*, VIII,1.19, p. 81; *On Strategy*, 19, pp. 62–8; Leo VI, *Taktika*, IX.12–16, p. 158; XVII.9, p. 396.
43. *Praecepta Militaria*, V.23–6, p. 52; see also Leo VI, *Taktika*, IX.15, p. 158.
44. *Histoire de Yahya*, 23, p. 455; Ibn al-Qalanisi, p. 51; Bar Hebraeus, I, p. 180.
45. M. Canard (1961), 'Les sources arabes de l'histoire byzantine aux confins des Xe et XIe siècles', *Revue des études byzantines*, 19, pp. 298–300.
46. *Histoire de Yahya*, 23, p. 456.

47. Canard, 'Sources arabes', p. 299.
48. Al-Ansari, *Muslim Manual of War*, p. 95.
49. Constantine Porphyrogenitus, *Three Military Treatises*, C.474–8, p. 124.
50. P. Armstrong and A. McBride (2003), *Stirling Bridge and Falkirk 1297–98: William Wallace's Rebellion*, Oxford: Osprey; P. Armstrong (2002), *Bannockburn 1314: Robert Bruce's Great Victory*, Oxford: Osprey.
51. Smail, *Crusading Warfare*, pp. 172–3; France, *Victory in the East*, pp. 284–5.
52. *Praecepta Militaria*, II.150–75, pp. 30–2; *Sylloge Taktikorum*, 47.9–13, p. 88.
53. *Praecepta Militaria*, II.69–72, p. 26; II.124–6, p. 28.
54. John Skylitzes, *Synopsis of Byzantine History*, p. 292; *History of Leo the Deacon*, IX.viii, pp. 196–7.
55. Canard, 'Sources arabes', p. 299.
56. *Strategikon*, II.16, pp. 32–3; Leo VI, *Taktika*, XII.52, 105, pp. 246, 272; XIV.3, p. 290; XVIII.138–40, p. 492. In his *Gothic Wars*, Procopius speaks out against the general himself risking too much, in what Whately describes as the 'Homeric dichotomy' between the Achillean ethos (assumed by Totila) and the Odyssean ethos (assumed by Belisarius); see Whately, *Battles and Generals*, pp. 188–95.
57. Onasander, *Strategikos*, XXIII, p. 463. See also Frontinus, Στρατηγήματα [*Strategemata*], II.iv. 9, p. 129; Leo VI, *Taktika*, XIV.97, p. 344.
58. For this incident at Achelous, see John Skylitzes, *Synopsis of Byzantine History*, pp. 197–8. A similar incident from the years that followed Basil II's accession is also mentioned in *Synopsis of Byzantine History*, p. 310.
59. *Praecepta Militaria*, II.119–20, p. 28; IV.120–3, pp. 44–6.
60. McGeer, *Sowing the Dragon's Teeth*, pp. 306–7; *Sylloge Taktikorum*, 46.9, p. 79; *Praecepta Militaria*, IV.142–4, p. 46.
61. *History of Leo the Deacon*, IV.iii, p. 107.
62. *Sylloge Taktikorum*, 46.17, 24, pp. 81, 82; *Praecepta Militaria*, II.32–6, p. 24; II.62–6, p. 26; IV.160–2, p. 48.
63. *Praecepta Militaria*, IV.141–52, p. 46. The same tactical role is described in the *Sylloge Taktikorum*, 46.9, p. 79.
64. See previous note.
65. *Strategikon*, VII, p. 65.
66. C. E. Bosworth (1992), 'Abu ʿAmr ʿUthman al-Turtusi's *Siyar al-thugur* and the Last Years of Arab Rule in Tarsus (Fourth/Tenth Century)', *Graeco-Arabica*, 5, pp. 194–5. For the aḥdāth, see Bosworth, 'City of Tarsus', pp. 268–86; Y. Lev (1982), 'The Fatimids and the Aḥdāth of Damascus 386/996–411/1021', *Die Welt des Orients*, 13, pp. 97–106; Nicolle, *Crusader Warfare*, II, pp. 31–3.
67. Vasiliev, *Byzance et les Arabes*, II.1, pp. 342, 360; Canard, *Hamdanides*, pp. 763, 793; Garrood, 'Byzantine Conquest of Cilicia', pp. 136–40.
68. Aeneas Tacticus, *On Defence against Siege*, XVI, p. 83; Onasander, *Strategikos*, X.2, p. 415; Polyaenus, 8.23.11; Frontinus, Στρατηγήματα [*Strategemata*], II.v.35, p. 158; Vegetius, *Epitome of Military Science*,

III.10; *On Strategy*, 40, p. 118; *Strategikon*, X.2; Leo VI, *Taktika*, IX.25, XVII.5, XVII.37, XVII.54.
69. Cappel, 'Byzantine Response to the Arab', pp. 113–32.
70. *History of Leo the Deacon*, II.5, p. 75; John Skylitzes, *Synopsis of Byzantine History*, p. 182. Frontinus, Vegetius and the military treatise *On Strategy* also report this stratagem to escape one's pursuer: Frontinus, Στρατηγήματα [*Strategemata*], II.xiii.1–3; Vegetius, *Epitome of Military Science*, III.22, pp. 109–10; *On Strategy*, pp. 119–21.
71. The language of the treatises, no doubt, gives away a degree of caution rooted in decades of experience on the battlefields of Cilicia and Syria; see *Praecepta Militaria*, II.44–9, p. 24; II.75–9, p. 26; IV.151–3, p. 46; IV.163–6, p. 48; *On Tactics*, 25, pp. 313–15.

Summaries and Conclusions

> If you know the enemy and know yourself, you need not fear the result of a hundred battles. If you know yourself but not the enemy, for every victory gained you will also suffer a defeat. If you know neither the enemy nor yourself, you will succumb in every battle.
> —Sun-Tzu, *The Art of War*, III.18

War is a violent form of interaction that has dominated human activity since the dawn of mankind, but to explain the 'insanity' of going to war against one's own species, other sciences like psychology, sociology, evolutionary biology and anthropology have contributed their views in the quest to find answers as to why we are the only mammals deliberately killing our own kind. In order to explain this 'pathological behaviour', evolutionary biologists have put the blame on several factors ranging from a 'selfish gene' most eager to replicate, to excessive amounts of testosterone directly linked to aggressiveness. Psychological explanations put forward by James as early as 1910 have suggested that warfare is as prevalent as it is because of its positive psychological effects, both on the individual and on society as a whole.[1]

As any form of interaction between intelligent beings, war has been fluid and permutable over the two hundred millennia of our existence for which we have archaeological evidence, and over the roughly six millennia for which we have written records. Throughout recorded history, war has *not* developed linearly from a primitive to a more sophisticated, and hence deadlier, form of killing. As van Creveld has emphasised, there have been fluctuations in warfare but no real breakthroughs, with many factors remaining 'unaltered well into the age of gunpowder' or even the twentieth century.[2] In her book on the *Evolution of Strategy*, Heuser refers to 'fashions in warfare', by which she means the prevailing attitudes in both technological aspects of war but also in strategic and tactical thinking, where she has highlighted this rather obvious, but worth emphasising, idea:

> The conduct of war has rarely if ever been static . . . as hostile groups encountering each other always sought to maximise the advantage they could draw from any particularly successful ways of fighting they had developed or

Summaries and Conclusions

were developing. Differences of tactics, or weapons technology, then usually became subjects of great interest to the inferior party, and technology transfer was always seen as desirable.[3]

This idea was put to the test in this book and it was my intention to showcase the way competing states like the Byzantine Empire and the Hamdanid Emirate adapted to the strategies and tactics of their enemies in the operational theatres of eastern Asia Minor, northern Mesopotamia and northern Syria in the middle of the tenth century. My objective was to understand and clarify the mechanisms that lay behind the diffusion of 'military knowledge' between the competing military cultures of the region. This was brought about by scrutinising two different pools of evidence – the historians' accounts of the conflicts and the military treatises that proliferated in the tenth century – in an attempt to identify and trace the origin of any changes in the battlefield tactics of the armies that operated in the East in a period of great socio-political upheaval and military expansion that defined the region in the tenth century.

Therefore, this book is a comparative study of the *military cultures* that clashed in the region in the period in question, which culminated in the annexation of Cilicia and northern Syria by the imperial forces led by the famous campaigning emperors Nicephorus Phocas and John Tzimiskes. What we have come to identify as *military culture* is a system of beliefs and behavioural norms that influence what people think is (morally) right and wrong in war and conflict or, to put it plainly, how each peoples justified war, theorised about it, and developed different customs that shaped the planning and conduct of warfare. The people who would actually shape these beliefs in the Middle Ages were the elite classes.

In the case of the Byzantine Empire, the cultural traditions of the Hellenistic and imperial Roman past were largely intertwined with the early Christian distaste for the shedding of – primarily Christian – blood that produced a mentality over war and conflict that not only sanctioned, but encouraged the avoidance of battles and the use of craft, intelligence, wiles, bribery and 'other means' to bring a war to the quickest and most cost-efficient end. Nonetheless, the notion of honour in battle and the way it was perceived by different cultures varied greatly, and while the many 'cultures of war' that emerged in the medieval world shared some basic characteristics, what is more broadly comparable are the processes or dynamics that shaped military cultures around the world. I touched upon the different notions of bravery and honourable combat between the military cultures of the Christian East and West, and those of the Muslim East, only to reach a number of very interesting conclusions on the cultural boundaries that delineated and defined military practices. In the end, even the 'Western' fighters, imbued as they were in the heroic poetry of the *chansons de geste*, were not so far removed from the notions of trickery and cunning behaviour in war that their Byzantine and Muslim counterparts competing for booty and glory in the East engaged in.

This is a promising field of study that could yield compelling data on the intersection of ideas about war and the socio-economic, political and geographic background of various peoples. Of special interest are those regions where various military cultures have come into contact and conflict, in geographical areas that have been contested for millennia, which should be studied for further evidence not only within cultures, but between big and sub- cultures. I wish to expand more on this field in a future monograph, applying Morillo's typology of transcultural wars in Europe and its periphery to a specific region and period: Italy in the eleventh century, as the Italian Peninsula and the island of Sicily became a theatre of intense cultural interaction and transcultural warfare between the incoming Normans and the local Lombard, Greek and Muslim polities. How can we identify what kind of conflict we are faced with, based on an analysis of the accepted 'conventions of conflict' that can generate greater or lesser degrees of ritual associated with or as part of combat? Put simply, were these cultures different in terms of the conduct of warfare? And, if so, how did they perceive their enemies who came from different 'military cultures', and how did they react?

The culture of war and the Byzantine notion of chivalric and honourable battle was one of the three main considerations (or factors) that determined the empire's strategic thinking and planning, at least until the late twelfth century. The rest were the empire's – and, most importantly, the capital's – geopolitical location in the confluence of two continents, and the state's reliance on agriculture with the concomitant reaction of the economy to warfare. All three were interrelated and helped to develop a sort of 'grand strategy' (or rather 'political pragmatism', as I have explained in Chapter 1) that was based on the acute awareness of material considerations and the imbalance of resources between the empire and its enemies, considerations that can explain the commanders' reluctance to face enemies in different operational theatres at the same time. While the dominant approach in the empire's military thinking throughout its history was clearly defensive, recent studies have shown that the idea that the Byzantine Empire was a non-bellicose state must be discounted.

In the second chapter of this study, I defined in detail the different forms of warfare that dominated the eastern border regions of the empire, and I explained the strategy and strategic goals of all regional opponents leading up to the great 'reconquest' of territories in the mid-tenth century. In general terms, it seems that the Byzantine state up to the end of the twelfth century developed a number of variants on a system of defence in depth. This strategy encouraged two basic approaches to the Arab invasions, namely the following and harassing of the enemy on friendly ground, and the ambushing of foraging parties while keeping a close eye on the main invading army. All this was directed at maximising the enemy's logistical problems and flushing them out of the country, while themselves sustaining as few casualties as possible.

This was a decentralised system of defence that had evolved over centuries before it was put down in great detail for posterity by the anonymous author of the c. 960 treatise *On Skirmishing*. The author vividly portrays the anxieties of the defenders of Anatolia over the constant invasions of the Muslim emirs over the Taurus and Anti-Taurus mountain ranges that contributed to the political instability, militarisation and economic, commercial, agricultural and demographic decline of central and eastern Asia Minor. Byzantine commanders had always to tread a fine line between gathering intelligence and shadowing the invading forces and bringing them to battle, a balance that often led the defending forces to being outmanoeuvred by an experienced enemy. It is against this socio-political and geographical background that we can appreciate the key role that the three Muslim bastions for razzias (Tarsus, Melitene and Theodosiopolis) played in this conflict that reached a climax in the middle of the century, and why I have – repeatedly – called them a 'thorn' in the side of the empire's eastern frontiers.

There is a direct correlation between the empire's policy in the East against the Muslim emirs and the delicate way the emperors planned their foreign policy with the local *naxarars* of medieval Armenia in the north-eastern border regions. The main points that I attempted to elucidate in Chapter 3 focus on the political and strategic importance of Armenia proper – and more specifically the cantons of Taron and Vaspourakan – as the 'back door' for any enemy invasion routes into Anatolia. I described the political reasons behind the empire's involvement in Armenia and northern Mesopotamia in the first half of the tenth century, the wars with the Muslims, the empire's delicate diplomatic negotiations with the Armenian princes and the emergence of a new enemy in the East. At precisely this junction lies the paradox that I have tried to disentangle: if Armenia was strategically far more important than Cilicia and Syria, and the government in Constantinople did not contemplate any territorial expansion in the region, how can we explain the escalation of violence that led to the extensive annexation of territories in Cilicia and Syria in the 950s–60s?

The crux of this discussion is that, if we take a closer look at the political background of both protagonists – Constantine VII and Sayf ad-Dawla – and their place within their respective courts, we see that this war had turned into an 'all-out' conflict by the end of the 950s to which none of them could (politically) afford to succumb. Eventually, it was the seemingly infinite resources the empire was able to pour into the operational theatre in the East that had proved decisive by 962.

A key aspect of the information flow between cultures was the knowledge and the impression each nation had for its neighbours. Therefore, in Chapter 4, I tried to put the spotlight on another fascinating aspect of the primary sources' information on the Byzantine–Muslim conflicts of the tenth century. This revolves around the stereotypical portrayal of the 'other' as warrior, whereby the 'other' is the main enemy of the Byzantines on the battlefields of the period – the Eastern Muslims. A constant motif in all the sources of the period is the conflict between a 'real' (or 'ideal') soldier and a 'false' one, between the 'moral' and the 'immoral', where the spiritual and material motives of the state and individuals are contrasted with the ideals of *just/holy war* and *jihad*. As any comments of this kind, they must be read within the socio-political context of the centuries of warfare that had fundamentally shaped both the Christian and Muslim polities, and should be interpreted not as a rhetorical exercise or state propaganda, but rather as some sort of response to a particular set of Muslim doctrines and institutions with which Byzantine society was confronted.

Another motif that emerges in the secular and ecclesiastical sources of the period about the Muslim invaders of Christian lands and seas is that of a cunning and opportunistic enemy immersed in the pursuit of money, plundering and booty, unchivalric and uncivilised. No wonder that any description of this kind would inspire contempt and deeply rooted hatred. Nevertheless, it seems that the same sources were highly impressed by the fighting abilities of the Muslims, who were portrayed as intelligent and furiously resilient with high morale and ready to die for their cause. This shows the clear dichotomy that had emerged in the minds of the authors over what was (morally) right and what was (morally) wrong: it is the Byzantines who held the moral high ground; it is they who would achieve eternal glory with the help of God.

As Sun-Tzu highlighted more than two and a half millennia ago, 'an army without secret agents [bringing intelligence] is exactly like a man without eyes or ears'.[4] Hence, Chapters 5 and 6 of my study dealt extensively with the importance of intelligence for a medieval state and its procurement both before and after the outbreak of hostilities – a point in time used to determine where, exactly, espionage ends and reconnaissance

Summaries and Conclusions

begins. Modern analysts have neatly arranged intelligence into three categories: strategic intelligence (concerned with broad issues such as economics, political assessments, military capabilities and the intentions of foreign nations); operational intelligence (providing support to the army commander and attached to the formation's headquarters); and, thirdly, tactical intelligence (focused on supporting the operations on a tactical level). A distinction like this would not have been as easy for a medieval commander, although he would have used numerous intelligence channels to gather as detailed a report as possible about several questions regarding the geography of the enemy territory, information about roads, bridges and the countryside, and issues that focused on the character of the enemy sovereign, the political situation in the enemy country and the socio-religious demography of the target region.

Koutrakou has pointed out that the Byzantine military manuals do not always provide us with a clear-cut picture of intelligence-gathering, and both the technical terms and more common words that are used in Byzantine sources to refer to spies, special agents, commandos or, more generally, troops dispatched to conduct reconnaissance on the enemy 'offer interesting possibilities for discussion'.[5] Therefore, in Chapter 6 I sought to determine the existing cross-border channels used by governments to obtain intelligence about the enemy, as well as which professions, groups of people and places were considered ideal for the procuring of intelligence: from ambassadors, envoys and their staff, along with reports from travelling laymen and ecclesiastics, to social groups such as the famous *akritai*, whose relationship with the central government was ambiguous and whose movement across the fluid 'no-man's land' was flexible and informal.

A key question that emerged was whether reliable, detailed and fast tactical intelligence relayed back to the commander of an invading or defending army could bring about long-term change of tactics and/or strategy? My view on this is that, because war is not an intellectual activity but, rather, savage and primitive mayhem, a clash which could last for hours, days and even months, a general has always been required to come up with the best possible strategy and apply the best possible tactics to emerge victorious at a lesser cost than his adversary. Intelligence could provide him with the best tactical background information to make better decisions during battle. Therefore, intelligence should be considered one of dozens of different parameters that can greatly influence decision-making, strategy and battle tactics. Unlike other factors, however, such as the weather or pure luck, intelligence was something a general could control.

The richest and most useful sources for identifying tactical changes in the Byzantine army are the military treatises of the tenth century, the so-called *strategika* or *taktika* – a literary category of works written (primarily) for officers of the army that contained constitutions and treatises of military nature, such as the *Praecepta Militaria* of Nicephorus Phocas (c. 969), the anonymous *Sylloge Taktikorum* (c. 930) and the *Taktika* attributed to Emperor Leo VI (written c. 895–908). Several studies have revealed important links between the written sources that the authors used to compile their works, especially regarding their access to their first-century counterparts like Onasander or Frontinus, and the structure and contents of the *taktika* through the centuries, from their notion of strategy to the different battle formations and the employment of stratagems to defeat an enemy. Even though these manuals served to preserve the knowledge of the ancient Greeks and Romans in the art of war, these were not just copied texts taken from ancient authorities on the subject. Byzantine strategic thinking was rather the conscious imitation and adaptation of ancient dicta to the current geopolitical situation and, although these treatises were more prescriptive than descriptive in nature, they offered a great degree of discretion on the battlefield and encouraged adaptability and improvisation.

Therefore, military historians of the period were fortunate – compared with other historical periods of the pre-modern era or with other cultures like the medieval 'West' – in having a considerable number of military treatises from which they could draw information on the strategy and tactics of the imperial armies in the tenth century. The fact that four of them were compiled within the first six decades of the century, and with just a few decades between them, provides fertile ground for comparison and for historians to point out any tactical changes – small or large – that could have occurred in the intervening period between the writing of the treatises, especially when significant geopolitical events had transpired, such as a war.

As a result, based on the close study of the military treatises of the tenth century, I have concluded that there is clear evidence of a number of significant changes in the tactical structure and deployment of the Byzantine armies – both infantry and cavalry – in the operational theatres of eastern Anatolia, northern Mesopotamia and northern Syria in the middle of the tenth century. These range from the hollow double-ribbed infantry square, the depth and deployment of the men in each file of an infantry phalanx, the position of the cavalry in a mixed army formation, changes in the battle formation of the *kataphraktoi*, the appearance of a new heavy infantry unit armed with long pikes – the *menavlatoi* – and the addition of a third line of cavalry identified by its Arabic name, *saqah*. Finally, in the eighth chapter of this study, I tried to pinpoint the *catalyst* (be it a battle, an encounter with an enemy nation in battle and so on) that may have provided the Byzantines with the impetus to develop these tactical changes. Upon concluding my research, it is my firm conviction that the differences I have outlined should be considered as a response to the tactics employed by the Arabs and experienced by the great officers in the East such as John Curcuas, Nicephorus Phocas, John Tzimiskes and Nicephorus Uranus.

A final key question that has been addressed is whether 'theory translated into practice' on the battlefields of this period. What I mean by that is whether the primary sources of the period provide adequate and clear information to back up the assertions made about any tactical changes in the imperial armies of the period, based on the comparative reading of the tenth-century *taktika*. My study is based on primary material written by people from various social strata, including a palace cleric (Leo the Deacon), an army officer and court official (John Skylitzes), an educated physician (Yahya ibn Said), a poet (al-Mutanabbi), a Muslim prince (Abu Firas) and a madrassa professor (Ibn Zafir). In the final section of this study (Chapters 9 and 10), I focused on the chroniclers' social, religious and educational backgrounds, the dates and places of the compilation of their work, their own sources and the way they gleaned their information from them, their biases and sympathies and thus their level of impartiality as historians. Furthermore, I engaged in a comparative analysis of the sources from a military perspective, asking questions such as 'To what extent are the figures they provide for army sizes reliable?', 'What is their knowledge of the local geography, and to what extent – if at all – were they familiar with the terrain of the operations (battles, sieges or the campaign routes of armies) they describe?', 'How far do their narratives permit the accurate reconstruction of a chain of events, especially regarding the battlefield manoeuvres of armies in action?'.

First, I have noted the great disparity of the historians and chroniclers of the period regarding the focus of their work. The Arab and Syriac sources are much more detailed on the achievements of their patrons against the Byzantines on the battlefields of Cilicia and Syria than the Byzantines, who concentrate rather on the Balkan theatre of war. Deducing any kind of reliable data on the numbers involved in campaigns or battles and reconstructing any battlefield tactics and/or battlefield manoeuvres from the descriptions provided is very problematic, despite the fact that two of the sources in question were eyewitnesses of the wars in the East (al-Mutanabbi and Abu Firas). In addition, the same sources do not give us anything about the nature of the ground or the topography of the battlefields – factors which played a vital role in the outcome of any battle – nor do they analyse the battle tactics employed by the opposing armies. Instead, they choose to focus on the achievements of individual protagonists, namely their patron Sayf ad-Dawla.

However, before we dismiss the accounts of the sources in terms of their description of campaigns and combat episodes as insignificant or – frankly – useless, we should bear in mind that describing a military campaign in the form of what we would call today 'war correspondence' was not their objective; hence, we should refrain from criticising any medieval historian for a tendency to romanticise and idealise and, on occasion, to reduce battles to a series of conventional *classicising* images. In many cases, these were mere attempts to produce works that were didactic and would appeal to the peculiar nature of their audience, which was society as a whole or the political and/or intellectual elite. The latter, in particular, were people who were largely interested in the glorious side of war, the heroic actions of leading generals, battle exhortations and duels between champions. In a sense, we should appreciate *how* the medieval historians and their audience understood battle, *not* whether their descriptions fit our understanding.

Following that assumption, we can draw a great deal of important conclusions about the fighting tactics, training, morale and *esprit de corps* of the – predominantly Eastern – armies that operated in the region of eastern Anatolia and Cilicia, from the study of the major pitched battles of the second half of the century. The Battles of Hadath (954) and Apamea (998) highlight the key role of the commander for the morale of the soldiers and for the outcome of the battle, a fact that was appreciated by the author of the *Praecepta*, who emphasised the importance of targeting the enemy leader in an all-out cavalry attack. At Tarsus (965) and Dorystolon (971), there is clear evidence that the Byzantine army was deployed in a very different way than the one recommended by the contemporary military manuals, more specifically regarding the place of the commander in battle and the deployment of the heavy cavalry of the *kataphraktoi* within the mixed formation. These changes can, indeed, offer hints at the ability of Byzantine commanders to react and adapt to the situation in hand according to the composition, deployment and fighting tactics of the enemy. Finally, the battles at Arkadiopolis (970) and Alexandretta (971) are two very good examples of how to lure an enemy into a trap using a combination of good intelligence and experience against familiar enemies, while emphasising the significance placed upon the regulation of the pursuit of a retreating foe, and the rigorous battle discipline (τάξις) and professionalism that was instilled in the armies of the period. In the end, it should have become clear just how important it was for the Byzantine commanders of the tenth century to learn their lessons from their experiences in the operational theatres of the East. The study of the existing evidence leaves little doubt over what good students they proved themselves to be.

Notes

1. W. James (2013), *The Moral Equivalent of War*, New York: Read Books.
2. M. van Creveld (1989), *Technology and War from 2000 BC to the Present*, New York: Free Press, pp. 25–50.
3. Heuser, *Evolution of Strategy*, p. 40.
4. Sun-Tzu, *Art of War*, p. 175.
5. Koutrakou, 'Spies of Towns', p. 246.

Primary Bibliography

Abu Yusuf Ya^{c}qub (1979), *Kitab al-kharaj* [Islamic Revenue Code], trans. A. A. Ali, Lahore: Islamic Book Centre.
Aeneas Tacticus (1928), *On Defence against Siege*, in The Illinois Classical Club (ed.), *Aeneas Tacticus, Asclepiodotus, Onasander*, The Loeb Classical Library, London: Heinemann.
Amatus of Montecassino (2004), *The History of the Normans*, trans. P. Dunbar, Woodbridge: Boydell & Brewer.
Ambroise (2003), *The History of the Holy War*, trans. M. Ailes, Woodbridge: Boydell & Brewer.
Ammianus Marcellinus (1948), *History*, trans. J. C. Rolfe, London: Heinemann.
Anna Comnena (2000), *The Alexiad*, trans. E. A. S. Dawes, Cambridge, ON: In Parentheses Publications.
al-Ansari (1961), *A Muslim Manual of War*, ed. G. T. Scanlon, Cairo: American University at Cairo Press.
Asclepiodotus (1928), *Tactics*, in The Illinois Classical Club (ed.), *Aeneas Tacticus, Asclepiodotus, Onasander*, The Loeb Classical Library, London: Heinemann.
Asochik (1883–1917), *Histoire Universelle*, trans. E. Dulaurier and F. Macler, Paris: Les presses universelles.
Asolik of Taron (1917), *Histoire Universelle*, trans. F. Macler, Paris: Leroux.
Caesar (1917), *The Gallic Wars*, trans. by H. J. Edwards, London: Heinemann.
de Charny, G. (1996), *A Knight's Own Book of Chivalry*, trans. E. Kennedy, Philadelphia: University of Pennsylvania Press.
Chatzelis, G., and J. Harris (eds) (2017), *A Tenth-Century Byzantine Military Manual: The Sylloge Tacticorum*, London: Routledge.
Constantine Porphyrogenitus (1990), *Three Treatises on Imperial Military Expeditions*, ed. J. F. Haldon, Vienna: Verlag der Österreichischen Akademie der Wissenschaften.
Constantine Porphyrogenitus (1985), *De Administrando Imperio=DAI*, ed. (Greek text) G. Moravcsik, trans. R. J. H. Jenkins, Washington, DC: Dumbarton Oaks Texts.
Constantine Porphyrogenitus (1829–30), *De Ceremoniis Aulae Byzantinae*, Corpus Scriptorum Historiae Byzantinae, vols 5–6, ed. I. Reiski, Bonn: Webber.
Dennis, G. T. (2008), 'The Anonymous Byzantine Treatise *On Skirmishing* by the Emperor Lord Nicephoros', in G. T. Dennis (ed.), *Three Byzantine Military Treatises*, Washington, DC: Dumbarton Oaks, pp. 143–239.

Primary Bibliography

Dennis, G. T. (ed.) (1985), *Three Byzantine Military Treatises*, Corpus Fontium Historiae Byzantinae, 25, Washington, DC: Dumbarton Oaks.

Dimitroukas, I. (ed.) (2005), *Ναυμαχικά, Λέοντος ς', Μαυρικίου, Συριανού Μαγίστρου, Βασιλείου Πατρικίου, Νικηφόρου Ουρανού* [Naumachika of Leo VI, Maurice, Syrianus Magister, Basil the Patrician, Nicephorus Ouranus], Athens: Kanaki.

Fulcherius Carnotensis (1913), *Historia Hierosolymitana*, ed. H. Hagenmeyer, Heidelberg: Carl Winters.

Georgius Monachus (1838), in *Theophanes Continuates, Ioannes Cameniata, Symeon Magister, Georgius Monachus, Οι μετά Θεοφάνην* [The Continuators of Theophanes], Corpus Scriptorum Historiae Byzantinae, vol. 33, ed. I. Bekker, Bonn: Webber.

Grammaticus, Leo (1842), *Chronographia*, Corpus Scriptorum Historiae Byzantinae, vol. 34, ed. I. Bekker, Bonn: Webber.

Guillaume de Pouille [William of Apulia] (1963), *La geste de Robert Guiscard*, ed. M. Mathieu, Palermo: Bruno Lavagnini.

al-Harthami (1964), *Mukhtasar siyasat al-hurub*, ed. ᶜAbd al-Raʾuf ᶜAwn, Cairo: Silsilat kutub al-turath.

Ibn Hawqal (1964), *Configuration de la terre (Kitāb surat al-arḍ)*, trans. J. H. Kramers and G. Wiet, Paris: G. P. Maisonneuve et Larose.

Ibn Khaldun (1969), *The Muqaddimah: An Introduction to History*, trans. by Franz Rosenthal, 3 vols, Princeton, NJ: Princeton University Press.

Ibn Khurradadhbeh (1889), *Kitāb al-masālik wa'l-mamālik*, ed. M. J. de Goeje, Bibliotheca Geographorum Arabicorum 6, Leiden: Brill.

Ibn Rustah, A. (1892), *Kitāb al-aᶜlāk an-nafīsa*, ed. M. J. de Goeje, Bibliotheca Geographorum Arabicorum 7, Leiden: Brill.

Joannes Scylitzes (1973), *Ioannis Scylitzae Synopsis historiarum*, ed. I. Thurn, Corpus Fontium Historiae Byzantinae, vol. 5, Berlin: Walter De Gruyter.

John Skylitzes (2010), *A Synopsis of Byzantine History, 811–1057*, trans. J. Wortley, Cambridge: Cambridge University Press.

de Joinville, J. (1963), 'Life of St-Louis', trans. M. R. B. Shaw, *Joinville and Villehardouin: Chronicles of the Crusades*, London: Penguin.

de Jomini, Baron A. H. (2008), *The Art of War*, restored edition, Kingston, ON: Legacy Books Press.

Kaminiatis, I. (2000), *John Kaminiates: The Capture of Thessaloniki*, ed. D. Frendo and A. Fotiou, Perth: Australian Association of Byzantine Studies.

Kaminiatis, I. (2000), *Για την άλωση της Θεσσαλονίκης* [Regarding the Sack of Thessalonike], Athens: Kanaki.

Leo VI (2010), *The 'Taktika' of Leo VI*, Corpus Fontium Historiae Byzantinae, vol. 49, trans. G. T. Dennis, Washington, DC: Dumbarton Oaks.

Leonis diaconi (1828), *De Velitatione bellica Nicephori augusti*, ed. C. B. Hass, Corpus Scriptorum Historiae Byzantinae, 4, Bonn: Weber.

Liudprand of Cremona (1993), *The Embassy to Constantinople and Other Writings*, ed. J. J. Norwich, London: J. M. Dent.

Malaterra, G. (2005), *The Deeds of Count Roger of Calabria and Sicily and of his Brother Duke Robert Guiscard*, ed. K. B. Wolf, Ann Arbor: University of Michigan Press.
al-Malik Farra, M. A. (1981), *A Critical Edition of 'Kitab Al-Buldan' by Al Ya'qubi, Ahmad Ibn Abi Ya'qub (Ishaq) B. Ja'far B. Wahb B. Wadih Al-Kitab Al-Abbasi*, London: British Library Research and Development Department .
al-Masᶜudi (1989), *The Meadows of Gold: The Abbasids*, trans. and ed. P. Lunde and C. Stone, London: Routledge.
al-Masᶜudi (1967), *al-Tanbih wa'l-ishraf*, 2nd edition, Baghdad.
Maurice's Strategikon: Handbook of Byzantine Military Strategy (1984), trans. G. T. Dennis, Philadelphia: University of Pennsylvania Press.
Michael Psellus (1899), *Chronographia*, ed. C. Sathas, London: Methuen.
Michaelis Pselli Scripta minora: magnam partem adhuc inedita (1941), ed. E. Kurtz and F. Drexl, Milan: Società editrice 'Vita e pensiero'.
al-Mutanabbi (1950), *Poem on Hadath*, in A. A. Vasiliev (ed.) and M. Canard (trans.), *Byzance et les Arabes, 867–959*, Brussels: Institut de philologie et d'histoire orientales.
Nicolas I, *Letters*, CFHB 2.
Nizam al-Mulk (2012), *The Book of Government or Rules for Kings, The Siyar al-Muluk of Siyasat-nama of Nizam al-Mulk*, trans. H. Darke, London: Routledge.
Odo of Deuil (1949), *La croisade de Louis VII, roi de France*, 4 vols, ed. H. Waquet, Paris: Académie des inscriptions et belles lettres.
Odorico, P. (2010), *Ιωάννης Καμινιάτης, Ευστάθιος Θεσσαλονίκης, Ιωάννης Αναγνώστης: Χρονικά των αλώσεων της Θεσσαλονίκης*, [Ioannes Kaminiates, Eustathios of Thessaloniki, Ioannes Anagnostes: Chronicles of the Conquests of Thessaloniki], trans. Ch. Messis, Athens: AGRA.
Onasander (1928), *Strategikos*, in The Illinois Classical Club (ed.), *Aeneas Tacticus, Asclepiodotus, Onasander*, The Loeb Classical Library, London: Heinemann.
Orderic Vitalis (1854), *Ecclesiastical History of England and Normandy*, trans. T. Forester, London: Henry G. Bohn.
Peters, E. (ed.) (1998), *The First Crusade: 'The Chronicle of Fulcher of Chartres' and Other Source Materials*, Philadelphia: University of Pennsylvania Press.
Nicephoras Uranus (1995), *Presentation and Composition on Warfare of the Emperor Nicephoros*, in E. McGeer (ed. and trans.), *Sowing the Dragon's Teeth: Byzantine Warfare in the Tenth Century*, Washington, DC: Dumbarton Oaks.
Procopius (1916), *Histories*, trans. H. E. Dewing, 6 vols, London: Heinemann.
Procopius (1924), *History of the Wars, Book VI: The Gothic War* [= *De Bello Gothico*], trans. H. B. Dewing, London: Heinemann.
Qudama ibn Jaᶜfar (1889), *Kitab al-kharaj*, ed. M. J. de Goeje, Bibliotheca Geographorum Arabicorum 6, Leiden: Brill.
al-Sarraf, Shihab (1996), 'Furusiyya Literature of the Mamluk Period', in D. Alexander (ed.), *Furusiyya*: vol. 1: *The Horse in the Art of the Near East*, Riyadh: King Abdulaziz Public Library.

Primary Bibliography

Sextus Julius Frontinus (1950), *The Stratagems and the Aqueducts of Rome*, ed. M. B. McElwain, trans. C. E. Bennett, London: Heinemann.

Σέξτος Ιούλιος Φροντίνος [Sextus Julius Frontinus] (2015), *Στρατηγήματα* [*Strategemata*], ed. G. Theotokis, trans. V. Pappas, Athens: Hellenic Army Press.

Sun-Tzu (1963), *The Art of War*, trans. S. B. Griffith, Oxford: Oxford University Press.

Sylloge Taktikorum = Sylloge Tacticorum, quae olim 'inedita Leonis Tactica' dicebatur (1938), ed. A. Dain, Paris: Société d'édition les belles lettres.

Symeon Magister and Logothete (2006), *Chronicon*, Corpus Fontium Historiae Byzantinae, vol. 44.1, ed. Staffan Wahlgren, Berlin: Walter de Gruyter.

al-Tabari (1985), *History: The Crisis of the Abbasid Caliphate*, trans. G. Saliba, New York: State University of New York Press.

Talbot, C. H. (1954), *The Anglo-Saxon Missionaries in Germany, Being the Lives of SS. Willibrord, Boniface, Leoba and Lebuin Together with the Hodoepericon of St. Willibald and a Selection from the Correspondence of St. Boniface*, London: Sheed and Ward.

al-Tarsusi (1948), *Tabsira arbab al-lubab*, in C. Cahen (trans.), 'Un traité d'armurerie composé pour Saladin', *Bulletin d'études orientales* 12, pp. 103–63.

Tabulae Imperii Byzantini (1976–2004), 10 vols, Vienna: Verlag der Österreichischen Akademie der Wissenschaften.

Taybugha al-Baklamishi al-Yunani (1970), *Saracen Archery*, trans. J. D. Latham and W. F. Paterson, London: Holland Press.

The Anonymous Byzantine Treatise 'On Strategy' (2008), in Dennis (ed.), *Three Byzantine Military Treatises*.

Kekaumenus (1965), *Strategikon*, ed. B. Wassiliewsky, V. Jernstedt, Amsterdam: Hakkert.

Κεκαυμένος (1996), *Στρατηγικόν*, Greek trans. D. Tsougarakes, Athens: Kanaki.

The Chanson d'Antioche: An Old-French Account of the First Crusade (2011), trans. S. Edgington and C. Sweetenham, Aldershot: Ashgate.

La Chanson d'Antioche (1977), ed. S. Duparc-Quioc, Paris: Librairie Orientaliste Paul Geuthner.

The Chronicle of Theophanes Confessor (1997), ed. C. Mango and R. Scott, Oxford: Oxford University Press.

The Chronography of Gregory Abû'l Faraj, the Son of Aaron, the Hebrew Physician, commonly known as Bar Hebraeus: Being the First Part of his Political History of the World (2003), trans. E. A. Wallis Budge, 2 vols, London: Gorgias.

The Ecclesiastical History of Orderic Vitalis (1969–80), ed. M. Chibnall, 6 vols, Oxford: Oxford University Press.

The History of Leo the Deacon: Byzantine Military Expansion in the Tenth Century (2005), trans. A. M. Talbot, Washington, DC: Dumbarton Oaks.

The History of Menander the Guardsman (1985), ed. R. C. Blockley, Cambridge: Cambridge University Press.

The History of Theophylact Simocatta: An English Translation with Introduction (1986), ed. and trans. by M. Whitby and M. Whitby, Oxford: Oxford University Press.
The History of William the Marshal (2016), trans. N. Bryant, Woodbridge: Boydell & Brewer.
The Russian Primary Chronicle, Laurentian Text (1953), ed. S. H. Cross and O. P. Sherbowitz-Wetzor, Cambridge, MA: Medieval Academy of America.
Theodosius Diaconus, *Expugnatio Cretae.*
Theophanes Continuates (1838), in *Theophanes Continuates, Ioannes Cameniata, Symeon Magister, Georgius Monachus, Οι μετά Θεοφάνην* [The Continuators of Theophanes], Corpus Scriptorum Historiae Byzantinae, vol. 33, ed. I. Bekker, Bonn: Webber.
Usamah ibn Munqidh (1929), *Memoirs of an Arab-Syrian Gentleman. Or, An Arab Knight in the Crusades: Memoirs of Usāmah ibn-Munqidh (Kitāb al-i'tibār)*, trans. P. Khuri Hitti, New York: Columbia University Press.
Vegetius: Epitome of Military Science (2001), trans. N. P. Milner, Liverpool: Liverpool University Press.
Vita Basilii [*Chronographiae Quae Theophanis Continuati Nomine Fertur Liber Quo Vita Basilii Imperatoris Amplectitur*] (2011), Corpus Fontium Historiae Byzantinae, vol. 42, ed. and trans. I. Ševčenko, Berlin: Walter de Gruyter.
Vita Euthymii (1888), ed. C. de Boor, Berlin.
Vita St-Pauli Junioris (1913 [1892]), ed. H. Delehaye, *Annalecta Bollandiana*. Reprinted in T. Wiegand (ed.), *Der Latmos, Milet: Ergebnisse der Ausgrabungen und Untersuchungen seit dem Jahre 1899*, Berlin: G. Reimer, vol III/1.
Widukind of Corvey (1935), *Res Gestae Saxonicae Sive Annalium Libri Tres*, ed. P. Hirsch, Hanover: MGH Scriptores rerum Germanicarum in usum scholarum 60.
Widukind of Corvey (2014), *Deeds of the Saxons*, trans. B. S. Bachrach and D. S. Bachrach, Washington, DC: Catholic University of America Press.
William of Apulia (n.d.) *William of Apulia, The Deeds of Robert Guiscard*, trans. G. A. Loud, Medieval History Texts in Translation series, available at: www.leeds.ac.uk/arts/downloads/file/1049/the_deeds_of_robert_guiscard_by_william_of_apulia (accessed 25 June 2018).
Xenophon (1914), *Cyropaedia*, trans. W. Miller, 2 vols, London: Heinemann.
Yahya al-Antaki Cronache dell'Egitto Fatimide e dell' Impero Bizantino 937–1033 (1998), trans. B. Pirone, Milan: Jaca Book.
Yahya ibn Said al-Antaki (1932–57), *Histoire de Yahya-ibn-Saʿīd d'Antioche, continuateur de Saʿīd-ibn-Bitriq*, ed. I. Krachkovskii and A. A. Vasiliev, *Patrologia Orientalis*, Paris: Firmin-Didot, vol. 18, fasc. 5.

Secondary Bibliography

Abels, R. (2008), 'Cultural Representation and the Practice of War in the Middle Ages', *Journal of Medieval Military History*, 6, pp. 1–31.
Abels, R., and S. Morillo (2005), 'A Lying Legacy? A Preliminary Discussion of Images of Antiquity and Altered Reality in Medieval Military History', *Journal of Medieval Military History*, 3, pp. 1–13.
Afinogenov, D. (2002), 'A Lost 8th-Century Pamphlet against Leo III and Constantine V?', *Eranos*, 100, pp. 15–17.
Agapitos, P. A. (1989), 'Η εικόνα του Αυτοκράτορα Βασιλείου Ά στη φιλομακεδονική γραμματεία 867–959' [The Image of Emperor Basil I in the Pro-Macedonian Literature 867–959], *Ελληνικά*, 40, pp. 285–322.
Agius, D. A. (2008), *Classic Ships of Islam: From Mesopotamia to the Indian Ocean*, Leiden: Brill.
Ahrweiler, H. (1975), *L'ideologie politique de l'Empire byzantine*, Paris: Presses Universitaires de France.
Ahrweiler, H. (1969), 'Recherches sur l'administration de l'Empire byzantin aux IXe–XIe siècles', *Bulletin de Correspondance hellénique*, 84.
Ahrweiler, H. (1965), 'La frontière et les frontières de Byzance en Orient', in H. Ahrweiler (ed.), *Byzance: les pays et les territoires*, London: Variorum.
Ahrweiler, H. (1965), *L'idéologie de l'Empire byzantin*, Paris: Presses Universitaires de France, pp. 9–24.
Ahrweiler, H. (1962), 'L'Asie mineure et les invasions arabes', *Revue historique*, 227, pp. 1–32.
Allmand, C. (1988), *The Hundred Years War*, Cambridge: Cambridge University Press.
Anastasiadis, M. P. (1994), 'On Handling the *Menavlion*', *Byzantine and Modern Greek Studies*, 18, pp. 1–10.
Anderson, J. G. C. (1897), 'The Road-System of Eastern Asia Minor with the Evidence of Byzantine Campaigns', *Journal of Hellenic Studies*, 17, pp. 22–44.
Anderson, J. G. C. (1896), 'The Campaign of Basil I against the Paulicians in 872 A.D.', *Classical Review*, 10, pp. 136–40.
Angelidi, C. (2013), 'Designing Receptions in the Palace (*De Cerimoniis* 2.15)', in A. Beihammer, S. Constantinou and M. Parani (eds), *Court Ceremonies and Rituals of Power in Byzantium and the Medieval Mediterranean: Comparative Perspectives*, Leiden: Brill, pp. 465–86.

Arberry, A. J. (1967), *Poems of Al-Mutanabbi*, Cambridge: Cambridge University Press.
Armstrong, P. (2002), *Bannockburn 1314: Robert Bruce's Great Victory*, Oxford: Osprey.
Armstrong, P., and A. McBride (2003), *Stirling Bridge and Falkirk 1297–98: William Wallace's Rebellion*, Oxford: Osprey.
Arthurson, I. (1991), 'Espionage and Intelligence from the Wars of the Roses to the Reformation', *Nottingham Medieval Studies*, 35, pp. 134–54.
Asa Eger, A. (2015), *The Islamic–Byzantine Frontier*, London: I. B. Tauris.
Ayalon, D. (1978), *Gunpowder and Firearms in the Mamluk Kingdom: A Challenge to a Mediaeval Society*, London: Frank Cass.
Bacharach, J. L. (1981), 'African Military Slaves in the Medieval Middle East: The Cases of Iraq (869–955) and Egypt (868–1171)', *International Journal of Middle East Studies*, 13, pp. 471–95.
Bachrach, B. S. (2007), '"A Lying Legacy" Revisited: The Abels–Morillo Defense of Discontinuity', *Journal of Medieval Military History*, 5, pp. 153–93.
Bachrach, B. S. (2006), 'Crusader Logistics: From Victory at Nicaea to Resupply at Doryleon', in J. H. Pryor (ed.), *Logistics of Warfare in the Age of the Crusades*, Aldershot: Ashgate, pp. 43–62.
Bachrach, B. S. (1999), 'Early Medieval Military Demography: Some Observations on the Methods of Hans Delbruck', in Kagay and Villalon (eds), *Circle of War in the Middle Ages*, pp. 3–20.
Beihammer, A. (2012), 'Strategies of Diplomacy and Ambassadors in Byzantine–Muslim Relations on the Tenth and Eleventh Centuries', in A. Becker and N. Drocourt (eds), *Ambassadeurs et ambassades au coeur des relations diplomatiques: Rome, Occident médiéval, Byzance, VIIIe s. avant J.-C.–XIIe s. après J.-C*, Metz: Centre de recherche universitaire lorrain d'histoire, Université de Lorraine, pp. 371–400.
Bellinger, A. R. (1956), 'The Coins and Byzantine Imperial Policy', *Speculum*, 31, pp. 70–81.
Bennett, M. (2006), 'Amphibious Operations from the Norman Conquest to the Crusades of St. Louis, c. 1050–c. 1250', in D. J. B. Trim and M. C. Fissel (eds), *Amphibious Warfare 1000–1700*, Leiden: Brill, pp. 51–68.
Bennett, M. (1998), 'The Myth of the Military Supremacy of Knightly Cavalry', in M. J. Strickland (ed.), *Armies, Chivalry and Warfare: Proceedings of the 1995 Harlaxton Symposium*, Stamford: Paul Watkins.
Berza, M., and E. Stănescu (eds), *Actes du XIVe Congrès international des études byzantines: Bucarest, 6–12 septembre 1971*, Bucharest: Editura Academiei Republicii Socialiste România.
Beshir, B. J. (1978), 'Fatimid Military Organization', *Islam*, 55, pp. 37–56.
Bikhazi, R. J. (1981), 'The Hamdanid Dynasty of Mesopotamia and North Syria 868–1014', PhD dissertation, University of Michigan.
Bivar, A. D. H. (1972), 'Cavalry Equipment and Tactics on the Euphrates Frontier', *Dumbarton Oaks Papers*, 26, pp. 271–91.

Secondary Bibliography

Blankinship, K. Y. (1994), *The End of the Jihad State: The Reign of Hisham Ibn Abd Al-Malik and the Collapse of the Umayyads*, New York: State University of New York Press.

Blysidu, V. (2001), *Αριστοκρατικές οικογένειες και εξουσία (9ος–10ος αι.). Έρευνες πάνω στα διαδοχικά στάδια αντιμετώπισης της Αρμενο-Παφλαγονικής και Καππαδοκικής αριστοκρατίας* [Aristocratic Families and Power (9th–10th c.): Investigations on the Successive Stages of Dealing with the Armenian-Paphlagonian and Cappadocian Aristocracy], Thessaloniki: Vanias.

Bonner, M. D. (2006), *Jihad in Islamic History, Doctrines and Practice*, Princeton, NJ: Princeton University Press.

Bonner, M. D. (1996), *Aristocratic Violence and Holy War: Studies in the Jihad and the Arab–Byzantine Frontier*, New Haven, CT: American Oriental Society.

Bonner, M. D. (1994), 'The Naming of the Frontier: *Awāṣim*, *Thughūr*, and the Arab Geographers', *Bulletin of the School of Oriental and African Studies*, 57, pp. 17–24.

Bonner, M. D. (1987), 'The Emergence of the *Thugur*: The Arab–Byzantine Frontier in the Early Abbasid Age', PhD dissertation, University of Princeton.

Bosworth, C. E. (1992), 'The City of Tarsus and the Arab–Byzantine Frontiers in Early and Middle Abbasid Times', *Oriens*, 33, pp. 268–86.

Bosworth, C. E. (1992), 'Abu 'Amr 'Uthman al-Turtusi's *Siyar al-thugur* and the Last Years of Arab Rule in Tarsus (Fourth/Tenth Century)', *Graeco-Arabica*, 5, pp. 183–95.

Bosworth, C. E. (1975), 'Recruitment, Muster and Review in Medieval Islamic Armies', in V. J. Parry and M. E. Yapp (eds), *War, Technology and Society in the Middle East*, London: Oxford University Press, pp. 59–77.

Bosworth, C. E. (1965–6), 'Military Organization under the Buyids of Persia and Iraq', *Oriens*, 18–19, pp. 143–67.

Bosworth, C. E. (1961), 'Ghaznavid Military Organization', *Der Islam*, 36, pp. 37–77.

Bradbury, J. (1985), *The Medieval Archer*, Woodbridge: Boydell & Brewer.

Brault, G. J. (2010), *Song of Roland: An Analytical Edition; Introduction and Commentary*, Philadelphia: University of Pennsylvania Press.

Brooks, E. W. (1901), 'Arabic Lists of the Byzantine Themes', *Journal of Hellenic Studies*, 21, pp. 67–77.

Browning, R. (1978), 'Literacy in the Byzantine World', *Byzantine and Modern Greek Studies*, 4, pp. 39–54.

Buckler, G. (1929), *Anna Comnena: A Study*, London: Oxford University Press.

Buckler, J., and H. Beck (2008), *Central Greece and the Politics of Power in the Fourth Century BC*, Cambridge: Cambridge University Press.

Burns, R. (2017), *Aleppo: A History*, London: Routledge.

Bury, J. B. (1958), *The Imperial Administrative System in the Ninth Century: With a Revised Text of the 'Kletorologion' of Philotheos*, New York: Burt Franklin.

Bury, J. B. (1909), 'Mutasim's March through Cappadocia in A.D. 838', *Journal of Hellenic Studies*, 29, pp. 120–29.
Cahen, C. (1972), 'L'administration financière de l'armée fatimide d'après al-Makhzūmī', *Journal of the Economic and Social History of the Orient*, 15, pp. 163–82.
Cahen, C. (1953), 'Evolution de l'iqta du IXe au XIIIe siècle', *Annales d'histoire économique et sociale*, 8, pp. 25–62.
Cameron, A. (1987), 'The Construction of Court Ritual: The Byzantine Book of Ceremonies', in D. Cannadine and S. Price (eds), *Rituals of Royalty, Power and Ceremonial in Traditional Societies*, Cambridge: Cambridge University Press.
Cameron, A. (1970), *Agathias*, Oxford: Oxford University Press.
Cameron, A. (1969), 'Agathias on the Sasanians', *Dumbarton Oaks Papers*, 23, pp. 67–183.
Campagnolo-Pothitou, M. (1995), 'Les échanges de prisonniers entre Byzance et l' Islam aux IXe et Xe siècles', *Journal of Oriental and African Studies*, 7, pp. 1–56.
Canard, M. (1964), 'Les relations politiques et sociales entre Byzance et les Arabes', *Dumbarton Oaks Papers*, 18, pp. 33–56.
Canard, M. (1961), 'Les sources arabes de l'histoire byzantine aux confins des Xe et XIe siècles', *Revue des études byzantines*, 19, pp. 284–314.
Canard, M. (1953), *Histoire de la dynastie des Hamdanides*, Paris: Presses Universitaires de France.
Canard, M. (1951), 'Le cérémonial fatimite et le cérémonial byzantin, essai de comparaison', *Byzantion*, 21, pp. 355–420.
Canard, M. (1949–51), 'Deux épisodes des relations diplomatiques Arabo-Byzantines au Xe siècle', *Bulletin d'études orientales*, 13, pp. 51–69.
Canard, M. (1948), 'Les Hamdanides et l'Armenie', *Annales de l'Institut d'études orientales d'Alger*, 7, pp. 77–94.
Canard, M. (1936), 'Mutanabbi et la guerre byzantino-arab: intérêt historique de ses poésies', in *Al Mutanabbi: recueil publiee à l'occasion de son millènaire*, Beirut: Imprimerie catholique, pp. 99–114.
Canard, M. (1936), 'Une lettre de Muhammad ibn Tugj al-Ihsid émir d'Égypte à l'empereur romain Lecapene', *Annales de l'Institut d'études orientales de la Faculté des lettres d'Alger*, 2, pp. 189–209.
Canard, M., and N. Adontz (1936), 'Quelques noms des personnages byzantins dans une pièce du poète Abû-Firâs', *Byzantion*, 11, pp. 451–60.
Cappel, A. J. (1994), 'The Byzantine Response to the Arab (10th–11th Century)', *Byzantinische Forschungen*, 20, pp. 113–32.
Cathers, K. (2002), '"Markings on the Land" and Early Medieval Warfare in the British Isles', in P. Doyle and M. R. Bennett (eds), *Fields of Battle: Terrain in Military History*, London: Kluwer, pp. 9–17.
Chamberlain, M. (2008), 'The Crusader Era and the Ayyubid Dynasty', in C. F. Petry (ed.), *The Cambridge History of Egypt*, vol. 1: *Islamic Egypt, 640–1517*, Cambridge: Cambridge University Press, pp. 227–9.

Secondary Bibliography

Charanis, P. (1975), 'Cultural Diversity and the Breakdown of Byzantine Power in Asia Minor', *Dumbarton Oaks Papers*, 29, pp. 1–20.

Charanis, P. (1963), *The Armenians in the Byzantine Empire*, Lisbon: Fundação Calouste Gulbenkian.

Charanis, P. (1961), 'The Transfer of Population as a Policy in the Byzantine Empire', *Comparative Studies in Society and History*, 3, pp. 140–54.

Charanis, P. (1959), 'Ethnic Changes in the Byzantine Empire in the Seventh Century', *Dumbarton Oaks Papers*, 13, pp. 28–43.

Charles, Archduke of Austria (1893–4), *Ausgewählte Schriften weiland seiner kaiserlichen Hoheit des Erzherzogs Carl von Oesterreich*, Vienna: W. Braumüller.

Chatzelis, G., and J. Harris (2017), *A Tenth-Century Byzantine Military Manual: The Sylloge Tacticorum*, London: Routledge.

Cheynet, J.-C. (2014), 'Les Arméniens dans l'armée byzantine au Xe siècle', *Travaux et mémoires*, 18, pp. 175–92.

Cheynet, J.-C. (2013), 'Réflexions sur le "pacifisme byzantin"', in C. Gastberger, C. Messis, D. Mureşan and F. Ronconi (eds), *Pour l'amour de Byzance: hommage à Paolo Odorico*, Frankfurt am Main: Peter Lang, pp. 63–73.

Cheynet, J.-C. (2008), *La société byzantine: l'apport des sceaux*, 2 vols, Paris: Association des amis du Centre d'histoire et civilisation de Byzance.

Cheynet, J.-C. (2003), 'Basil II and Asia Minor', in P. Magdalino (ed.), *Byzantium in the Year 1000*, Leiden: Brill, pp. 82–96.

Cheynet, J.-C. (2003), 'Les transferts de population sous la contrainte à Byzance', in L. Feller (ed.), *Travaux et recherches de l'UMLV. Littératures. Sciences humaines. Les déplacements contraints de population*, Marne-la-Vallée: Université de Marne-la-Vallée, pp. 45–70.

Cheynet, J.-C. (2002), 'Les limites du pouvoir à Byzance: une forme de tolérance?' in A. Nikolaou (ed.), *Ανοχή και καταστολή στους μέσους χρόνους: Μνήμη Λένου Μαυρομάτη* [Tolerance and Suppression in the Middle Ages: In Memory of Lenos Mavromatis], International Symposia 10, Athens: National Research Foundation, pp. 15–28.

Cheynet, J.-C. (2001), 'La conception militaire de la frontière orientale', in A. Eastmond, *Eastern Approaches to Byzantium*, Aldershot: Ashgate, pp. 57–69.

Cheynet, J.-C. (1998), 'Theophile Thèophobe et les Perses', in S. Lampakis (ed.), *Byzantine Asia Minor*, Athens: National Research Institute, pp. 39–50.

Cheynet, J.-C. (1996), 'Les Arméniens de l'empire en Orient de Constantin X à Alexis Comnène', in N. G. Garsoïan, *L'Armenie et Byzance: histoire et culture*, Paris: Publications de la Sorbonne, pp. 67–78.

Cheynet, J.-C. (1996), *Pouvoir et contestations à Byzance (963–1210)*, Paris: Publications de la Sorbonne.

Cheynet, J.-C. (1995), 'Les effectifs de l'armée byzantine aux Xe–XIIe siècles', *Cahiers de civilisation médiévale*, 38e année, 152, pp. 319–35.

Cheynet, J.-C. (1991), 'Fortune et puissance des grandes familles (Xe–XIIe siècle)', in V. Kravari, J. Lefort and C. Morrisson (eds), *Hommes et richesses*, II, pp. 199–213.

Cheynet, J.-C. (1986), *Études prosopographiques*, Paris: Publications de la Sorbonne.
Cheynet, J.-C. (1986), 'Les Phocas', in G. Dagron and H. Mihaescu (eds), *Le traité sur la guérilla (De velitatione) de l'Empereur Nicéphore Phocas (963–969)*, Paris: Editions du Centre national de la recherche scientifique, pp. 289–315.
Christides, V. (1997), 'Military Intelligence in Arabo-Byzantine Naval Warfare', in Tsiknakes (ed.), *Byzantium at War*, pp. 269–81.
Christides, V. (1984), *The Conquest of Crete by the Andalusians (ca. 824–961)*, Athens: Academy of Athens.
Christides, V. (1984), *The Conquest of Crete by the Arabs (ca. 824): A Turning Point in the Struggle between Byzantium and Islam*, Athens: Academy of Athens.
Christides, V. (1982), 'Two Parallel Naval Guides of the Tenth Century: Qudama's Document and Leo VI's *Naumachica*: A Study on Byzantine and Moslem Naval Preparedness', *Graeco-Arabica*, 1, pp. 52–103.
Christides, V. (1981), 'Once Again Kaminiates' Capture of Thessalonica', *Byzantinische Zeitschriftt*, 74, pp. 7–10.
Christides, V. (1981), 'The Raids of the Moslems of Crete in the Aegean Sea: Piracy and Conquest', *Byzantion*, 51, pp. 76–111.
Christides, V. (1979), 'Arabic Influence on the Akritic Cycle', *Byzantion*, 49, pp. 94–109.
Christides, V. (1962), 'An Arabo-Byzantine Novel, Umar b. Al-Nu'man Compared with Digenes Akritas', *Byzantion*, 32, pp. 549–604.
Christophilopoulou, E. (1935), *Το Επαρχικόν Βιβλίον του Λέοντος του Σοφού και αι Συντεχνίαι εν Βυζαντίω* [The Book of the Eparch by Leo the Wise and the Guilds in Byzantium], Athens: Pournara Publications.
Chrysos, E. (2005), 'Το Βυζάντιο και η διεθνής κοινωνία του Μεσαίωνα' [Byzantium and the International Community of the Middle Ages], in E. Chrysos (ed.), *Το Βυζάντιο ως Οικουμένη* [Byzantium as Oikumene], Athens: National Research Institute, pp. 59–78.
Chrysos, E. (2003), 'Ο πόλεμος έσχατη λύση' [War as the Ultimate Solution], in A. Avramea, A. Laiou and E. Chrysos (eds), *Βυζάντιο: Κράτος και Κοινωνία – Μνήμη Νίκου Οικονομίδη / Byzantium: State and Society – In Memory of Nikos Oikonomides*, Athens: National Research Institute, pp. 543–63.
Chrysos, E. (1993), 'Η Βυζαντινή διπλωματία ως μέσο επικοινωνίας' [Byzantine Diplomacy as a Means of Communication], in N. Moschonas (ed.), *Η Επικοινωνία στο Βυζάντιο* [Communication in Byzantium], Athens: National Research Foundation, pp. 399–407.
Chrysos, E. (1992), 'Byzantine Diplomacy, A.D. 300–800: Means and Ends', in Shepard and Franklin (eds), *Byzantine Diplomacy*, pp. 25–39.
Chrysostomides, J. (2001), 'Byzantine Concepts of War and Peace', in A. V. Hartmann and B. Heuser (eds), *War, Peace and World Orders in European History*, London: Routledge, pp. 91–102.

Secondary Bibliography

von Clausewitz, C. (1984), *On War*, ed. M. Howard and P. Paret, Princeton, NJ: Princeton University Press.

von Clausewitz, C. (2007), *On War*, ed. and trans. M. Howard and P. Paret, Oxford: Oxford University Press.

Constable, O. R. (2003), *Housing the Stranger in the Mediterranean World: Lodging, Trade, and Travel in Late Antiquity and the Middle Ages*, Cambridge: Cambridge University Press.

Cook, R. F. (1980), *'Chanson d'Antioche', chanson de geste: le cycle de la croisade est-il epique?*, Amsterdam: John Benjamins B. V.

Cosentino, S. (2009), 'Writing about War in Byzantium', *Revista de Història das Ideias*, 30, pp. 83–99.

Cosentino, S. (2000), 'Syrianos' *Strategikon* – A Ninth-Century Source?', *Bizantinistica: Rivista di studi bizantini e slavi*, 2, pp. 243–80.

Coulston, J. C. N. (1985), 'Roman Archery Equipment', in M. C. Bishop (ed.), *The Production and Distribution of Roman Military Equipment*, Oxford: Oxford University Press, pp. 220–366.

Cox, R. (2012), 'Asymmetric Warfare and Military Cconduct in the Middle Ages', *Journal of Medieval History*, 38, pp. 100–25.

van Creveld, M. (2013), *Wargames: From Gladiators to Gigabytes*, Cambridge: Cambridge University Press.

van Creveld, M. (2008), *The Culture of War*, New York: History Press Limited.

van Creveld, M. (1991), *The Transformation of War*, London: Free Press.

van Creveld, M. (1989), *Technology and War from 2000 BC to the Present*, New York: Free Press.

Crone, P. (2007), 'Quraysh and the Roman Army: Making Sense of the Meccan Leather Trade', *Bulletin of the School of Oriental and African Studies*, 70, pp. 63–88.

Crone, P. (1987), *Meccan Trade and the Rise of Islam*, Oxford: Oxford University Press.

Crouch, D. (2002), *William Marshal: Knighthood, War and Chivalry, 1147–1219*, London: Routledge.

Dagron, G. (1987), '"Ceux d'en face": les peoples étrangers dans les traits militaires byzantins', *TravMém*, 10, pp. 207–32.

Dagron, G. (1983), 'Byzance et le modèle islamique au Xe siècle, à propos des *Constitutions tactiques* de l'empereur Léon VI', *Comptes rendus des séances de l'année de l'Académie des inscriptions et belles-lettres*, pp. 219–43.

Dagron, G. (1976), 'Minorités ethniques et religieuses: l'immigration syrienne', *Travaux et mémoires*, 6, pp. 177–216.

Dagron, G., and H. Mihaescu (1986), *Le traité sur la guérilla ('De velitatione') de l'empereur Nicéphore Phocas (963–969)*, Paris: Editions du Centre national de la recherche scientifique.

Dahan, S. (1944), *Le 'Diwan' d'Abu Firas al-Hamdani (poète arabe du IVe siècle de l'hégire)*, Beirut: Institut français de Damas.

Dahmus, J. (1983), *Seven Decisive Battles of the Middle Ages*, Chicago, IL: Nelson-Hall.
Dain, A. (1967), 'Les stratégistes byzantins', *Travaux et mémoires*, 2, pp. 317–92.
Dain, A. (1940), *Appellations grecques du feu grégeois*, Paris: Mélanges Ernout.
Dawson, T. (2007), '"Fit for the Task": Equipment Sizes and the Transmission of Military Lore, Sixth to Tenth Centuries', *Byzantine and Modern Greek Studies*, 31, pp. 1–12.
Dawson, T., (2002), 'Suntagma Hoplon: Equipment of Regular Byzantine Troops, c. 950–c.1204', in D. Nicolle (ed.), *A Companion to Medieval Arms and Armour*, Woodbridge: Boydell & Brewer, pp. 81–96.
Dédéyan, G. (2002), 'Reconquête territoriale et immigration arménienne dans l'aire cilicienne sous les empereurs macédoniens (de 867 à 1028)', in M. Balard and A. Ducellier (eds), *Migrations et diasporas méditerranéennes: Xe–XVIe siècles. Actes du colloque de Conques, Octobre 1999*, Paris: Publications de la Sorbonne, pp. 11–32.
Dédéyan, G. (1993), 'Les Arméniens sur la frontière sud-orientale de Byzance, fin IXe–fin XIe siècles', *La frontière, travaux de la maison de l'Orient*, 21, pp. 67–85.
Dédéyan, G. (1975), 'Immigration arménienne en Cappadoce au XIe siecle', *Byzantion*, 45, pp. 41–117.
Delbruck, H. (1913), *Numbers in History*, London: Wentworth.
Dennis, G. T. (1997), 'The Byzantines in Battle', in Tsiknakes (ed.), *Byzantium at War*, pp. 165–78.
DeVries, K. (2004), 'The Use of Chroniclers in Recreating Medieval Military History', *Journal of Medieval Military History*, 2, pp. 1–17.
DeVries, K. (1999), 'God and Defeat in Medieval Warfare: Some Preliminary Thoughts', in Kagay and Villalon (eds), *Circle of War in the Middle Ages*, pp. 87–97.
DeVries, K. (1997), 'Catapults Are Not Atomic Bombs: Towards a Redefinition of "Effectiveness" in Premodern Military Technology', *War in History*, 4, pp. 454–70.
Dodgeon, M. H., and S. N. C. Lieu (eds) (2002), *The Roman Eastern Frontier and the Persian Wars, AD 226–363: A Documentary History*, London: Routledge.
Doyle, P., and M. R. Bennett (2002), 'Terrain in Military History: An Introduction', in P. Doyle and M. R. Bennett (eds), *Fields of Battle: Terrain in Military History*, London: Kluwer, pp. 1–7.
Drocourt, N. (2008), 'La diplomatie médio–byzantine et l'antiquité', *Anabases*, 7, pp. 57–87.
Duby, G. (1990), *The Legend of Bouvines: War, Religion and Culture in the Middle Ages*, trans. C. Tihanyi, Cambridge: Cambridge University Press.
Durak, K. (2008), 'Commerce and Networks of Exchange between the Byzantine Empire and the Islamic Near East from the Early Ninth Century to the Arrival of the Crusaders', PhD dissertation, Harvard University.
Dvornik, F. (1926), *La vie de Saint Grégoire le décapolite et les slaves macédoniens au IXe siècle*, Paris: Champion.

Dyson, S. B. (2014), 'Origins of the Psychological Profiling of Political Leaders: The US Office of Strategic Services and Adolf Hitler', *Intelligence and National Security*, 29, pp. 654–74.

Ekkebus, B. (2009), 'Heraclius and the Evolution of Byzantine Strategy', *Constructing the Past*, 10, pp. 73–96.

El Cheikh, N. M. (2004), *Byzantium Viewed by the Arabs*, Cambridge, MA: Harvard University Press.

Ellis-Davidson, H. R. (1973), 'The Secret Weapon of Byzantium', *Byzantinische Zeitschrift*, 66, pp. 61–74.

Fahmy, A. M. (1956), *Muslim Naval Organisation in the Eastern Mediterranean from the Seventh to the Tenth Century AD*, Cairo: National Publication and Print House.

Flori, J. (1993), 'Un problème de méthodologie: la valeur des nombres chez les chroniquers du Moyen Age, à propos des effectifs de la première croisade', *Le Moyen Age*, 119, pp. 399–422.

Forsyth, J. H. (1977), 'The Byzantine–Arab Chronicle (938–1034) of Yahya B. Said Al-Antaki', 2 vols, PhD dissertation, University of Michigan.

France, J. (2005), 'Close Order and Close Quarter: The Culture of Combat in the West', *International History Review*, 27, pp. 498–517.

France, J. (1999), *Victory in the East: A Military History of the First Crusade*, Cambridge: Cambridge University Press, pp. 122–42.

Frendo, D. (2000), *John Kaminiates: The Capture of Thessaloniki*, Perth: Australian Association of Byzantine Studies.

Gabriel, R. A., and D. W. Boose (1994), *The Great Battles of Antiquity: A Strategic and Tactical Guide to Great Battles that Shaped the Development of War*, Westport, CT: Greenwood.

Galatariotou, C. (1993), 'Travel and Perception in Byzantium', *Dumbarton Oaks Papers*, 47, pp. 221–41.

Gamber, O. (1968), '*Kataphrakten*, Clibanarier, Normanreiter', *Jahrbuch der Kunsthistorischen Sammlungen in Wien*, 64, pp. 7–44.

Garrood, W. (2008), 'The Byzantine Conquest of Cilicia and the Hamdanids of Aleppo, 959–965', *Anatolian Studies*, 58, pp. 127–40.

Garsoïan, N. G. (1998), 'The Problem of the Armenian Integration into the Byzantine Empire', in H. Ahrweiler and A. E. Laiou (eds), *Studies on the Internal Diaspora of the Byzantine Empire*, Washington, DC: Dumbarton Oaks Research Library, pp. 53–124.

Gillingham, J. (2004), '"Up with Orthodoxy!" In Defence of Vegetian Warfare', *Journal of Medieval Military History*, 2, pp. 149–58.

Gillingham, J. (1999), 'An Age of Expansion', in M. Keen (ed.), *Medieval Warfare*, Oxford: Oxford University Press.

Gillingham, J. (1989), 'William the Bastard at War', in C. Harper-Hill et al. (eds), *Studies in Medieval History Presented to R. Allen Brown*, Woodbridge: Boydell & Brewer, pp. 141–58.

Gillingham, J. (1988), 'War and Chivalry in the History of William the Marshal', in P. R. Coss and S. D. Lloyd (eds), *Thirteenth Century England II: Proceedings* of the Newcastle Upon Tyne Conference, 1987, Woodbridge: Boydell & Brewer, pp. 251–63.

Gillingham, J. (1984), 'Richard I and the Science of War in the Middle Ages', in J. Gillingham and J. C. Holt (eds), *War and Government in the Middle Ages*, Woodbridge: Boydell & Brewer, pp. 78–91.

Goldsworthy, A. (2000), *Roman Warfare*, London: Cassell.

Greatrex, G. (1996), 'Stephanus, the Father of Procopius of Caesarea?', *Medieval Prosopography*, 17, pp. 125–45.

Greatrex, G., and S. N. C. Lieu (eds) (2002), *The Roman Eastern Frontier and the Persian Wars, AD 363–628: A Narrative Sourcebook*, London: Routledge.

Grubbs, J. T. (2010), 'The Mongol Intelligence Apparatus: The Triumphs of Genghis Khan's Spy Network', *International Association for Intelligence Education*, pp. 3–14.

Gürkan, E. S. (2012), 'The Efficacy of Ottoman Counter-intelligence in the 16th Century', *Acta Orientalia Academiae Scientiarum Hungaricae*, 65, pp. 1–38.

Haldon, J. F. (2016), T*he Empire That Would Not Die: The Paradox of Eastern Roman Survival, 640–740*, Cambridge, MA: Harvard University Press.

Haldon, J. F. (2014), *A Critical Commentary on the 'Taktika' of Leo VI*, Washington, DC: Dumbarton Oaks Research Library and Collection.

Haldon, J. F. (2013), 'Information and War: Some Comments on Defensive Strategy and Information in the Middle Byzantine Period (ca. A.D. 660–1025)', in A. Sarantis and N. Christie (eds), *War and Warfare in Late Antiquity*, 2 vols, Leiden: Brill, pp. 373–93.

Haldon, J. F. (2006), 'The Organisation and Support of an Expeditionary Force: Manpower and Logistics in the Middle Byzantine Period', in N. Oikonomides (ed.), *Το εμπόλεμο Βυζάντιο* [Byzantium at War], Athens: National Hellenic Research Foundation, pp. 111–51.

Haldon, J. F. (2001) *The Byzantine Wars: Battles and Campaigns of the Byzantine Era*, Stroud: Tempus.

Haldon, J. F. (2000), 'Theory and Practice in Tenth-Century Military Administration: Chapters II, 44 and 45 of the *Book of Ceremonies*', *Travaux et mémoires*, 13, pp. 201–352.

Haldon, J. F. (1999), *Warfare, State and Society in the Byzantine World, 565–1204*, London: UCL Press.

Haldon, J. F. (1997), *Byzantium in the Seventh Century: The Transformation of a Culture*, Cambridge: Cambridge University Press.

Haldon, J. F. (1995), 'Strategies of Defence, Problems of Security: The Garrisons of Constantinople in the Middle Byzantine Period', in G. Dagron and C. Mango (eds), *Constantinople and Its Hinterland*, Aldershot: Ashgate, pp. 143–55.

Haldon, J. F. (1995), '"Fighting for Peace": Justifying Warfare and Violence in the Medieval East Roman World', in R. W. Kaeuper, D. G. Tor and H. Zurndorfer (eds), *The Cambridge World History of Violence*, vol. 2: AD 500–AD 1500, Cambridge: Cambridge University Press, forthcoming.

Secondary Bibliography

Haldon, J. F. (1993), 'Military Service, Military Lands, and the Status of Soldiers: Current Problems and Interpretations', *Dumbarton Oaks Papers*, 47, pp. 1–67.

Haldon, J. F. (1992), 'Blood and Ink: Some Observations on Byzantine Attitudes towards Warfare and Diplomacy', in Shepard and Franklin (eds), *Byzantine Diplomacy*, pp. 281–95.

Haldon, J. F. (1990), *Byzantium in the Seventh Century*, Cambridge: Cambridge University Press.

Haldon, J. F. (1975), Some Aspects of Byzantine Military Technology from the Sixth to the Tenth Centuries, *Byzantine and Modern Greek Studies*, 1, pp. 11–47.

Haldon, J. F., and M. Byrne (1977), 'A Possible Solution to the Problem of Greek Fire', *Byzantinische Zeitschrift*, 70, pp. 91–9.

Haldon, J. F., and H. Kennedy (1980), 'The Arab–Byzantine Frontier in the Eighth and Ninth Centuries: Military Organization and Society in the Borderlands', *Zbornik Radova Vizantoloski Institut*, 19, pp. 79–116.

Haldon, J. F., V. Gaffney, G. Theodoropoulos and P. Murgatroyd (2013), 'Marching across Anatolia: Medieval Logistics and Modeling the Mantzikert Campaign', *Dumbarton Oaks Papers*, 65/66, pp. 1–27.

Haldon, J. F., N. Roberts, A. Izdebski et al. (2014), 'The Climate and Environment of Byzantine Anatolia: Integrating Science, History, and Archaeology', *Journal of Interdisciplinary History*, 45, pp. 113–61.

Hamblin, W. J. (1985), 'The Fatimid Army during the Early Crusades', PhD dissertation, University of Michigan.

Hamidullah, M. (1961), *Muslim Conduct of State*, Lahore: Islamic Book Centre.

Hanson, V. D. (2001), *Carnage and Culture: Landmark Battles in the Rise of Western Power*, New York: First Anchor.

Hanson, V. D. (1989), *The Western Way of War: Infantry Battle in Classical Greece*, New York: A. Knopf.

Harari, Y. N. (2007), 'The Concept of "Decisive Battles" in World History', *Journal of World History*, 18, pp. 251–66.

Harmon, R. S., F. H. Dillon III and J. B. Garver Jr (2004), 'Perspectives on Military Geography', in D. R. Caldwell, J. Ehlen and R. S. Harmon (eds), *Studies in Military Geography and Geology*, London: Kluwer, pp. 7–20.

al-Hassan, A. Y., and D. R. Hill (1992), *Islamic Technology: An Illustrated History*, Cambridge: Cambridge University Press.

Hendy, M. F. (1985), *Studies in the Byzantine Monetary Economy, c. 300–1450*, Cambridge: Cambridge University Press.

Heuser, B. (2016), 'Theory and Practice, Art and Science in Warfare: An Etymological Note', in D. Marston and T. Leahy (eds), *War, Strategy and History: Essays in Honour of Professor Robert O'Neill*, Canberra: Australian National University Press, pp. 179–96.

Heuser, B. (2010), *The Evolution of Strategy: Thinking War from Antiquity to the Present*, Cambridge: Cambridge University Press.

Heuser, B. (2002), *Reading Clausewitz*, London: Pimlico.
Hewitt, H. J. (1966), *The Organization of War under Edward III, 1338–62*, Manchester: Manchester University Press.
Hillenbrand, C. (1999), *The Crusades: Islamic Perspectives*, Edinburgh: Edinburgh University Press.
Holmes, C. (2010), 'Provinces and Capital', in L. James (ed.), *A Companion to Byzantium*, Oxford: Blackwell, pp. 55–66.
Holmes, C. (2005), *Basil II and the Governance of the Empire (975–1025)*, Oxford: Oxford University Press.
Holmes, C. (2002), 'The Byzantine Eastern Frontier in the Tenth and Eleventh Centuries', in D. Abulafia and N. Berend (eds), *Medieval Frontiers: Concepts and Practices*, Aldershot: Ashgate, pp. 83–104.
Howard-Johnston, J. (2010), *Witnesses to a World Crisis: Historians and Histories of the Middle East in the Seventh Century*, Oxford: Oxford University Press.
Howard-Johnston, J. (1999), 'Heraclius' Persian Campaigns and the Revival of the East Roman Empire, 622–630', *War in History*, 6, pp. 1–44.
Howard-Johnston, J. (1995), 'Crown Lands and the Defence of Imperial Authority in the Tenth and Eleventh Centuries', *Byzantinische Forschungen*, 21, pp. 75–100.
Hull, I. V. (2005), *Absolute Destruction: Military Culture and the Practices of War in Imperial Germany*, Ithaca, NY: Cornell University Press.
Hunger, H. (1978), *Die hochsprachliche profane Literatur der Byzantiner*, 2 vols, Munich: Beck.
Husayn, F. (2012), 'The Participation of Non-Arab Elements in the Umayyad Army and Administration', in F. Donner (ed.), *The Articulation of Early Islamic State Structures*, London: Routledge, pp. 279–80.
Huxley, G. L. (1982), 'Topics in Byzantine Historical Geography', *Proceedings of the Royal Irish Academy. Section C: Archaeology, Celtic Studies, History, Linguistics, Literature*, 82C, pp. 89–110.
Huxley, G. L. (1975), 'The Emperor Michael III and the Battle of Bishop's Meadow (A.D. 863)', *Greek, Roman, and Byzantine Studies*, 16, pp. 443–50.
Hyland, A. (1996), *The Medieval Warhorse from Byzantium to the Crusades*, Conshohocken, PA: Combined Books.
Isaac, B. (1990), *The Limits of Empire: The Roman Army in the East*, Oxford: Oxford University Press.
Isaac, B. (1988), 'The Meaning of the Terms *Limes* and *Limitanei* in Ancient Sources', *Journal of Roman Studies*, 78, pp. 125–47.
Ivanov, S. (2002), 'Casting Pearls before Circe's Swine: The Byzantine View of Mission', *Travaux et mémoires*, 14, pp. 295–301.
Jacobi, R. (1996), 'The Origins of the Qasida Form', in Sperl and Shackle (eds), *Qasida Poetry in Islamic Asia and Africa*, pp. 21–35.
James, W. (2013), *The Moral Equivalent of War*, New York: Read Books.

Secondary Bibliography

Janin, R. (1936), 'Un ministre arabe à Byzance: Samonas', *Echos d'Orient*, 36, pp. 307–18.

Jayyusi, S. K. (1996), 'The Persistence of the Qasida Form', in Sperl and Shackle (eds), *Qasida Poetry in Islamic Asia and Africa*.

Jeffreys, E. (2000), 'Akritis and Outsiders', in D. C. Smythe (ed.), *Strangers to Themselves: The Byzantine Outsider*, Aldershot: Ashgate, pp. 189–202.

Jeffreys, E., J. Haldon and R. Cormack (eds) (2008), *The Oxford Handbook of Byzantine Studies*, Oxford: Oxford University Press.

Jenkins, R. J. (1948), 'The "Flight" of Samonas', *Speculum*, 23, pp. 217–35.

John, S. (2012), 'The Use of Oral Evidence in the Twelfth-Century Historical Writing of the First Crusade', *The Crusades and the Latin East Seminar Series*, London: Institute of Historical Research (12 March).

Kaegi, W. E. (2003), *Heraclius, Emperor of Byzantium*, Cambridge: Cambridge University Press.

Kaegi, W. E. (2000), *Byzantium and the Early Islamic Conquests*, Cambridge: Cambridge University Press.

Kaegi, W. E. (1991), 'Challenges to the Late Roman and Byzantine Military Operations in Iraq (4th–9th Ccenturies)', *Klio*, 73, pp. 586–94.

Kaegi, W. E. (1989), 'Changes in Military Organisation and Daily Life on the Eastern Frontier', in *He kathemerine zoe sto Byzantio*, Athens: National Institute of Byzantine Research, pp. 507–21.

Kaegi, W. E. (1986), 'The Frontier: Barrier or Bridge?', in G. Vikan (ed.), *17th International Byzantine Congress. Major Papers*, New York: A. D. Caratzas.

Kaegi, W. E. (1983), 'Some Thoughts on Byzantine Military Strategy', *The Hellenic Studies Lecture*, Brookline, MA: Hellenic College Press, pp. 1–18.

Kaegi, W. E. (1981), 'Constantine and Julian's Strategies of Strategic Surprise against the Persians', *Athenaum*, 69, pp. 209–13.

Kaegi, W. E. (1967), 'Some Reconsiderations on the Themes: Seventh–Ninth Centuries', *Jahrbuch der Österreichischen Byzantinistik*, 16, pp. 39–53.

Kaegi, W. E. (1964), 'The Contribution of Archery to the Turkish Conquest of Anatolia', *Speculum*, 39, pp. 96–108.

Kagan, K. (2006), *The Eye of Command*, Ann Arbor: University of Michigan Press.

Kagay, D. J., and L. J. A. Villalon (eds), *The Circle of War in the Middle Ages: Essays on Medieval Military and Naval History*, Woodbridge: Boydell & Brewer.

Kaldellis, A. (2017), 'Did the Byzantine Empire have "Ecumenical" or "Universal" Aspirations?', in C. Ando and S. Richardson (eds), *Ancient States and Infrastructural Power: Europe, Asia, and America*, Philadelphia: University of Pennsylvania Press, pp. 272–300.

Kaldellis, A. (2015), *The Byzantine Republic: People and Power in New Rome*, Cambridge, MA: Harvard University Press.

Karales, B. (2000), Λέων Διάκονος. Ιστορία [Leo the Deacon. History], Athens: Kanaki.

Karpozilos, A. (2002), *Βυζαντινοί Ιστορικοί και Χρονογράφοι (8ος–10ος αιώνας)* [Byzantine Historians and Chroniclers (8th–10th Centuries)], vol. 2, Athens: Kanaki.

Kay, S. (2005), *The Chansons de Geste in the Age of Romance: Political Fictions*, Oxford: Oxford University Press.

Kazhdan, A. (1992), 'The Notion of Byzantine Diplomacy', in Shepard and Franklin (eds), *Byzantine Diplomacy*, pp. 3–21.

Kazhdan, A. (1978), 'Some Questions Addressed to the Scholars, who Believe in the Authenticity of Kaminiates' Capture of Thessalonika', *Byzantinische Zeitschrift*, 71, pp. 301–14.

Kazhdan, A. (1961), 'Iz istorii vizantiiskoi khronografii X v. 2. Istochniki L'va D'iakona i Skilitsy dlia istorii tretei chtverti X stoletiia', *VizVrem*, 20, pp. 106–28.

Keegan, J. (2004), *The Face of Battle: A Study of Agincourt, Waterloo and the Somme*, London: Pimlico.

Keegan, J. (2003), *Intelligence in War: Knowledge of the Enemy from Napoleon to Al-Qaeda*, New York: Pimlico.

Keegan, J. (1993), *A History of Warfare*, New York: Vintage.

Keen, M. (2004), *Chivalry*, New Haven, CT: Yale University Press.

Kelsay, J., and J. T. Johnson (eds) (1991), *Just War and Jihad: Historical and Theoretical Perspectives on War and Peace in Western and Islamic Traditions*, Westport, CT: Greenwood.

Kennedy, H. (2004), 'Byzantine–Arab Diplomacy in the Near East from the Islamic Conquests to the Mid Eleventh Century', in M. Bonner (ed.), *Arab–Byzantine Relations in Early Islamic Times*, Aldershot: Ashgate, pp. 81–91.

Kennedy, H. (2004), *The Prophet and the Age of the Caliphates*, London: Longman.

Kennedy, H. (2001), *The Armies of the Caliphs: Military and Society in the Early Islamic State Warfare and History*, London: Routledge.

Khadduri, M. (1955), *War and Peace in the Law of Islam*, Baltimore, MD: Johns Hopkins University Press.

Khalilieh, H. S. (1999), 'The *Ribât* System and its Role in Coastal Navigation', *Journal of the Economic and Social History of the Orient*, 42, pp. 212–25.

Khouri al Odetallah, R. A. (1983), 'Άραβες και Βυζαντινοί. Τό Πρόβλημα τών Αιχμαλώτων Πολέμου' [Arabs and Byzantines: The Problem of Prisoners of War], PhD thesis, Aristotle University of Thessaloniki.

Kiapidou, E. S. (2010), *Η Σύνοψη Ιστοριών του Ιωάννη Σκυλίτζη και οι Πηγές της (811–1057): Συμβολή στη Βυζαντινή Ιστοριογραφία κατα τον 11ο αιώνα* [The Synopsis Historion of Ioannes Skylitzes and its Sources (811–1057): Contribution to the Byzantine Historiography of the Eleventh Century], Athens: Kanaki.

Kiapidou, I. S. (2003), 'Battle of Dazimon, 838', *Online Encyclopaedia of the Hellenic World, Asia Minor*, Athens: Foundation of the Hellenic World.

Klopsteg, P. E. (1987), *Turkish Archery and the Composite Bow*, London: Butler & Tanner.

Secondary Bibliography

Kochly, H. and Rustow, W. (1853–5), *Griechische Kriegsschriftsteller*, Leipzig: Engelmann.

Koder, J., and I. Stouraitis (eds) (2012), *Byzantine War Ideology between Roman Imperial Concept and Christian Religion: Akten Des Internationalen Symposiums (Wien, 19.–21. Mai 2011)*, Vienna: Austrian Academy of Sciences Press.

Koder, J., and I. Stouraites (2012), 'Byzantine Approaches to Warfare (6th–12th Centuries): An Introduction', in Koder and Stouraites (eds), *Byzantine War Ideology*, pp. 9–15.

Kolia-Dermitzaki, A. (1991), *Ὁ βυζαντινός «ἱερός πόλεμος». Ἡ ἔννοια καί ἡ προβολή τοῦ θρησκευτικοῦ πολέμου στο Βυζάντιο* [Byzantine 'Holy War': The Concept and Evolution of Religious Warfare in Byzantium], Athens: Historikes Monografies 10.

Kolias, G. (1939), *Léon Choerosphactès, magistre, proconsul et patrice*, Athens: Verlag der 'Byzantinisch-neugriechischen jahrbücher.

Kolias, T. G. (1997), 'Ἡ πολεμική τακτική των Βυζαντινών: θεωρία και πράξη' [The War Tactic of the Byzantines: Theory and Practice], in Tsiknakes (ed.), *Byzantium at War*, pp. 153–64.

Kolias, T. G. (1993), *Νικηφόρος Β΄ Φωκάς (963–969). Ο Στρατηγός Αυτοκράτωρ και το Μεταρρυθμιστικό του Ἔργο* [Nikephoros Phokas the Second (963–969): General Imperator and his Reforms], Athens: Vasilopoulos.

Kolias, T. G. (1988), *Byzantinische Waffen. Ein Beitrag zur byzantinischen Waffenkunde von den Anfängen biz zur lateinischen Eroberung*, Vienna: Verlag der Österreichische Akademie der Wissenschaften.

Kolias, T. G. (1984), 'The *Taktika* of Leo VI and the Arabs', *Graeco-Arabica*, 3, pp. 129–35.

Korres, K. (1995), 'Ὑγρόν Πῦρ', Ένα Όπλο της Βυζαντινής Ναυτικής Τακτικής ['Greek Fire', a Weapon in Byzantine Naval Tactics], Thessaloniki: Vanias.

Kortüm, H. H. (ed.) (2006), *Transcultural Wars: From the Middle Ages to the 21st Century*, Berlin: Akademie Verlag.

Koutrakou, N. (2005), 'Βυζαντινή διπλωματική παράδοση και πρακτικές. Μια προσέγγιση μέσω της ορολογίας' [Byzantine Diplomatic Tradition and Practices: A Terminology Approach], in S. Patoura-Spanou (ed.), *Διπλωματία και Πολιτική, Ιστορικές Προσεγγίσεις* [Diplomacy and Politics: Historical Approaches], Athens: National Research Institute, pp. 92–5.

Koutrakou, N. (2000), '"Spies of Towns": Some Remarks on Espionage in the Context of Arab–Byzantine Relations (VIIth–Xth Centuries)', *Graeco-Arabica*, 7–8, pp. 243–66.

Koutrakou, N. (1995), 'Diplomacy and Espionage: Their Role in Byzantine Foreign Relations, 8th–10th Centuries', *Graeco-Arabica*, 6, pp. 125–44.

Koutrakou, N. (1995), '*Logos* and *Pathos* between Peace and War: Rhetoric as a Tool of Diplomacy in the Middle Byzantine Period', *Thesaurismata*, 25, pp. 7–20.

Krawczyk, J. L. (1985), 'The Relationship between Pastoral Nomadism and Agriculture: Northern Syria and the Jazira in the Eleventh Century', *JUSUR*, 1, pp. 1–22.

Kumin, B. (1999), 'Useful to Have, but Impossible to Govern: Inns and Taverns in Early Modern Bern and Vaud', *Journal of Early Modern History*, 3, pp. 153–203.

Laiou, A., and C. Morrisson (2007), *The Byzantine Economy*, Cambridge: Cambridge University Press.

Lambton, A. K. S. (1965), 'Reflections on the *Iqṭāʿ*', in G. Makdisi (ed.), *Arabic and Islamic Studies in Honor of Hamilton A. R. Gibb*, Leiden: Brill, pp. 358–76.

Landau-Tasseron, E. (2006), 'The Status of Allies in Pre-Islamic and Early Islamic Arabian Society', *Islamic Law and Society*, 13, pp. 6–32.

Langer, W. C. (1972), *The Mind of Adolf Hitler: The Secret Wartime Report*, New York: Basic Books.

Larkin, M. (2008), *Al-Mutanabbi: Voice of the ʿAbbasid Poetic Ideal*, Oxford: Oxford University Press.

Lassner, J. (1970), *The Topography of Baghdad in the Early Middle Ages: Text and Studies*, Detroit, MI: Wayne State University Press.

Latham, J. D. (1979), 'Towards a Better Understanding of al-Mutanabbī's Poem on the Battle of al-Hadath', *Journal of Arabic Literature*, 10, pp. 1–22.

Lee, A. D., and J. Shepard (1991), 'A Double Life: Placing the Peri Presbeon', *Byzantinoslavica*, 52, pp. 15–39.

Lemerle, P. (1971), *Le premier humanisme byzantin*, Paris: Presses Universitaires de France.

Lemerle, P. (1965), 'Thomas le slave', *Travaux et mémoires*, 1, pp. 255–97.

Lendon, J. E. (1999), 'The Rhetoric of Combat: Greek Military Theory and Roman Culture in Julius Caesar's Battle Descriptions', *Classical Antiquity*, 18, pp. 273–329.

Letsios, D. (1992), 'Die Kriegsgefangenschaft nach Auffassung der Byzantiner', *Byzantinoslavica*, 53, pp. 213–27.

Lev, Y. (2006), 'Infantry in Muslim Armies during the Crusades', in J. H. Pryor (ed.), *Logistics of Warfare in the Age of the Crusades*, Aldershot: Ashgate, pp. 185–207.

Lev, Y. (ed.) (1997), *War and Society in the Eastern Mediterranean, 7th–15th Centuries*, Leiden: Brill.

Lev, Y. (1991), 'The Evolution of the Tribal Army', in Lev (ed.), *State and Society*, pp. 81–92.

Lev, Y. (1991), *State and Society in Fatimid Egypt*, Leiden: Brill.

Lev, Y. (1988), 'The Fatimids and Egypt 301–358/914–969', *Arabica*, 35, pp. 186–96.

Lev, Y. (1987), 'Army, Regime, and Society in Fatimid Egypt, 358–487/968–1094', *Journal of Middle East Studies*, 19, pp. 337–65.

Secondary Bibliography

Lev, Y. (1982), 'The Fatimids and the *Aḥdāth* of Damascus 386/996–411/1021', *Die Welt des Orients*, 13, pp. 97–106.

Lev, Y. (1980), 'The Fatimid Army, A.H. 358–427/968–1036 C.E.: Military and Social Aspects', *Asian and African Studies*, 14, pp. 165–92.

Leverage, P. (2010), *Reception and Memory: A Cognitive Approach to the Chansons de Geste*, Amsterdam: Editions Rodopi.

Liddell, H. G., and R. Scott (eds) (1953), *A Greek–English Lexicon*, Oxford: Clarendon Press.

Lilie, R.-J. (1984), 'Des Kaisers Macht und Ohnmacht. Zum Zerfall der Zentralgewalt in Byzanz vor dem vierten Kreuzzug', in R.-J. Lilie and P. Speck (eds), *Varia* I (Poikila Byzantina 4), Bonn: R. Habelt, pp. 9–120.

Lilie, R.-J. (1976), *Die Byzantinische Reaktion auf die Ausbreitung der Araber: Studien zur Strukturwandlung des byzantinischen Staates im 7. und 8. Jhd*, Munich: Institut für Byzantinistik und Neugriechische Philologie der Universität München.

Ljubarskij, J. (1991), 'Writer's Intrusion in Early Byzantine Literature', in I. Ševčenko, G. G. Litavrin and W. K. Hanak (eds), *XVIIIth International Congress of Byzantine Studies: Selected Papers, Main and Communications*, Moscow: Byzantine Studies Press, pp. 433–56.

Lounghis, T. (2010), *Byzantium in the Eastern Mediterranean: Safeguarding East Roman Identity (407–1204)*, Nicosia: Cyprus Research Center.

Lounghis, T. (1999), 'La théorie de l'oecumène limité et la revision du *Constitutum Constantini*' [The Theory of the Limited *Oecumene* and the Revision of the *Constitutum Constantini*], in A. Dzhurova and G. Bakalov (eds), *Obshchoto i spetsifichnoto v balkanskite kulturi do kraya na XIX vek. Sbornik v chest na prof. Vasilka Tupkova-Zaimova*, Sofia: Ivanovo State University Press, pp. 119–22.

Lounghis, T. (1997), 'Επιθεώρηση ενόπλων δυνάμεων πριν από εκστρατεία' [Inspection of Armed Forces before a Campaign], in Tsiknakes (ed.), *Byzantium at War*, pp. 93–110.

Lounghis, T. (1995), 'Die byzantinische Ideologie der "begrenzten Ökumene" und die römische Frage im ausgehenden 10. Jahrhundert', *Byzantinoslavica*, 56, pp. 117–28.

Lounghis, T. (1993), *Η Ιδεολογία της Βυζαντινής Ιστοριογραφίας* [The Ideology of Byzantine Historiography], Athens: Herodotos.

Lounghis, T. (1990), *Κωνσταντίνου Ζ΄ Πορφυρογέννητου De Administrando Imperio (Προς τον ίδιον υιόν Ρωμανόν). Μια μέθοδος ανάγνωσης* [Constantine VII Porphyrogenitus' *De Administrando Imperio* (To my own son Romanus): A Reading Method], Thessaloniki: Vanias.

Luttwak, E. (2009), *The Grand Strategy of the Byzantine Empire*, Cambridge, MA: Harvard University Press.

Lykaki, M. (2016), 'Οι Αιχμάλωτοι Πολέμου στη Βυζαντινή Αυτοκρατορία (6ος–11ος αι.): Εκκλησία, Κράτος, Διπλωματία και Κοινωνική Διάσταση'

[Prisoners of War in the Byzantine Empire (6th–11th c.): Church, State, Diplomacy and Social Dimensions], unpublished PhD thesis, University of Athens.

Lynn, J. A. (2003), 'Chivalry and *Chevauchée*: The Ideal, the Real, and the Perfect in Medieval European Warfare', in J. A. Lynn (ed.), *Battle: A History of Combat and Culture*, Philadelphia, PA: Westview, pp. 73–110.

McCormick, M. (1986), *Eternal Victory: Triumphal Rulership in Late Antiquity: Byzantium, and the Early Medieval West*, Cambridge: Cambridge University Press.

MacDonald, J. (1988), *Great Battlefields of the World*, New York: Michael Joseph.

McGeer, E. (2003), 'Two Military Orations of Constantine VII', in J. W. Nesbitt (ed.), *Byzantine Authors: Literary Activities and Preoccupations*, Leiden: Brill, pp. 111–38.

McGeer, E. (1995), *Sowing the Dragon's Teeth: Byzantine Warfare in the Tenth Century*, Washington, DC: Dumbarton Oaks.

McGeer, E. (1995), 'The Legal Decree of Nikephoros II Phokas Concerning Armenian *Stratiotai*', in Miller and Nesbitt (eds), *Peace and War in Byzantium*, pp. 123–37.

McGeer, E. (1992), 'The Syntaxis Armatorum Quadrata: A Tenth-Century Tactical Blueprint', *Revue des études byzantines*, 50, pp. 219–29.

McGeer, E. (1986–7), 'Μεναύλιον – Μεναύλατοι', *ΔΙΠΤΥΧΑ*, 4, pp. 53–8.

McGrath, S. (1995), 'The Battles of Dorostolon (971): Rhetoric and Reality', in Miller and Nesbitt (eds), *Peace and War in Byzantium*, pp. 152–64.

Madgearu, A. (2013), *Byzantine Military Organization on the Danube, 10th–12th Centuries*, Leiden: Brill.

Magdalino, P. (2010), 'Byzantium = Constantinople', in L. James (ed), *A Companion to Byzantium*, Oxford: Blackwell, pp. 43–54.

Magdalino, P. (2004), 'In Search of the Byzantine Courtier: Leo Choirosphaktes and Constantine Manasses', in H. Maguire (ed.), *Byzantine Court Culture from 829 to 1204*, Washington, DC: Dumbarton Oaks.

Magdalino, P. (1989), 'Honour among *Romaioi*: The Framework of Social Values in the World of Digenes Akrites and Kekaumenos', *Byzantine and Modern Greek Studies*, 13, pp. 183–218.

Mahan, A. T. (1892), *The Influence of Sea Power Upon History, 1660–1783*, London: Little, Brown.

Mango, C. (1985), 'On the Re-reading of the Life of St. Gregory the Decapolite', *Βυζαντινά*, 13, pp. 633–46.

Mango, C. (1973), 'Eudocia Ingerina, the Normans, and the Macedonian Dynasty', *Zbornik Radova Vizantološkog Instituta*, 14–15, pp. 17–27.

Mango, C. (1965), 'Byzantinism and Romantic Hellenism', *Journal of the Warburg and Courtauld Institutes*, 28, pp. 29–43.

Maniati-Kokkini, T. (1997), 'Η επίδειξη ανδρείας στον πόλεμο κατά τους ιστορικούς του 11ου και 12ου αι.' [Demonstrating Bravery in War According to the Historians of the 11th and 12th c.] in Tsiknakes (ed.), *Byzantium at War*, pp. 239–59.

Mann, M. (2012), *The Sources of Social Power: A History of Power from the Beginning to AD 1760*, Cambridge: Cambridge University Press.

Markopoulos, A. (2012), 'The Ideology of War in the Military Harangues of Constantine VII Porphyrogennetos', in Koder and Stouraites (eds), *Byzantine War Ideology*, pp. 47–57.

Markopoulos, A. (2003), 'Byzantine History Writing at the End of the First Millennium', in P. Magdalino (ed.), *Byzantium in the Year 1000*, Leiden: Brill, pp. 183–97.

Markopoulos, A. (2000), 'Ζητήματα κοινωνικού φύλου στον Λέοντα τον Διάκονο' [Issues of Gender in Leo the Deacon], in S. Kaklamanes and A. Markopoulos (eds), *Ενθύμησις Νικολάου Μ. Παναγιωτάκη* [Studies in Memory of Nikolaos M. Panagiotakes], Heraklion: Crete University Press.

Markopoulos, A. (1985), 'Theodore Daphnopates et la continuation de Theophane', *Jahrbuch der Österreichischen Byzantinistik*, 35, pp. 171–82.

Markopoulos, A. (2008), 'Education', in Jeffreys et al. (eds), *Oxford Handbook of Byzantine Studies*, pp. 785–95.

May, T. (2007), *The Mongol Art of War: Chinggis Khan and the Mongol Military System*, Barnsley: Pen & Sword Military.

Mayor, A. (2006), *Υγρόν Πυρ, Δηλητηριώδη Βέλη και Σκορπιοί–Βόμβες, Βιολογικά και Χημικά Όπλα στον Αρχαίο Κόσμο* [Greek Fire, Poisonous Arrows and Scorpion Bombs: Biological and Chemical Weapons in the Ancient World], trans. Annita Gregoriadou, Athens: Enalios.

Mercier, M. (1952), *Le feu gregéois, les feux de guerre depuis l'antiquité, la poudre à canon*, Paris: Paul Geuthner.

Metcalfe, A. (2009), *The Muslims of Medieval Italy*, Edinburgh: Edinburgh University Press.

Miles, G. C. (1964), 'Byzantium and the Arabs: Relations in Crete and the Aegean Area', *Dumbarton Oaks Papers*, 18, pp. 1–32.

Millar, F. (1988), 'Government and Diplomacy in the Roman Empire during the First Three Centuries', *International History Review*, 10, pp. 345–77.

Miller, T., and J. Nesbitt (eds), *Peace and War in Byzantium*, Washington, DC: Catholic University of America Press.

Miotto, M. (2015), 'Ααουάσιμ και Θουγούρ, το στρατιωτικό σύνορο του Χαλιφάτου στην ανατολική Μικρά Ασία' [*Awāṣim* and *Tuġūr*: The Military Frontier of the Islamic State (Caliphate) in Eastern Anatolia], *Byzantiaka*, 32, pp. 133–56.

Moffatt, A. (1979), 'Early Byzantine School Curricula and a Liberal Education', in I. Dujcev (ed.), *Byzance et les slaves: mélanges Ivan Dujcev*, Paris: Association des amis des études archéologiques, pp. 275–88.

Montagu, J. D. (2006), *Greek and Roman Warfare: Battles, Tactics, and Trickery*, London: Greenhill.

Moore, R. L. (2002), 'The Art of Command: The Roman Army General and His Troops, 135 BC–AD 138', PhD dissertation, University of Michigan.
Moosa, M. (1969), 'The Relation of the Maronites of Lebanon to the Mardaites and Al-Jarājima', *Speculum*, 44, pp. 597–608.
Morillo, S. (2013), 'Justifications, Theories and Customs of War', in D. Graff (ed.), *The Cambridge History of War*, Cambridge: Cambridge University Press, pp. 1–24.
Morillo, S. (2013), *What is Military History?*, Cambridge: Cambridge University Press.
Morillo, S. (2006), 'Expecting Cowardice: Medieval Battle Tactics Reconsidered', *Journal of Medieval Military History*, 4, pp. 65–73.
Morillo, S. (2006), 'A General Typology of Transcultural Wars: The Early Middle Ages and Beyond', in H. H. Kortüm (ed.), *Transcultural Wars: From the Middle Ages to the 21st Century*, Berlin: Akademie Verlag, pp. 29–42.
Morillo, S. (2003), 'Battle Seeking: The Context and Limits of Vegetian Strategy', *Journal of Medieval Military History*, 1, pp. 21–41.
Morillo, S. (2001), 'Cultures of Death: Ritual Suicide in Medieval Europe and Japan', *The Medieval History Journal*, 4, pp. 241–57.
Morillo, S. (2001), 'Milites, Knights, and Samurai: Medieval Military Terminology and the Problem of Translation', in R. P. Abels and B. S. Bachrach (eds), *The Normans and Their Adversaries at War: Essays in Memory of C. Warren Hollister*, Woodbridge: Boydell & Brewer, pp. 167–84.
Morillo, S. (1999), 'The "Age of Cavalry" Revisited', in Kagay and Villalon (eds), *Circle of War in the Middle Ages*, pp. 45–58.
Morillo, S. (1996), *The Battle of Hastings: Sources and Interpretations*, Woodbridge: Boydell & Brewer.
Morillo, S., J. Black and P. Lococo (2009), *War in World History: Society, Technology and War from Ancient Times to the Present*, 2 vols, New York: McGraw-Hill Education.
Morris, R. (1994), 'Succession and Usurpation: Politics and Rhetoric in the Late Tenth Century', in P. Magdalino (ed.), *New Constantines: The Rhythm of Imperial Renewal in Byzantium, 4th–13th Centuries*, Aldershot: Ashgate.
Munir, M. (2010), 'Debates on the Rights of Prisoners of War in Islamic Law', *Islamic Studies*, 49, pp. 463–92.
Murphy, R. (1999), *Ottoman Warfare, 1500–1700*, London: UCL Press.
Murray, H. A. (1943), 'Analysis of the Personality of Adolph Hitler with Predictions of his Future Behavior and Suggestions for Dealing with him Now and After Germany's Surrender', Report delivered to the OSS, October 1943.
Naval Intelligence Division (1942), *Geographic Handbook Series: Syria*, London.
Naval Intelligence Division (1942), *Geographical Handbook Series: Turkey*, London.
Nerlich, D. (1999), *Diplomatische Gesandtschaften zwischen Ost- und Westkaisern 756–1002*, Bern: P. Lang.

Secondary Bibliography

Newth, M. A. (2005), *Heroes of the French Epic: A Selection of* Chansons de Geste, Woodbridge: Boydell & Brewer.

Nicolle, D. (2007), *Crusader Warfare, Muslims, Mongols and the Struggle against the Crusades*, London: Hambledon Continuum.

Nicolle, D. (1997), 'Arms of the Umayyad Era: Military Technology in a Time of Change', in Lev (ed.), *War and Society*, pp. 9–100.

Nicolle, D. (1993), *Armies of the Muslim Conquests*, London: Osprey.

Nielsen, J. S. (1991), 'Between Arab and Turk: Aleppo from the 11th till the 13th Centuries', *Byzantinische Forschungen*, 16, pp. 323–40.

Norris, H. T. (2009), 'The Sacred Sword of Maslamah B. "Abd Al-Malik"', *Oriente Moderno/Studies on Islamic Legends*, 89, pp. 389–406.

Odahl, C. M. (1976), 'Constantine and the Militarization of Christianity: A Contribution to the Study of Christian Attitudes toward War and Military Service', unpublished DPhil dissertation, University of California, San Diego.

Οντορίκο, Π. [Odorico, P.] (2010), *Ιωάννης Καμινιάτης, Ευστάθιος Θεσσαλονίκης, Ιωάννης Αναγνώστης: Χρονικά των αλώσεων της Θεσσαλονίκης*, trans. Χ. Μεσσής, Athens: AGRA.

Odorico, P. (2005), *Jean Caminiatès, Eustathe de Thessalonique, Jean Anagnostès – Thessalonique: chroniques d'une ville prise*, Toulouse: Anacharsis.

Oikonomides, N. (1997), 'Το όπλο του χρήματος' [Money as a Weapon], in Tsiknakes (ed.), *Byzantium at War*, pp. 261–8.

Oikonomides, N. (1995), 'The Concept of "Holy War" and Two Tenth-Century Byzantine Ivories', in Miller and Nesbitt (eds), *Peace and War in Byzantium*, pp. 62–86.

Oikonomides, N. (1979), 'L'épopée de Digénes et la frontière orientale de Byzance aux Xe et Xie siècles', *Travaux et mémoirs*, 7, pp. 375–97.

Oikonomides, N. (1974), 'L'organisation de la frontière orientale de Byzance aux Xe–XIe siècles et le Taktikon de l'Escorial', in Berza and Stănescu (eds), *Actes du XIVe Congrès international des études byzantines*, I, pp. 285–302.

Oikonomides, N. (1972), *Les listes de préséance byzantines des neuvième et dixième siècles*, Paris: Ed. du Centre national de la recherche scientifique.

Okwess-O'Bweng, K. (1988), 'Le portrait du soldat noir chez les Arabes et les Byzantins d'après l'anonyme "Foutouh al-Bahnasâ" et "De Expugnatione Thessalonicae" de Jean Caminiatès', *Βυζαντινός Δόμος*, 2, pp. 41–7.

Ostrogorsky, G. (1956), 'The Byzantine Emperor and the Hierarchical World Order', *Slavonic and East European Review*, 35, pp. 1–14.

Panagiotakes, N. M. (1996), 'Fragments of a Lost Eleventh-Century Byzantine Historical Work', in C. Constantinides, N. M. Panagiotakes, E. Jeffreys and A. D. Angelou (eds), *Philhellen: Studies in Honour of Robert Browning*, Venice: Istituto ellenico di studi bizantini e postbizantini di Venezia, pp. 321–57.

Panagiotakes, N. M. (1965), *Leon o Diakonos*, Athens: Association of Byzantine Studies.

Parry, V. J. (1970), 'Warfare', in P. M. Holt, A. K. S. Lambton and B. Lewis (eds) *The Cambridge History of Islam*, Cambridge: Cambridge University Press.

Parry, V. J., and M. E. Yapp (eds) (1975), *War, Technology and Society in the Middle East*, London: Oxford University Press.

Partington, J. R. (1999), *A History of Greek Fire and Gunpowder*, London: Johns Hopkins University Press.

Paterson, W. F. (1966), 'The Archers of Islam', *Journal of the Economic and Social History of the Orient*, 9, pp. 69–87.

Patoura, S. (1994), *Οι Αιχμάλωτοι ως Παράγοντες Επικοινωνίας και Πληροφόρησης* [Prisoners as Agents of Communication and Information], Athens: National Research Institute.

Patoura-Spanou, S. (2005), 'Όψεις της Βυζαντινής διπλωματίας' [Facets of Byzantine Diplomacy], in S. Patoura-Spanou (ed.), *Διπλωματία και Πολιτική, Ιστορικές Προσεγγίσεις* [Diplomacy and Politics: Historical Approaches], Athens: National Research Institute, pp. 131–64.

Pattenden, P. (1983), 'The Byzantine Early Warning System', *Byzantion*, 53, pp. 258–99.

Pertusi, A. (1974), 'Tra storia e leggenda: Akritai e Ghazi sulla frontiera orientale di Bisanzio', in Berza and Stănescu (eds), *Actes du XIVe Congrès international des études byzantines*, I, pp. 285–382.

Polemis, D. I. (1965), 'Some Cases of Erroneous Identification in the Chronicle of Skylitzes', *Byzantinoslavica*, 26, pp. 74–81.

Pritchett, W. K. (1971), *Ancient Greek Military Practices*, Berkeley: University of California Press.

Pryor, J. H. (2000), *Geography, Technology, and War: Studies in the Maritime History of the Mediterranean, 649–1571*, Cambridge: Cambridge University Press.

Pryor, J. H. (1984), 'Transportation of Horses during the Era of the Crusades, Eighth Century to 1285, Part I: to c. 1285', *Mariner's Mirror*, 70, pp. 9–27.

Pryor, J. H., and E. M. Jeffreys (2006), *The Age of the ΔΡΟΜΩΝ: The Byzantine Navy c. 500–1204*, Leiden: Brill.

Rahe, P. A. (1981), 'The Annihilation of the Sacred Band at Chaeronea', *American Journal of Archaeology*, 85, pp. 84–7.

Ramsay, W. M. (1972), *The Historical Geography of Asia Minor*, New York: John Murray.

Ramsay, W. M. (1903), 'Cilicia, Tarsus, and the Great Taurus Pass', *The Geographical Journal*, 22, pp. 357–410.

Rance, P. (2007), 'The Date of the Military Compendium of Syrianus Magister (formerly the Sixth-Century *Anonymous Byzantinus*)', *Byzantinische Zeitschrift*, 100, pp. 701–37.

Rance, P. (2004), 'Drungus, δρούγγος, and δρουγγιστί: A Gallicism and Continuity in Late Roman Cavalry Tactics', *Phoenix*, 58, pp. 96–130.

Reisch, G. A. (1991), 'Chaos, History, and Narrative', *History and Theory*, 30, pp. 1–20.

Secondary Bibliography

Richmond, J. A. (1988), 'Spies in Ancient Greece', *Greece & Rome*, 45, pp. 1–18.
Richter, M. (1994), *The Oral Tradition in the Early Middle Ages*, Turnhout: Stroud.
Rogers, C. J. (2008), 'The Battle of Agincourt', in L. J. A. Villalon and D. J. Kagay (eds), *The Hundred Years War (Part II): Different Vistas*, Leiden: Brill, pp. 37–131.
Rogers, C. J. (2006), 'Strategy, Operational Design, and Tactics', in J. C. Bradford (ed.), *International Encyclopaedia of Military History*, New York: Routledge.
Rogers, C. J. (2003), 'The Vegetian "Science of Warfare" in the Middle Ages', *Journal of Medieval Military History*, 1, pp. 1–19.
Rogers, C. J. (2000), *War Cruel and Sharp: English Strategy under Edward III, 1327–1360*, Woodbridge: Boydell & Brewer.
Rogers, C. J. (1999), 'Edward III and the Dialectics of Strategy, 1327–1360', in C. J. Rogers (ed.), *The Wars of Edward III: Sources and Interpretations*, Woodbridge: Boydell & Brewer, pp. 83–102.
Roland, A. (2008), 'Secrecy, Technology, and War: Greek Fire and the Defense of Byzantium, 678–1204', in J. France and K. DeVries (eds), *Warfare in the Dark Ages*, Aldershot: Ashgate, pp. 655–79.
Rolington, A. (2013), *Strategic Intelligence for the 21st Century: The Mosaic Method*, Oxford: Oxford University Press.
Rosser, J. (1976), 'John the Grammarian's Embassy to Baghdad and the Recall of Manuel', *Byzantinoslavica*, 37, pp. 168–71.
Roueché, C. (2000), 'Defining the Foreign in Kekaumenos', in D. C. Smythe (ed.), *Strangers to Themselves: The Byzantine Outsider*, Aldershot: Ashgate, pp. 203–14.
Runciman, S. (1977), *The Byzantine Theocracy*, Cambridge: Cambridge University Press.
Runciman, S. (1929), *The Emperor Romanus Lecapenus and his Reign: A Study of Tenth-Century Byzantium*, Cambridge: Cambridge University Press.
Russel, F. S. (1999), *Information Gathering in Classical Greece*, Ann Arbor: University of Michigan Press.
Rydén, L. (1984), 'The Portrait of the Arab Samonas in Byzantine Literature', *Graeco-Arabica*, 3, pp. 101–8.
Safi, O. (2006), *The Politics of Knowledge in Premodern Islam: Negotiating Ideology and Religious Inquiry*, Chapel Hill: University of North Carolina Press.
Sears, M. A., and C. Willekes (2016), 'Alexander's Cavalry Charge at Chaeronea, 338 BCE', *Journal of Military History*, 80, pp. 1017–35.
Setton, K. M. (1954), 'On the Raids of the Moslems in the Aegean in the Ninth and Tenth Centuries and their Alleged Occupation of Athens', *American Journal of Archaeology*, 58, pp. 311–19.
Ševčenko, I. (1979–80), 'Constantinople Viewed from the Eastern Provinces in the Middle Byzantine Period', *Harvard Ukrainian Studies*, 3/4, pp. 726–46.

Ševčenko, I. (1969–70), 'Poems on the Deaths of Leo VI and Constantine VII in the Madrid Manuscript of Scylitzes', *Dumbarton Oaks Papers*, 23/24, pp. 185–228.

Shean, J. F. (2010), *Soldiering for God: Christianity and the Roman Army*, Leiden: Brill.

Shepard, J. (2002), 'Emperors and Expansionism: From Rome to Middle Byzantium', in D. Abulafia and N. Berend (eds), *Medieval Frontiers: Concepts and Practices*, Aldershot: Ashgate, pp. 55–82.

Shepard, J. (2001), 'Constantine VII, Caucasian Openings and the Road to Aleppo', in A. Eastmond (ed.), *Eastern Approaches to Byzantium*, Aldershot: Ashgate, pp. 19–40.

Shepard, J. (1995), 'Imperial Information and Ignorance: A Discrepancy', *Byzantinoslavica*, 56, pp. 107–16.

Shepard, J. (1992), 'Byzantine Diplomacy, A.D. 800–1204: Means and Ends', in Shepard and Franklin (eds), *Byzantine Diplomacy*, pp. 41–71.

Shepard, J. (1985), 'Information, Disinformation and Delay in Byzantine Diplomacy', *Byzantinische Forschungen*, 10, pp. 233–93.

Shepard, J., and S. Franklin (eds) (1992), *Byzantine Diplomacy: Papers from the Twenty-Fourth Spring Symposium of Byzantine Studies*, Aldershot: Ashgate.

Shepherd Creasy, E. (1863), *The Fifteen Decisive Battles of the World: From Marathon to Waterloo*, New York: A. L. Burt.

Sidnell, P. (2006), *Warhorse: Cavalry in Ancient Warfare*, London: Continuum.

von Sievers, F. (1979), 'Military, Merchants and Nomads: The Social Evolution of the Syrian Cities and Countryside during the Classical Period, 780–969/164–358', *Der Islam*, 56, pp. 212–44.

Simeonova, L. V. (2000), 'Foreigners in Tenth-Century Byzantium: A Contribution to the History of Cultural Encounter', in D. C. Smythe (ed.), *Strangers to Themselves: The Byzantine Outside*, Aldershot: Ashgate, pp. 229–44.

Simeonova, L. V. (1998), 'In the Depths of Tenth-Century Byzantine Ceremonial: The Treatment of Arab Prisoners of War at Imperial Banquets', *Byzantine and Modern Greek Studies*, 22, pp. 75–104.

Siuziumov, M. I. A. (1916), 'Ob istochnikakh L'va D'iakona', *Vizantiiskoe obozrenie*, 2, pp. 106–66.

von Sivers, P. (1982), 'Taxes and Trade in the Abbāsid *Thughūr*, 750–962/133–351', *Journal of the Economic and Social History of the Orient*, 25, pp. 71–99.

Snyder, J. (2002), 'Anarchy and Culture: Insights from the Anthropology of War', *International Organization*, 56, pp. 7–45.

Soeters, J. L., D. J. Winslow and A. Wibull (2003), 'Military Culture', in G. Caforio (ed.), *Handbook of the Sociology of the Military*, New York: Springer, pp. 237–54.

Spence, I. G. (1993), *The Cavalry of Classical Greece: A Social and Military History with Particular Reference to Athens*, Oxford: Oxford University Press.

Sperl, S., and C. Shackle (eds) (1996), *Qasida Poetry in Islamic Asia and Africa*, Leiden: Brill.

Secondary Bibliography

Stenkevych, S. P. (1996), 'Abbasid Panegyric and the Poetics of Political Allegiance: Two Poems of Al-Mutanabbi on Kafur', in Sperl and Shackle (eds), *Qasida Poetry in Islamic Asia and Africa*, pp. 36–63.
Stephenson, I. P. (2006), *Romano-Byzantine Infantry Equipment*, Stroud: Tempus.
Stephenson, P. (2007), 'Imperial Christianity and Sacred War in Byzantium', in J. K. Wellman (ed.), *Belief and Bloodshed: Religion and Violence Across Time and Tradition*, Plymouth: Rowman & Littlefield.
Stern, S. M. (1950), 'An Embassy of the Byzantine Emperor to the Fatimid Caliph Al-Mu'izz', *Byzantion*, 20, pp. 239–53.
Stouraites, I. (2012), 'Conceptions of War and Peace in Anna Comnena's *Alexiad*', in Koder and Stouraitis (eds), *Byzantine War Ideology*, pp. 69–80.
Stouraites, I. (2012), '"Just War" and "Holy War" in the Middle Ages: Rethinking Theory through the Byzantine Case-Study', *Jahrbuch der Österreichischen Byzantinistik*, 62, pp. 227–64.
Stouraites, I. (2011), '*Jihād* and Crusade: Byzantine Positions towards the Notions of "Holy War"', *Byzantina Symmeikta*, 21, pp. 11–63.
Stouraites, I. (2009), *Krieg und Frieden in der politischen und ideologischen Wahrnehmung in Byzanz*, Vienna: Fassbaender.
le Strange, G. (1897), 'A Greek Embassy to Baghdad in 917 A.D.', *Journal of the Royal Asiatic Society of Great Britain and Ireland*, pp. 35–45.
Strickland, M. (1996), *War and Chivalry: The Conduct and Perception of War in England and Normandy, 1066–1217*, Cambridge: Cambridge University Press.
Strickland, M., and R. Hardy (2005), *From Hastings to the Mary Rose: The Great Warbow*, Stroud: Sutton Publishing.
Sumberg, L. A. M. (1968), *La Chanson d'Antioche, étude historique et littéraire*, Paris: Picard.
Tantum, G. (1979), 'Muslim Warfare: A Study of a Medieval Muslim Treatise on the Art of War', in R. Elgood (ed.), *Islamic Arms and Armour*, London: Scolar Press.
Tarn, W. W. (1948), *Alexander the Great*, Cambridge: Cambridge University Press.
Teall, J. L. (1959), 'The Grain Supply of the Byzantine Empire, 330–1025', *Dumbarton Oaks Papers*, 13, pp. 87–139.
Theotokis, G. (2015), 'Promoting the Newcomer: Myths, Stereotypes, and Reality in the Norman Expansion in Italy during the XIth Century', *Porphyra*, 24, pp. 28–38.
Theotokis, G. (2014), *The Norman Campaigns in the Balkans, 1081–1108 AD*, Woodbridge: Boydell & Brewer.
Theotokis, G. (2014), 'From Ancient Greece to Byzantium: Strategic Innovation or Continuity of Military Thinking?', in B. Kukjalko, I. Rūmniece and O. Lāms (eds), *Antiquitas Viva 4: Studia Classica*, Riga: University of Latvia Press, pp. 106–18.
Theotokis, G. (2012), 'Rus, Varangian and Frankish Mercenaries in the Service of the Byzantine Emperors (9th–11th c.) – Numbers, Organisation and Battle

Tactics in the Operational Theatres of Asia Minor and the Balkans', *Byzantina Symmeikta*, 22, pp. 125–56.

Theotokis, G. (2010), 'The Norman invasion of Sicily, 1061–1072: Numbers and Military Tactics', *War in History*, 17, pp. 381–402.

Tlusty, A. (2002), 'The Public House and Military Culture in Germany, 1500–1648', in A. Tlusty and B. Kumin (eds), *The World of the Tavern: Public Houses in Early Modern Europe*, Aldershot: Ashgate, pp. 136–59.

Tobias, N. (2007), *Basil I, Founder of the Macedonian Dynasty: A Study of the Political and Military History of the Byzantine Empire in the Ninth Century*, New York: Edwin Mellen Press.

Tor, D. G. (2007), *Violent Order, Religious Warfare, Chivalry and the Ayyar Phenomenon in the Medieval Islamic World*, Würzburg: Ergon.

Tougher, S. (1998), 'The Imperial Thought-World of Leo VI, the Non-campaigning Emperor of the Ninth Century', in L. Brubaker (ed.), *Byzantium in the Ninth Century: Dead or Alive?*, Aldershot: Ashgate, pp. 51–60.

Tougher, S. (1997), *The Reign of Leo VI (886–912): Politics and People*, Leiden: Brill.

Toumanoff, C. (1971), 'Caucasia and Byzantium', *Traditio*, 27, pp. 111–58.

Toynbee, A. (1973), *Constantine Porphyrogenitus and His World*, London: Oxford University Press.

Treadgold, W. (2015), 'The Formation of a Byzantine Identity', in M. B. P. Maleon and A. E. Maleon (eds), *Studies in Byzantine Cultural History*, Bucharest: Editura Academiei Române, pp. 315–37.

Treadgold, W. (2013), *The Middle Byzantine Historians*, Basingstoke: Palgrave Macmillan.

Treadgold, W. (2006), 'Byzantium, the Reluctant Warrior', in N. Christie and M. Yazigi (eds), *Noble Ideals and Bloody Realities: Warfare in the Middle Ages, 378–1492*, Leiden: Brill, pp. 209–33.

Treadgold, W. (2005), 'Standardized Numbers in the Byzantine Army', *War in History*, 12, pp. 1–14.

Treadgold, W. (1997), *A History of the Byzantine State and Society*, Stanford, CA: Stanford University Press.

Treadgold, W. (1995), *Byzantium and its Army, 284–1081*, Stanford, CA: Stanford University Press.

Treadgold, W. (1992), 'The Army in the Works of Constantine Porphyrogenitus', *Rivista di Studi Bizantini e Neoellenici*, 29, pp. 77–162.

Treadgold, W. (1989), 'On the Value of Inexact Numbers', *Byzantinoslavica*, 50, pp. 57–61.

Treadgold, W. (1983), 'Remarks on the Work of al-Jarmī on Byzantium', *Byzantinoslavica*, 44, pp. 205–12.

Treadgold, W. (1980), 'Notes on the Numbers and Organization of the Ninth-Century Byzantine Army', *Greek, Roman and Byzantine Studies*, 21, pp. 269–77.

Secondary Bibliography

Trombley, F. R. (2002), 'Military Cadres and Battle during the Reign of Heraclius', in G. J. Reinink and B. H. Stolte (eds), *The Reign of Heraclius (610–641): Crisis and Confrontation*, Leuven: Peeters, pp. 241–59.
Tsaras, J. (1988), 'Η αυθεντικότητα του Χρονικού του Ιωάννου Καμινιάτη' [The Authenticity of the Chronicle of Ioannes Kaminiates], *Βυζαντιακά*, 8, pp. 43–58.
Tsiknakes, K. (ed.), *Το Εμπόλεμο Βυζάντιο (9ος–12ος αι.) = Byzantium at War (9th–12th c.)*, Athens: National Research Foundation.
Tsougarakes, D. (1988), *Byzantine Crete: From the 5th Century to the Venetian Conquest*, Athens: St. D. Basilopoulos.
Tyldesley, J. (2000), *Ramesses II: Egypt's Greatest Pharaoh*, London: Penguin.
The U.S. Army & Marine Corps Counterinsurgency Field Manual, U.S. Army Field Manual No. 3-24, Marine Corps Warfighting Publication No. 3–33.5 (2007), Chicago, IL: The University of Chicago Press.
Vasiliev, A. A. (1932), 'Härün-ibn Yahya and his Description of Constantinople', *Seminarium Kondakovianum*, 5, pp. 149–63.
Vasiliev, A. A. (ed.), and M. Canard (trans.) (1935), *Byzance et les Arabes, 867–959*, 2 vols, Madison, WI: Institut de philologie et d'histoire orientales.
Verbruggen, J. F. (1997), *The Art of Warfare in Western Europe during the Middle Ages from the Eighth Century to 1340*, Woodbridge: Boydell & Brewer.
Vest, B. A. (2007), *Geschichte der Stadt Melitene und der umliegenden Gebiete: Vom Vorabend der arabischen bis zum Abschluss der türkischen Eroberung (um 600–1124)*, Hamburg: Armenian Research Center.
Walter, B. (2011), 'Urban Espionage and Counterespionage during the Burgundian Wars (1468–1477)', *Journal of Medieval Military History*, 9, pp. 132–43.
Walker, P. E. (1972), 'A Byzantine Victory over the Fatimids at Alexandretta (971)', *Byzantion*, 42, pp. 431–40.
van Wees, H. (2004), *Greek Warfare, Myths and Realities*, London: Bloomsbury Academic.
Whately, C. (2016), *Battles and Generals: Combat, Culture, and Didacticism in Procopius' Wars*, Leiden: Brill.
Wheatley, P. (2000), *The Places Where Men Pray Together: Cities in Islamic Lands, Seventh Through the Tenth Centuries*, Chicago, IL: University of Chicago Press.
Wheeler, E. L. (1988), 'The Modern Legality of Frontinus' Stratagems', *Militargeschichtliche Mitteilungen*, 44, pp. 7–29.
Wheeler, E. L. (1988), 'Πολλά τα κενά του πολέμου: The History of a Greek Proverb', *Greek, Roman and Byzantine Studies*, 29, pp. 153–84.
Whetham, D. (2008), 'The English Longbow: A Revolution in Technology?', in L. J. A. Villalon and D. J. Kagay (eds), *The Hundred Years War*, part 2: *Different Vistas*, Leiden: Brill, pp. 213–30.
Whittow, M. (1996), *The Making of Byzantium, 600–1025*, Berkeley: University of California Press.

Whittow, M. (2009), 'The Political Geography of the Byzantine World: Geographical Survey', *OHBS*, pp. 219–31.
Wiita, J. E. (1977), 'The Ethnika in Byzantine Military Treatises', PhD dissertation, University of Minnesota.
Wilson, P. H. (2008), 'Defining Military Culture', *Journal of Military History*, 72, pp. 11–41.
Wojnowski, M. (2012), 'Periodic Revival or Continuation of the Ancient Military Tradition? Another Look at the Question of the *Katafraktoi* in the Byzantine Army', *Studia Ceranea*, 2, pp. 195–220.
Worley, L. J. (1994), *Hippeis: The Cavalry of Ancient Greece*, Oxford: Oxford University Press.
Zuckerman, C. (1990), 'The Military Compendium of Syrianus Magister', *Jahrbuch der Österreichischen Byzantinistik*, 40, pp. 209–24.

Index

Abasgia, 71
Abbasid(s), 72, 74, 87, 149, 163, 164, 166, 170, 174, 177, 204, 243, 244, 246
  administration, 172
  army(ies), 56, 61, 250
  -Byzantine peace treaty of 938, 96
  Caliphate, 4, 14, 55, 154, 156, 165
Abu Firas, 17, 91, 96 (n15), 103 (n83), 168, 188 (n117), 236, 244, 245, 246, 254, 255, 259, 261, 265, 274 (n139), 277, 305
Abu Kafur, 244
Abu'l Asair, 93
Adata (Hadath), 60, 61, 62, 71, 81, 89, 277
  Battle of Hadath, 3, 17, 86, 89, 91, 192, 254, 255, 260, 265, 276, 277, 278, 289, 306
  Leo Phocas' siege of, 260
Aelian, 13, 25, 197, 199, 208
Aeneas Tacticus, 12, 25, 129, 151, 152, 160
Afshin, 61, 167
Agathias of Myrina, 176, 237, 251
aḥdāth, 292
Akroinon, Battle of, 56
Alakasseus, John, 263, 280, 281

al-Ansari, 36, 134, 144 (n26), 230, 232 (n19), 288
al-Aziz, 202, 228
al-Dawadari, Abu Bakr ibn, 280, 293
Aleppo, city, 1, 10, 12, 52, 74, 85, 87, 88, 93, 94, 95, 121, 166, 180, 180 (n2), 244, 245, 286, 287
  emirate, 2, 12, 86, 88, 90, 95, 118
Alexander, emperor, 71, 102 (n60)
Alexander the Great, 134, 142 (n10), 155, 213 (n29), 218 (n105)
Alexandretta, Battle of, 3, 17, 192, 276, 279, 280, 285, 293, 306
al-Hakim, 242
al-Maqrizi, 279, 286
al-Muᶜtamid, 71, 156
al-Mutanabbi, 17, 73, 87, 88, 89, 90, 91, 236, 244, 245, 246, 247, 259, 260, 261, 265, 277, 305
al-Qalanisi, 286, 287, 289
al-Samsana, Jaysh ibn, 287, 288
Amatus of Montecassino, 34, 39, 161
Amisos, 56, 62
Amorion, 56, 140, 156

Anazarbos, 61, 257
Ancyra, 56
Anemas, 37, 174, 262, 290
Anonymous Sylloge Taktikorum
    (c. 930), 3, 16, 105, 174, 192,
    193, 196, 197, 198, 199, 200,
    204, 206, 207, 209, 210, 219,
    223, 224, 228, 229, 230, 249,
    290, 291, 304
Apamea, city, 172, 285, 287
  Battle of, 3, 17, 192, 276,
    290, 306
Araxes River, 58
Arian, 197
Arkadiopolis, Battle of, 241, 253,
    256, 258, 263, 276, 281, 282,
    293, 294, 306
Armeniakon, theme, 56, 65 (n23),
    77, 139, 173
Armenian chapters, 75, 188 (n114)
Armenian (Ἀρμενι[α]κά) themes,
    201
Armosata, 61, 73, 173
arrada, 204
Arzes, 69, 72, 76
Ascalon, 135, 169, 231 (n7)
Asclepiodotus, 197, 208, 212
    (n22), 223, 224
Ashot I, 71, 72, 73
Athanasius of Lavra, 239
Attaleia, 56, 61, 158, 169, 174
Attaleiates, 222
ᶜAwāṣim, 52

bandum, 206
Bangratids, 71, 75, 77
Banu Taghlib, 87
Bar Hebraeus, 285, 286

Basil I, emperor, 58, 68 (n51),
    69, 70, 78, 79, 80, 83, 93,
    97 (n33), 98 (n34), 100 (n46),
    101 (n60), 102 (n63),
    109, 166
Basil II, emperor, 28, 40, 96
    (n19), 117, 202, 227, 237,
    238, 240, 241, 243, 248,
    285, 296 (n58)
Basil Hexamilites, 94
Basil the Nothos, 237
Basil Parakoimomenos, 255
Bithynia, 56, 57, 59, 62
Bitlis, 72, 73, 76, 79
Bohemond of Taranto, 136, 137,
    161, 175
Boulgarofygon, Battle of, 167,
    292
Bourgtheroulde, Battle of, 226
Bourtzes Michael, 285, 286, 287
Brémule, Battle of, 31, 226
Bryennius Nicephorus, 252
buffer zone(s), 7, 8, 10, 55, 56,
    69, 76, 87, 171, 173
Buyids of Persia and Iraq, 4, 74

Caesarea (modern Kayseri), 59,
    61, 62, 259
Calabria, 38, 39, 40, 70
calcatio, 93
Cappadocia, 52, 57, 58, 59, 60,
    61, 62, 63, 77, 93, 98 (n34),
    138, 157, 166, 167, 173, 254,
    256, 257, 259
catalyst [in war and/or in battle],
    219, 305
Caucasus, 7, 59, 71, 72, 76, 203
Chaldea, 1, 76

Index

Chandax, 81, 82, 113, 117, 137, 252, 256, 261, 266
Chanson d'Antioche by Richard le Pèlerin, 30
Charsianon, 57, 61, 62, 65 (n23), 157, 167, 168, 254, 260
   castle, 259
cheiromaggana [χειρομάγγανα], 203
chelandia, 117
chevauchée, 9, 21 (n29)
chiliarch (taxiarch), 194, 198, 212 (n12)
chronicle of Pseudo-Symeon, 238
Chrysocheir, 70
Cilician Gates, 61, 78
clibanarii, 7
Coloneia, 73
Comnena, Anna, 34, 48 (n68), 102 (n63), 175, 251
Comnenus, Alexius, 136, 137, 175, 228, 239
Comnenus, Isaac, 33
Constantine V, emperor, 149, 154
Constantine VII, emperor, 2, 13, 15, 33, 69, 70, 71, 72, 74, 75, 76, 79, 80, 81, 83, 85, 86, 87, 88, 90, 92, 93, 94, 95, 101 (n60), 115, 116, 118, 121, 122, 129, 153, 157, 161, 162, 169, 170, 176, 222, 227, 232 (n8), 240, 243, 251, 302
   De Administrando Imperio, 15, 33, 69, 71, 75, 76, 84, 115, 122, 161
Constantine IX, emperor, 35, 240
Cordoba, embassy, 159
   Umayyad Caliph of, 83, 162

Crete, 26, 81, 82, 83, 85, 88, 101 (n54), 113, 137, 150, 162, 175, 201, 214 (n54), 252, 261
Curcuas, John, 58, 72, 73, 74, 75, 76, 85, 92, 193, 201, 211, 221, 248, 254, 257, 305
Curcuas, Theophilus, 58, 72, 73
Cyprus, 77, 82, 101 (n54), 153, 181 (n18), 201
Cyropaedia, 24

Dalassinos, Damianos, 287, 288, 289, 290
Dalmatia, theme, 70, 82
Dār al-Islām, 52, 159
Daylami, 73, 202, 204, 205, 215 (n57), 228, 244, 279, 285, 286, 287, 288, 289, 295 (n39)
Dazimon, 56, 70
de Charny, Geoffrey, 30
decisive battle, 2, 6, 7, 8, 9, 10, 39, 53, 279, 283, 302
Digenes Akritas, 78, 139, 141, 171
dikaios polemos (just war), 41, 112, 114, 120
Dorylaion (Dorylaeum), 55, 62
Dorystolon, Battle of, 37, 174, 192, 247, 250, 253, 257, 258, 262, 264, 266, 276, 282, 290, 291, 292, 293, 306
doukatores (δουκάτορες) *see minsouratores* (μινσουράτορες)
Dyrrachium, Battle of, 136, 161, 228
   siege of, 175

Euphrates, 60, 61, 68 (n47), 70, 72, 78, 79, 84, 85, 88, 102, 259, 260
excubitai, 80
expilatores (εκσπηλατόρες), 138

Fatimids of Egypt and North Africa, 2, 4, 10, 83, 116, 123 (n20), 162, 202, 220, 228, 252, 279, 280, 286, 288
Fertile Crescent, 9, 60
fire signals, 133
formation *cuneus*, 208
formation double-faced, 196, 197, 206, 207, 219, 222, 223, 226, 229
formation rhomboid (*ρομβοειδεῖ*), 208
formation wedge (*τρίγωνος παράταξις*), 198, 207, 208, 211, 218 (n105), 227
formation *αμφίστομος*, 196, 219
formation *αντίστομος*, 219
formation *φοσσατικώς*, 222
Frontinus, Sextus Julius, 25, 136, 160, 162, 205, 213 (n29), 224, 233 (n25), 290, 304
Futuh, 279, 280

Gagic Arsdrouni, 71
Genesius, Joseph, 240
gens Normannorum, 37
Germanikeia, 60, 61, 71, 93, 156
ghazis, 292
ghulam, 73, 225, 228, 254, 255, 265, 277, 287, 288
Greek Fire – liquid fire, 203, 206

Hadath *see Adata*
Hamdanid
  armies, 4, 10, 95, 260, 265, 286
  dynasty, 3, 14, 15, 58, 75, 76, 86, 87, 92, 93, 115, 122, 164, 168, 173, 220, 231, 244, 245, 299
hand-pump [*χειροσίφουνα*], 203
Hauteville, Robert 'Guiscard' (the Cunning), 37, 38, 50 (n85)
Hauteville, Roger, 37, 38, 39, 40, 50 (n85)
Heraclius, 7, 8, 26, 78
Herodotus, 5, 236, 237, 238
Hexamilites, Basil, 94

Ibn Zafir, 17, 52, 73, 91, 236, 245, 246, 254, 255, 259, 265, 277
Ibn Khaldun, 36, 221, 230, 272 (n94)
  Muqaddimah, 220
Ibn Miskawaih, 243
Ikhshidids of Egypt, 81, 87, 88, 202
Ikonion (modern Konya), 59, 140
iqta, 220
iron caltrops (*τρίβολος*), 137

Jafar ibn Falah, 279
jihad, 14, 29, 52, 74, 82, 88, 90, 114, 120, 302
Joinville, *Life of St Louis*, 205
Just War *see dikaios polemos*

Kalbite Muslims of Sicily, 35
Kalikala, Umayyad Emirate of, 58
Kaminiates, Ioannes, 15, 112, 113, 116, 117, 121, 170

Index

karr wa farr, 221, 232 (n11)
kataphraktoi, 91, 92, 108, 174,
   199, 207, 211, 217 (n94), 227,
   229, 230, 248, 249, 262, 265,
   270 (n71), 277, 278, 282, 289,
   291, 304, 306
Kekaumenus, 65 (n13), 78,
   139–41, 150, 160
   Strategikon, 78, 140–1, 171
Khliat, 61, 69, 72, 75, 76
kleisourai, 55, 56, 65 (n23), 71,
   75, 141, 201, 257
Kritorikios of Taron, 75
Kutama Berber, 287

Leo IV, emperor, 55, 78
Leo VI 'the Wise', emperor, 3, 6,
   14, 25, 31, 66 (n32), 71, 77,
   79, 101 (n2), 118, 132, 151,
   163, 166, 167, 170, 174, 176,
   177, 201, 204, 247, 290
   Επαρχικόν Βιβλίον, 151
   Taktika, 3, 6, 11, 15, 31, 108,
   109–12, 118, 121, 143 (n13),
   166, 174, 178, 193, 194–200,
   206, 210, 219, 222, 224, 229,
   250, 291, 304
Leo of Tripoli, 82
Leo the Deacon, 14, 17, 37, 81,
   112, 113, 121, 201, 236, 240,
   241–61, 264, 265, 278, 282,
   283–90, 293, 305
limitanei, 55, 139, 145 (n51)
Livre de Chevalerie, by Geoffrey
   de Charny, 30
Loulon, 61, 62
Luʾluʾ al-Kabir, 286
Lykandos, 60, 61, 71, 72

Magyars, 7, 33, 109, 253, 257
Malagina, 56
Malaterra, Geoffrey, 37–40
Maniakes, George, 35
manjaniq or *mangonel*, 204
Manjutakin, 285–7
Mansourah, Battle of, 205
Manzikert, Battle of, 6, 7, 176
   city, 58, 69, 73, 75
Maurice, emperor, 6, 13, 106, 251
   Strategikon, 6, 13, 25, 106,
   107–11, 130, 135, 175, 178,
   193, 196–9, 206, 207–10, 212
   (n21), 219, 224, 227, 229, 273
   (n95), 291, 292
Mayyafariqin, 69, 73, 74, 157
Melitene (modern Malatya), 13,
   26, 57, 58–63, 70, 72, 73–5,
   80, 120, 163, 173, 176, 201,
   257, 259, 260, 285, 301
   emir of, 56, 72, 125 (n47), 167
Menander the Guardsman, 251
menavlatoi, 198, 199, 211, 227,
   228, 261, 304
Michael III, emperor, 56, 80, 100
   (n46), 166, 167
military culture(s), 2, 5, 9, 10,
   11, 17, 31, 41, 46 (n40), 171,
   299, 300
minsouratores (μινσουράτορες),
   133, 134
Mirdasids, 202
Mleh, 72, 73, 201
monokoursa, 54
Mopsuestia, 10, 90, 95, 258
Mosul, city, 73, 78, 87, 88,
   172, 173
   emirate, 4, 87, 288

mubarizan, 37
Mysticus, Nicolas, 72, 123 (n18), 187 (n106)

Nadja al-Kasaki, 255
naffatun, 204, 205
Nasir ad-Dawla, 14, 73, 75, 87, 88, 173, 245
New Military History, 11
Nicaea, 62, 70
Nicephorus the Deacon, 240
Nicetas the Paphlagonian, 240
Nicomedia, 70, 157
Nizam al-Mulk, 130, 135, 161, 164, 165
   The Book of Government or Rules for Kings, 130
normanitas, 37

Oikumene, 26, 34, 42 (n17), 84 ['limited Oikumene']
Onasander, 13, 25, 197, 223, 224, 290, 304
oplitarches (οπλιτάρχης), 227
Orderic Vitalis, 31, 226, 247, 249
Orontes, Battle of, 3, 17, 88, 192, 276, 285–9

Parsakoutenoi clan, 238
Pastilas, 137, 261
Pechenegs, 33–4
Perkri, 69, 72, 75, 76
Phocas, Bardas, 89, 91, 92, 201, 255, 259, 276, 277, 289
Phocas, Constantine, 254
Phocas, Leo, 86, 118, 167, 247, 255, 259, 260

Phocas, Nicephorus, 2, 16, 40, 52, 55, 66 (n32), 81, 85, 88, 91, 105, 110, 118, 136–8, 168, 174, 192, 194, 197–201, 203, 205, 207, 211, 219, 220, 226–31, 237, 239, 241, 243, 245, 249, 252, 254, 257, 261, 263, 277, 278, 284, 290, 291, 299, 304, 305
   Praecepta Militaria (c. 969), 2, 15, 16, 88, 91, 119, 120, 137, 192–206, 220–30, 256, 277, 281, 284, 286, 290, 291, 294, 304, 306
Pliska, Battle of, 282
Polyaenus, 13, 25
Procopius, 26, 45 (n34), 154, 181 (n6), 227, 236, 247, 251, 270 (n61), 296 (n56)
prokoursatores (προκουρσάτορες, lat. *procursor*), 134, 208, 230, 231, 235 (n48), 280
Psellus, Michael, 33, 64 (n6), 222, 227, 242

Qasida, 88, 244, 247

razzia (raid), 9, 13, 21 (n29), 58, 62, 74, 82, 139, 301
Romanus I Lecapenus, emperor, 13, 69, 70, 72, 73–9, 92, 98 (n34), 113, 129, 150, 173
Romanus II, emperor, 201, 237, 240, 243, 252
Romanus III, emperor, 243
Romanus IV 'Diogenes', 67 (n42)
Russian Primary Chronicle, 35

saqah [saka], 209, 211, 229, 230, 305
Scythian way of fighting, 108–9
Skleros, Bardas, 241, 243, 258, 263, 280, 285
skoulkatores (σκουλκάτορες), 133, 134
Skylitzes, John, 17, 92, 102 (n61), 118–22, 167, 236, 238, 240–43, 248, 250, 253–5, 257, 258, 263–5, 268 (n24), 279, 280–5, 305
Smbat, 71
Svyatoslav, 35, 136, 253, 257, 258, 283, 284, 292
Sygkellos, Georgios, 240
Symeon, Bulgarian tsar, 1, 71, 72, 167, 251
Symeon Logothetes, 238, 243
Syrianus Magistrus, 24, 249

Tarasius, patriarch, 251
Taron, 1, 14, 72, 73, 75, 76, 78, 95, 176, 301
Tarsus, 10, 13, 58, 61, 63, 82, 85, 90, 94, 95, 110, 112, 117, 150, 159, 163, 169, 201, 256, 258, 261, 263, 266, 278, 285, 290, 301
  Battle of, 3, 17, 136, 192, 252, 255, 276, 290, 292, 293, 306
tasinarioi (τασινάριοι), 138, 165, 280
taxiarchies, 193, 194, 195, 198, 199, 202, 211
Theodore Daphnopates, 240, 251
Theodore of Antioch, patriarch, 154

Theodore of Sebastea, 240, 241
Theodore of Side, 240
Theodosiopolis (Qaliqala), 13, 58, 63, 72, 73, 75, 78, 81, 260, 301
Theophanes, 55, 56, 78, 162, 172, 177, 232 (n8), 240, 251
Theophanes Continuatus, 85, 154, 162, 240, 242, 243
Theophilos, emperor, 56, 61
Thessaloniki, 112, 125 (n47), 137, 257
  sack of, 15, 82, 112–13, 117, 121, 174
thughūr (frontiers), 3, 29, 52, 58, 82, 124 (n36), 140, 157, 180 (n2), 232 (n8)
topoteretes, 230
Tornikioi, family, 75, 78
Trajan the Patrician, 251
Tyana, 61
Tzamandos, 61, 71, 157, 201, 254, 259
Tziniskes, John, 40, 58, 78, 118, 136, 151, 173, 211, 237, 238, 241, 243, 250, 252, 253, 255, 257, 258–65, 278, 282–5, 291, 292, 299, 305

Umar al-Aqta, 58
Uranus, Nicephorus, 174, 199, 200, 209, 211, 228, 229, 305

Van, Lake, 1, 60, 61, 69, 72, 73, 75–9, 176
Varangian Guard, 202, 228

Vaspourakan, 1, 14, 71–8, 95, 301
Vegetian (warfare/theory/strategy), 5, 6, 13, 19 (n16), 53, 120, 137

Yahya ibn Said al-Antaki, 17, 52, 81, 236, 242, 243–6, 251, 254, 255, 259, 260, 261, 265, 285–7, 295 (n39), 305

Index of Terms in Greek

άκρον (pl. άκρα), 139
άπλεκτον, 222
βιγλάτορες (Lat. vigilator), 132–3
βοηθός line, 210, 229
ίλαρχος, 208
καμινοβιγλάτορες (caminus, meaning the path, and βιγλάτορες), 132

κράσις, 107, 110
ουραγός, 197, 207, 226, 227
πανσιδήρους ιππότας, 278, 283
πλαγιοφύλακες, 208
πρόμαχος line, 210
τάξις (order, discipline), 227, 290, 294, 307
φοσσάτου, 54
χονσάριοι or χωσάριοι, 139

EU representative:
Easy Access System Europe
Mustamäe tee 50, 10621 Tallinn, Estonia
Gpsr.requests@easproject.com

www.ingramcontent.com/pod-product-compliance
Lightning Source LLC
Chambersburg PA
CBHW071758300426
44116CB00009B/1123